PALMS

THROUGHOUT THE WORLD

PALMS

THROUGHOUT THE WORLD

David L. Jones

Foreword by John Dransfield
Royal Botanic Gardens, Kew

Smithsonian Institution Press
Washington, D.C.

ACKNOWLEDGEMENTS

I would like to thank the many people who contributed in various ways to this book.

My wife Barbara deserves special mention for her numerous ideas, unstinting help, patience, competent word processing and cheerful companionship.

I sincerely thank John Dransfield for his solid support and encouragement. He read the entire text and particularly checked the botanical content, and especially the nomenclature. He also assisted with some derivations of obscure species names, wrote the foreword and provided a selection of photographs, both colour and black and white.

I am especially grateful to Catherine Jordan for locating numerous scientific texts, some of which proved to be very elusive.

Special mention must be made of Keith Thompson and John Bolger, who supplied details on the cultivation of some species and suggested a number of others which should be included. Other palm growers who contributed to the project were Stan and Jane Walkley, Peter Jones and Marty and Blue Bishop.

Martin Gibbons and Tobias Spanner commented on the *Trachycarpus* text and provided photographs of members of that genus.

Additional colour photographs for the book were provided by Greg Cuffe, Chris Goudey, Agnes Rinehart, Austen Bowden-Kerby and Ish Sharma.

I thank Don Evans, Superintendent, Fairchild Tropical Gardens, Florida, for cheerful guidance around the collection.

I would also like to thank Bill Templeman of Reed Books for his encouragement and support of my writing and Robert Taylor, also of Reed Books, for his help throughout the project.

Finally I also wish to thank all of those people who contributed in various ways to the success of my first palm book, *Palms in Australia*. Without that success, this book would not have happened.

DAVID L. JONES
Canberra, Australia
July 1994

Cover photo: *Licuala ramsayi,* Australian Fan Palm. D. Jones
Opp. title page: *Hyphaene petersiana* in habitat, Zimbabwe. C. Goudey
Page 8: *Livistona mariae* in its arid habitat. D. Jones

Published in the United States of America by Smithsonian Institution Press

ISBN 1-56098-616-6

Library of Congress Catalog Number 95-68615

This edition first published in 1998 by
New Holland Publishers Pty Ltd
Sydney • Melbourne • London • Cape Town

First Published in Australia by
Reed Books, Chatswood, N.S.W

Edited by Mary Halbmeyer
Proofread by Peter Meredith
Designed by Warren Penney

Typeset in Berkeley Old Style

Printed in Hong Kong

00 99 98 5 4 3 2

CONTENTS

FOREWORD

Palms are widely regarded as being second only to the grasses in economic importance and whole communities in the tropics and subtropics may depend on palms for their livelihood. They are wonderful plants. Not only are they so important but their great beauty and ornamental potential means that they are widely grown in tropical and subtropical gardens. As pot plants, they form a significant part of the horticultural trade in foliage plants, in temperate and tropical regions. Interest in palms seems never to have been greater and there is a real need for good introductory literature for the palm grower. *Palms in Australia,* by David Jones, was first published in 1984. It has proved to be a best-seller in the palm world. To my way of thinking, it has been the best of the popular introductions to palms and their cultivation and, despite its title, has enjoyed wide readership throughout the palm-growing world. When asked to recommend a popular introduction to palms, I have recommended *Palms in Australia* without hesitation.

Now in 1994, the range of palms available for the grower has increased dramatically and many more species are being introduced into cultivation every year. There are also many more palm growers. The International Palm Society has a worldwide membership and there are now a score of local palm societies. The demand for a good, popular, up-to-date introduction to palms, that provides practical information on cultivation and pests and diseases has never been so great. In the past decade there have also been significant changes in the naming of palms; changes due to the great increase in palm research. David Jones provides here just what is needed. *Palms Throughout the World* is much more than a new edition of *Palms in Australia.* One of the most exciting features of the book is that for some genera, we are provided with more than the usual one or two species—for the increasingly popular *Licuala* and *Pinanga* we are given accounts of substantial numbers of species that are not found in any popular books. With its beautiful illustrations and accessible text, *Palms Throughout the World* fills an important niche. I applaud David Jones for the role he plays in providing authoritative and reliable, yet easily comprehensible information on palms.

JOHN DRANSFIELD
Royal Botanic Gardens, Kew,
United Kingdom

PREFACE

The interest in palms is continuing to grow and in some countries these plants have a substantial following which even borders on the fanatical. They are now widely planted in many parts of the world and are used for a wide variety of purposes and products. Palms are valued for their ornamental qualities, as a commercial crop, as an aid to subsistence living, to add that tropical element to holiday venues and for decorating the foyers of public buildings and indoor rooms in private dwellings.

The inspiration to write this second book about palms was brought about by the tremendous popularity of *Palms in Australia*. Although primarily written for Australian conditions, that book became well accepted in many countries, especially in the USA. This new book is much more comprehensive and covers about 800 species of palms. Unfortunately even this substantial number is not exhaustive of all the palms cultivated by enthusiasts.

If this book has a bias it is directed towards those palms which are to be found in various countries between New Guinea and Malaysia. This region contains some special gems which offer tremendous prospects for cultivation. The wonderfully impressive species of *Johannesteijsmannia* are becoming familiar to palm enthusiasts, but many other exciting palms, especially dwarf to small palmlets, are to be found in other genera such as *Pinanga* and *Licuala*. Because of their out-standing prospects for cultivation these genera are given emphasis in this work.

I have endeavoured to keep abreast of current botanical literature and the names used in this publication are as up-to-date as possible. This has involved the perusal of numerous scientific journals, some of which are readily available in certain libraries, whereas others have a much more restricted distribution. On the botanical aspects I have been greatly aided by John Dransfield of the Royal Botanic Gardens, Kew, United Kingdom.

Common synonyms are included in the text and additionally as a separate entry in the index. As one aim of this work is to promote palms to the general public, the species entries are in alphabetical order and there has been no attempt at a taxonomic grouping. The use of botanical terms has been kept to a minimum and they are generally explained where first used. Common names are included only where they are well known. A detailed glossary is included as is a comprehensive bibliography for those who wish to learn more about these fascinating plants.

The book is illustrated by 350 colour photos, most of which were taken by the author. Others are credited where used.

DAVID L. JONES
Canberra, Australia
December 1994

Part One
THE NATURE OF PALMS

Pinanga patula (*Botanical Magazine*, Vol. 37 of the 3rd series, plate 6581)

CHAPTER 1

INTRODUCTION TO PALMS

Palms are woody monocotyledons in the family Arecaceae (alternative name Palmae) which is placed in the order Arecales. They are a natural group of plants with a characteristic appearance that enables most people to recognise them without great difficulty, although unrelated plants with a similar general appearance—such as Cycads, Cyclanths, Pandans, Cordylines—are often included with them by the inexperienced.

Although palms possess some features in common with the families Cyclanthaceae, Pandanaceae and even Araceae, these relationships are really only superficial and the palms stand out as an isolated and distinct group.

Palms are an ancient life form with fossil records from the late Cretaceous period. Being composed largely of durable materials, palms leave a relatively good fossil record and may have existed before this time. Changes have certainly occurred since those days because the pollen of the Mangrove Palm (*Nypa fruticans*), which today is only found in the tropics, has been recovered from Eocene sediments near London. These same sediments have also yielded pollen of a palm referred to as a *Phoenix* species, *Calamus daemonorops*, fruit of an *Oncosperma* and *Sabal* and seeds of a *Caryota*-like palm.

Palms are regarded as being 'Princes' among plants and indeed were labelled such (as 'Principes') by none other than the great Swedish botanist Carolus Linnaeus, the founder of the modern binomial system of plant nomenclature. He recognised palms as a distinct group of plants and described several species. Today, *Principes* is the name of the quarterly journal of the International Palm Society which was established in Florida in 1956. This is a very active society which combined forces with the L.H. Bailey Hortorium to print the most magnificent and comprehensive book on palms yet produced, this being based on the life's work of Harold E. Moore Jr. (See N.W. Uhl and J. Dransfield, 'Genera Palmarum', The L.H. Bailey Hortorium and The International Palm Society, 1987, Allen Press, Kansas.)

NUMBER OF SPECIES

In the literature, the estimate of the number of species of palms varies from 2500 to 3500 in 210 to 236 genera. A more accurate modern estimation is probably about 2600 species in 200 genera. Variations in the tallies basically arise because of disagreement between botanists on the delimitations of species and genera. More species await discovery and description as new areas are explored. Whatever the number of species, palms are a highly significant and extremely diverse group and rank fourth or fifth in size in the monocotyledons.

PALM CLASSIFICATION

Human beings like to categorise species and palms are no exception. Those plants or animals with a common set of characters which sets them apart from another group of closely related plants or animals are known as a species, while a whole group of closely related species comprise a genus. Related genera are then grouped into families; in the case of palms this is Arecaceae or Palmae (not Palmaceae as is sometimes written).

TABLE 1
Palm classification according to Moore (1973)

GROUP	LEAF FOLDING	LEAF SHAPE	NUMBER GENUS/ SPECIES	DISTRIBUTION	EXAMPLE
Coryphoid	induplicate	costa-palmate	32/322	Pantropical	*Corypha, Sabal, Thrinax*
Phoenicoid	induplicate	pinnate	1/17	Africa–Indo China	*Phoenix*
Borassoid	induplicate	palmate	6/56	Africa–New Guinea	*Borassus*
Caryotoid	induplicate	bipinnate	3/35	Asia	*Caryota*
Nypoid	reduplicate	pinnate	1/1	Ceylon–New Guinea	*Nypa*
Lepidocaryoid	reduplicate	pinnate (rarely palmate)	22/664	Pantropical	*Metroxylon Calamus*
Pseudophoenicoid	reduplicate	pinnate	1/4	Caribbean	*Pseudo-phoenix*
Ceroxyloid	reduplicate	pinnate	4/30	South America	*Ceroxylon*
Chamaedoroid	reduplicate	pinnate	6/146	Mascarenes–South America	*Chamaedorea*
Iriarteoid	reduplicate	pinnate	8/52	South America	*Iriartea Socratea*
Podococcoid	reduplicate	pinnate	1/2	West Africa	*Podococcus*
Arecoid	reduplicate	pinnate	88/760	Pantropical	*Areca, Linospadix, Cocos, Butia*
Cocosoid	reduplicate	pinnate	28/583	South America	*Syagrus*
Geonomoid	reduplicate	pinnate	6/92	South America	*Geonoma*
Phytelephantoid	reduplicate	pinnate	4/15	South America	*Phytelephas*

Palms are a complex group of plants which have numerous variations around a basic theme. Although there have been numerous attempts to classify them, most of the earlier efforts have been unsatisfactory and have caused unnecessary confusion. Fortunately over recent years there has been an intensive study of palms carried out by various researchers both in the laboratory and in the field. Chief among these researchers was the late Dr H.E. Moore Jr of the L.H. Bailey Hortorium at Cornell University, New York. Dr Moore devoted his life to the study of palms and in 1973 produced an informal classification which is reproduced in part above. (For a complete coverage see H.E. Moore, 'The major groups of palms and their distribution', Gentes Herbarium 11 (2) (1973): 27–141.)

More recent detailed studies by John Dransfield of the Royal Botanic Gardens, Kew and Natalie Uhl of the L.H. Bailey Hortorium, have refined and modified the informal classification of Moore and given formal nomenclatural status to the family categories above generic level. Each category at this level is typified by the name of the genus which forms the root of the name as designated in the rules of the International Code of Botanical Nomenclature. (See J. Dransfield and N.W. Uhl, 'An outline of a classification of palms', Principes 30 (1) (1986): 3–11.)

TABLE 2
Palm classification according to Dransfield and Uhl (1986)

SUBFAMILY	TRIBE	SUBTRIBE	NUMBER OF GENERA	EXAMPLES OF GENERA
Coryphoideae	Corypheae	Thrinacinae	14	*Thrinax, Chamaerops, Trachycarpus, Rhapis*
		Livistoninae	12	*Livistona, Licuala, Pritchardia, Brahea*
		Coryphinae	4	*Corypha, Chuniophoenix, Kerriodoxa, Nannorrhops*
		Sabalinae	1	*Sabal*
	Phoeniceae		1	*Phoenix*
	Borasseae	Lataniinae	4	*Latania, Borassus, Borassodendron, Lodoicea*
		Hyphaeninae	3	*Hyphaene, Medemia, Bismarckia*
Calamoideae	Calameae	Ancistrophyllinae	3	*Laccosperma, Eremospatha*
		Eugeissoninae	1	*Eugeissona*
		Metroxylinae	2	*Metroxylon, Korthalsia*
		Calamineae	8	*Calamus, Salacca, Daemonorops, Eleiodoxa*
		Plectocomiinae	3	*Plectocomia, Myrialepis, Plectocomiopsis,*
		Pigafettinae	1	*Pigafetta*
		Raphiinae	1	*Raphia*
		Oncocalaminae	1	*Oncocalamus*
	Lepidocaryeae		3	*Lepidocaryum, Mauritia, Mauritiella*
Nypoideae			1	*Nypa*
Ceroxyloideae	Cyclospatheae		1	*Pseudophoenix*
	Ceroxyleae		5	*Ceroxylon, Oraniopsis, Juania, Ravenea*
	Hyophorbeae		5	*Hyophorbe, Chamaedorea, Synechanthus, Gaussia*
Arecoideae	Caryoteae		3	*Caryota, Arenga, Wallichia*
	Iriarteeae	Iriarteinae	4	*Iriartea, Iriartiella, Dictyocaryum, Socratea*
		Wettiniinae	2	*Wettinia, Catoblastus*
	Podococceae		1	*Podococcus*
	Areceae	Oraniinae	2	*Orania, Halmoorea*
		Manicariinae	1	*Manicaria*
		Leopoldiniinae	1	*Leopoldinia*
		Malortieinae	1	*Reinhardtia*
		Dypsidinae	6	*Dypsis, Neodypsis, Chrysalidocarpus, Phloga*

SUBFAMILY	TRIBE	SUBTRIBE	NUMBER OF GENERA	EXAMPLES OF GENERA
		Euterpeinae	6	*Euterpe, Prestoa, Oenocarpus*
		Roystoneinae	1	*Roystonea*
		Archontophoenicinae	7	*Archontophoenix, Rhopalostylis, Hedyscepe, Chambeyronia*
		Cyrtostachydinae	1	*Cyrtostachys*
		Linospadicinae	4	*Linospadix, Calyptrocalyx, Laccospadix, Howea*
		Ptychospermatinae	9	*Ptychosperma, Veitchia, Balaka, Drymophloeus*
		Arecinae	8	*Areca, Pinanga, Hydriastele, Gulubia*
		Iguanurinae	27	*Iguanura, Dictyosperma, Pelagodoxa, Lepidorrhachis*
		Oncospermatinae	8	*Oncosperma, Deckenia, Nephrosperma, Verschaffeltia*
		Sclerospermatinae	2	*Sclerosperma, Marojejya*
	Cocoeae	Beccariophoenicinae	1	*Beccariophoenix*
		Butiinae	9	*Butia, Cocos, Syagrus, Jubaea*
		Attaleinae	4	*Attalea, Scheelea, Orbignya, Maximiliana*
		Elaeidinae	2	*Elaeis, Barcella*
		Bactridinae	6	*Bactris, Aiphanes, Acrocomia, Gastrococos*
		Geonomeae	6	*Geonoma, Calyptronoma, Calyptrogyne, Pholidostachys*
Phytelephantoideae			3	*Phytelephas, Ammandra Palandra*

GENERALISED GROWTH FEATURES

Palms are perennial, unisexual or bisexual woody plants, with a distinctive crown of leaves and which reproduce by seeds. Most species have a prominent trunk but a few have underground trunks and some have much reduced trunks and appear to be trunkless (acaulescent). A few are climbers with slender stems of indeterminate growth produced at intervals from an underground rhizome (or solitary). While most palms have a solitary trunk, a significant number produce multiple trunks by suckering from the base and are known as clumping, clustering or caespitose palms. The individual growth habit is usually characteristic but both habits may be present in the one genus (*Licuala, Hyphaene, Ptychosperma, Gronophyllum, Pinanga*) and sometimes the habit is variable within a species (*Laccospadix australasica, Chrysalidocarpus pembanus*).

Palm leaves are very prominent and have a characteristic shape. Most species are either palmate or pinnate but a very distinctive group has bipinnate leaves (*Caryota*). Each species of palm usually carries a characteristic number of leaves in its crown, whether numerous and dense, or few leaves in an open and sparse crown.

A dense, suckering clump of *Chrysalidocarpus lutescens*, the Golden Cane Palm or Butterfly Palm.

Palms flower when they are mature. In some species this may occur after three to five years while others may take more than forty years to mature. Most palms flower regularly throughout their life (termed pleonanthic) producing inflorescences from successive nodes up the trunk (termed acropetal flowering). Some specialised palms have restricted flowering processes. These palms flower down the stem producing inflorescences from each leaf axil (termed basipetal flowering) and the whole plant dies when the lowermost bunch of fruit ripens (*Arenga pinnata, Caryota urens*). Suckering palms can have this habit with the individual stem dying when the lowermost bunch of fruit ripens (termed hapaxanthic—*Arenga engleri, Caryota mitis*). Monocarpic palms flower but once in their lifetime and then die (*Corypha* spp., *Raphia australis*). In these palms the inflorescence terminates the trunk which dies after the fruit are mature. In *Nanorrhops ritchieana* each inflorescence is terminal on the branch of a stem and this flowering branch dies back to the next branch after maturing fruit.

Palm flowers are generally small and not spectacular by comparison with many garden flowers. They are, however, often borne in profusion, may be quite colourful (even showy) and are sometimes fragrant. Early literature recorded that most palms are wind-pollinated but today it is known that many species are pollinated by insects such as beetles, flies and bees. Palm flowers may be unisexual or bisexual and it is not uncommon for unisexual flowers of either sex to be distributed on the same inflorescence, often in close proximity to each other. In some genera, bisexual and unisexual flowers may occur on the one inflorescence.

Palm fruits may be small and borne in profusion, or large and carried in small numbers (the best example is the coconut). Frequently they are very colourful and are an additional decorative feature. The fruits contain one to three seeds (rarely more) which generally have a hard or fibrous covering. Further details on palm structure can be found in Chapter 2.

GENERALISED BIOLOGY

Palms are long-lived, perennial plants which produce unisexual or bisexual flowers on the one plant or on separate plants and reproduce by seeds. Most palms flower regularly each year but some have specialised limited flowering procedures.

In most species the transference of pollen from the stamens to the stigma is via insect vectors (beetles or small bees), but in certain species wind may be involved or perhaps both mechanisms operate. Prior to flowering, the buds of some palms may heat up. Many palm flowers release odours to attract pollinators.

The fruit of many palms are brightly coloured to attract dispersal agents and often there is a fleshy mesocarp which is eaten by various birds and animals. The seeds of most palms can germinate on maturity but some may have an after-ripening mechanism. For more details on palm biology see Chapter 4.

WORLD DISTRIBUTION

Palms are widely distributed in the well-watered zones of the world but are absent or rare in very dry or very cold regions. They are uncommon in the temperate zones but proliferate in the tropics. Not only do the vast majority of species occur here but palms are also a commonly encountered and sometimes dominant component of the vegetation. In the tropics they can be found from the seashore to inland districts and to high altitudes.

Latitudinal Range

Palms reach their greatest proliferation in the tropics and are much less prominent and diverse in the temperate regions. The most northerly palm species is *Chamaerops excelsa*, being found in Europe at about 44° north. In the United States of America, palms occur at about 33° north. These are *Washingtonia filifera* in California and *Rhapidophyllum hystrix* in South Carolina, Mississippi, Alabama and Georgia. The most southerly palm is found on Chatham Island off the east coast of New Zealand at 44° 18' south. This is a variant of *Rhopalostylis sapida*. On the west coast of the South Island of New Zealand, *Rhopalostylis sapida* occurs to about 43° south. In Australia *Livistona australis* extends to about 37° south.

Altitudinal Range

Palms are found from sea-level (see *Mangrove Palms* and *Littoral Palms*, pages 16, 17) to montane forests, mountain tops and high plateaux. The species which occurs at the highest altitude would appear to be *Ceroxylon utile*, which grows in the Andes of South America to about 4000 m altitude. Species of *Euterpe* in South America also grow at high altitudes (about 3000 m) and on Mount Kinabalu in Borneo, both *Calamus gibbsianus* and *Pinanga capitata* grow at about 3000 m. In the Himalayas, *Trachycarpus martianus* grows up to 2400 m altitude and in the Andes of South America *Geonoma* spp. occur at a similar altitude. Many species of palms occur in montane forests up to about 2000 m altitude, including *Caryota maxima*, *Chrysalidocarpus* spp., *Hyospathe* spp., *Plectocomia himalayana* and *Welfia regia*.

Generic Distribution

Despite the wide distribution and frequency of palms in the world's tropics, there are very few large and widespread genera. In fact, most palm genera contain five or fewer species and monotypic genera are common. The climbing genus *Calamus* is the largest with 370 species with a Pantropical distribution. Many small genera have a worldwide distribution—for example, *Borassus*, *Phoenix* and *Raphia*—while some larger genera have proliferated in relatively restricted areas. For example, *Licuala* and *Pinanga*—each containing more than 100 species—are largely found between Malaysia and New Guinea, while the 100 species of *Chamaedora* are restricted to Central and South America. As a general rule there are more differences than similarities between palms of the Old World and those of the New World.

Endemism

Many palm genera are localised in relatively small areas and there is considerable development of small genera and monotypic genera endemic to isolated areas and islands. This indicates the ability of palms to evolve and make use of specific niches in the environment. On Lord Howe Island, for example, there are four species of palms in three genera, all endemic. Two species are common on the coast and extend some way up the mountains; the third is found at intermediate altitudes and the fourth is restricted to the mountain tops. Fiji has about twenty-two species of palm in about ten genera. More than ninety per cent of the species are endemic while about half of the genera are not found elsewhere.

The proliferation of palms on some islands can be demonstrated by comparing Australia with the nearby islands of New Caledonia and New Guinea. Australia, with its large, but generally dry landmass, is poor in palms, with about twenty-two genera and sixty species. Of these, five genera and thirty-nine species are endemic (seventy-nine per cent). By contrast, New Caledonia, which is small by comparison, has thirty-nine species of palm in about seventeen genera, the vast majority of which are endemic. To the north of Australia, New Guinea has about thirty genera and 270 species, ninety-five per cent of which are endemic. A similar situation applies to the African continent when it is compared with the nearby island of Madagascar.

PALM HABITATS

It has been estimated that more than two-thirds of the world's palm species grow in rainforests. Here they may be emergent plants with their crowns well clear of the forest canopy; of intermediate size mingling with the canopy of the trees but not emerging; or growing as small understorey plants in the generally shady, dull conditions of the rainforest floor. Climbing palms are also common in rainforests and while these have their roots anchored in the soil of the forest floor, their uppermost leaves mingle with the outer foliage of the canopy.

A significant number of palms grow in open habitats such as savanna grassland, open woodlands and sparse forests. Many of these are hardy species which are capable of withstanding periods of dryness and occasional fires. A characteristic of palms growing in open sites is that they are frequently found in extensive colonies, usually of a single species—for example, *Borassus, Livistona, Phoenix, Raphia*. By contrast, palms found in rainforest are not often in colonies and many different species may exist together in a small area—for example, *Pinanga, Licuala*.

Some palms that grow in open areas tend to favour wet sites such as marshes and swamps, or the margins of permanent streams, lakes or lagoons. Often the palms growing in these sites occur in dense colonies or thickets (*Raphia taedigera, Phoenix paludosa*). Even sites subject to periodic inundation can support palms because their deep roots can tap ground water during the dry times. Although they may have excellent water-conserving devices, palms are absent from very dry habitats. Even those found in the desert—such as the Date Palm—will only survive where their roots can tap permanent underground water supplies.

A few palms have adapted to specialised habitats where they can compete successfully with other plants. These include situations along the seashore, in coastal estuaries, within the flood range of streams and even growing within the streams themselves. Few palms will tolerate snow but *Nannorrhops ritchiana* from the mountains of Afghanistan and species of *Trachycarpus* from the Himalayas are covered with snow regularly each winter. Some palms have adapted to unusual soil types and occur naturally on no other soil regime.

Emergent Palms

Palms commonly compete with trees to form the canopy of forests, and it is not uncommon for the taller palms to outgrow the trees and extend above the canopy as emergents. Palms impart a distinctive appearance to such forests and they are often highly noticeable when viewed from the air or when silhouetted on a ridge. Emergent palms are common in the forests of many countries and include *Beccariophoenix madagascariensis, Borassus* spp., *Ceroxylon* spp., *Clinostigma exorrhizum, Gulubia* spp., *Gronophyllum chaunostachys, Livistona saribus, Makeea magnifica* and *Neoveitchia storkii*.

Understorey Palms

Many palms are found growing naturally in the understorey of forests. These are shade-loving species which range in size from small acaulescent types to species of moderate dimensions which do not reach the forest canopy. Understorey palms are found in most countries where palms grow and in many different forest types. Examples include *Aiphanes* spp., *Asterogyne martiana, Calyptrocalyx* spp., *Chamaedorea* spp., *Dypsis* spp., *Geonoma* spp., many *Licuala* spp., *Linospadix* spp., *Livistona exigua, Rhapis* spp. and *Wallichia* spp.

Mangrove Palms

Only one species of palm is a true mangrove and that is the Mangrove Palm, *Nypa fruticans*. This species, found in Asia and the western Pacific, grows in estuaries often forming extensive colonies. The plants grow in soft mud within tidal influence where the water flow is slow. Seeds of this palm germinate while attached to the infructescence and the fruits float and are dispersed by currents. A few other palms, such as *Copernica gigas*, *Lucula paludosa* and *Phoenix paludosa*, grow on the landward fringes of mangrove communities but cannot be considered to be strictly mangroves.

Littoral Palms

A number of palms have adapted to coastal conditions and grow naturally in the littoral zone close to the sea. These palms usually grow in sandy soils—their roots can often tap reserves of ground water and they are able to withstand buffeting by salt-laden winds. The Coconut Palm (*Cocos nucifera*) is probably the best example, growing as it does right on the foreshore. Other littoral palms include *Allagoptera arenaria*, *A. brevicalyx*, *Arenga australasica*, *Copernicia ekmanii*, *Cyphosperma nucele*, *Pritchardia thurstonii*, *Pseudophoenix sargentii*, *Ptychosperma elegans* and *Thrinax radiata*.

Rheophytic Palms

A rheophyte is a plant which grows close to the edge of a stream and is often submerged during flooding. There are a number of rheophytic palms and it would seem that most of these have slender to very narrow leaflets, which presumably offer little resistance to the current when the plants are submerged. *Chamaedorea cataractarum* is a well-known rheophyte from Central America. In this region certain species of *Geonoma* are also rheophytic. Less well known are species of *Pinanga* from Borneo, including *P. rivularis* and *P. tenella*. *Areca rheophytica* from Borneo is also a rheophyte.

Aquatic Palms

Only one species of palm is truly aquatic, although a number of others may grow as rheophytes (see *Rheophytic Palms*, above). The only aquatic palm is *Ravenea musicalis* from Madagascar. This species grows in a fast-flowing stream in water to 2.5 m deep. The seeds germinate within the fruit which, when mature, falls into the water and floats until the flesh rots and the seeds sink to the bottom of the river and continue germination. The second and third scale leaves of the seedling have a hooked apex which catches on protuberances on the bed of the river, thus anchoring the seedling. Rootlets are produced in abundance from the seedling and grow upwards towards the light. The seedling leaves, which are completely submerged, are floppy and wave about in the current, offering little resistance to the water flow.

Desert Palms

Palms have few features which enable them to survive long droughts and hence are absent from arid and semi-arid regions, except for a few species which grow on the margins of permanent streams or in oases where ground water is close to the surface. The best-known oasis palm is undoubtedly the Date Palm, *Phoenix dactylifera*, which has been widely planted by humans. *Livistona carinensis* also grows in desert oases in Arabia and Somalia and *Medemia argun* occupied similar sites in Egypt and Sudan but is now possibly extinct. The Australian palm *Livistona mariae* is a relict confined to the margins of a permanent stream in Central Australia. In the USA and Mexico, including Baja California, two species of *Washingtonia* are confined to springs or streams in, or close to, gorges and canyons. Species of *Sabal* and *Brahea* may grow on similar sites.

Calciphiles

Calciphiles are plants which grow naturally on limey soils (basic not acidic) derived from limestone, chalk or, less commonly, coral. A large number of palms grow in the wild on limey soils—a selective adaptation for survival. It should be noted that such soils can become acidic by chemical reaction even though they are derived from a basic rock type. Calciphiles can obviously cope well with alkaline soils, but in cultivation this is not a strict requirement and they are mostly adaptable to a range of soil types.

Many species of palms originating in the Caribbean region are calciphiles, for limestone is the dominant rock on many of the islands. Included here are species of *Coccothrinax*, *Thrinax*, *Pseudophoenix*, *Gaussia* and *Calyptronoma rivalis*. Species of *Brahea* and *Sabal* may grow in similar habitats in adjacent countries. *Hyophorbe verschaffeltii* grows on limestone on the Mascarene

Islands. In the Pacific region and Asia, limestone outcrops are more sporadic but some unique palms are associated with them. Thus *Gronophyllum apricum* grows on limestone outcrops in New Guinea, *Licuala calciphila* in Vietnam and *Salacca rupicola* in Borneo. Two unique genera of fan palms are known from low scrub on limestone hills. These are *Guihaia* (two species) from southern China and north Vietnam and *Maxburretia* (three species) from Malaysia and Thailand. *Pritchardia thurstonii* is a fan palm well adapted to the coralline limestone of some Pacific islands.

Ultrabasic, Ultramafic and Serpentinic Soils

Soils of these types are rich in heavy metals such as chromium, iron, copper and manganese. Some of these soils are very acid (ultramafics—pH4) whereas many others are highly alkaline (ultrabasics—pH8). Serpentinic soils are rich in magnesium. Some soils of these types are also extremely porous with even heavy rainfall draining away quickly through the large pores.

Palms which grow on soils rich in heavy metals are frequently unique and unfortunately are also often very difficult to cultivate. Included in this category are about one-third of the palms which occur naturally in New Caledonia. These include desirable ornamentals such as *Actinokentia* spp., some *Basselinia* spp., *Brongniartikentia* spp., *Campecarpus fulcitus*, *Clinosperma bracteale*, *Cyphokentia macrostachya* and *Pritchardiopsis jeannenyi*. Ultrabasic soils also occur in Sabah and the Philippine island of Palawan.

HORTICULTURAL APPEAL

Palms are becoming increasingly popular for cultivation in streets, parks and gardens and as indoor and container plants. Because their shapes are more predictable than those of other trees, they can be used as lawn specimens, street trees and for avenues. They are especially popular in tropical and subtropical areas.

Because of the constant demand for them, palms are now an integral part of the nursery scene, and many nurseries specialise in their production. Some slower-growing species may take many years to become a saleable plant and are quite expensive. New palms are constantly being introduced by enthusiasts and as these prove themselves under various conditions, they may be taken up by the nursery trade and produced in quantity. There are many small-growing palms with tremendous horticultural potential that are hardly known because of the lack of propagating material or indeed await introduction to cultivation.

PALM HYBRIDS

Like many other groups of plants, palms can and do hybridise. Hybrids occur in nature but are much more frequent in cultivated plants. In some genera, such as *Latania*, *Phoenix*, *Ptychosperma*, *Chamaedorea* and *Copernicia*, hybrids are common and it is unwise to collect seed from plantings of mixed species. In many other genera, however, hybrids are virtually unknown. It should be realised that within an outcrossing genus even dissimilar species such as *Phoenix reclinata* and *P. roebelenii* may hybridise.

Intergeneric Hybrids

Hybrids between different genera are generally rare in nature and if this rule is broken, then it is probably more a reflection on the state of taxonomy of the genera involved rather than anything else. Within palms, intergeneric hybrids generally occur very sporadically and then usually between closely related genera. The hybrid progeny are mostly sterile but exceptions occur. Hybrid palms may be named as botanical entities and the names of intergeneric (nothogeneric) hybrids are often—but not always—a combination of the parents involved in the cross.

In the cocosoid palms, intergeneric hybrids occur sporadically between three closely related genera of the subtribe Butiinae. These are *Syagrus* x *Butia* (to form the hybrid genus X *Butiagrus*—formerly X *Butiarecastrum*) and *Jubaea* x *Butia*. X *Butiagrus nabonnandii* results from a cross between *Butia capitata* and *Syagrus romanzoffiana* (formerly *Arecastrum romanzoffianum*). The progeny of these hybrids are usually sterile.

Fertile hybrids are produced in the subtribe Attaleinae between the genera *Attalea* and *Orbignya*, resulting in a highly variable swarm of progeny. From this the hybrid genus *X Attabignya* was named (*X Attabignya minarum* originates from *Attalea compta* x *Orbignya oleifera*). Another hybrid genus, *X Markleya*, results from natural hybridisation between the genera *Orbignya* and *Maximiliana* (*X Markleya dahlgreniana* originates from *Orbignya phalerata* x *Maximiliana maripa*). The progeny of this intergeneric cross are also fertile. It is interesting to note that all of these intergeneric hybrids have originated in South America, particularly Brazil.

Interspecific Hybrids

Natural hybrids within a genus are generally uncommon, although perhaps as more groups are studied in detail more natural hybrids may come to light. The phenomenon of natural hybridisation indicates a close relationship between the parents, as well as other factors such as an overlap in flowering time and the sharing of at least one pollinating agent. Often the hybridisation occurs only where the parents grow in close proximity and the incidence of hybridisation may be enhanced on disturbed sites. It is common for the hybrid progeny to be sterile. Some of these natural hybrids are just recorded as novelties; others are given botanical status and are named. At least five hybrids in the genus *Syagrus* have been named. These are *Syagrus X camposportoana* (*S. romanzoffiana* x *S. coronata*), *S. X teixeiriana* (*S. romanzoffiana* x *S. oleracea*), *S. X costae* (*S. coronata* x *S. oleracea*), *S. X matafome* (*S. vagans* x *S. coronata*) and *S. X tostana* (*S. coronata* x *S. schizophylla*), all occurring in Brazil. *Syagrus X fairchildiana* was described from a plant cultivated at Fairchild Tropical Gardens in Florida, but this hybrid is now known to be identical with *X Butiagrus nabonnandii*.

About six natural hybrids have been named in the genus *Copernicia*. These are *Copernicia X burretiana* (*C. hospita* x *C. macroglossa*), *C. X occidentalis* (*C. curtissii* x *C. brittonorum*), *C. X shaferi* (*C. hospita* x *C. cowellii*), *C. X sueroana* (*C. hospita* x *C. rigida*), *C. X textilis* (*C. hospita* x *C. baileyana*) and *C. X vespertilionum* (*C. gigas* x *C. rigida*). One natural hybrid in the genus *Bactris* has also been named—*Bactris X moorei* (*B. oligoclada* x *B. humilis*).

Natural hybrids also occur in the genus *Orbignya* (*O. phalerata* x *O. eichleri* and *O. bacaba* x *O. minor*) and on Lord Howe Island the two species of *Howea* (*H. forsteriana* and *H. belmoreana*) are reputed to hybridise on occasion. Hybrids have also been noted in the genera *Areca* (*A. catechu* x *A. triandra*), *Chrysalidocarpus* (*C. lucubensis* x *C. lutescens*) and have been artificially produced in *Elaeis* (*E. oleifera* x *E. guineensis*). Hybrids also occur in arecoid palms in the subtribe Euterpeinae, between the closely related species of *Oenocarpus*.

PALM ODDITIES

As a group, palms contain some very unusual and interesting plants. It is perhaps not widely known that the largest seed, the largest inflorescence and the longest leaf in the plant kingdom are produced by palms. The seed is that of the Double Coconut (*Lodoicea maldivica*) of Seychelles. Individual seeds of this palm may weigh up to 20 kg and early this century were prized as a collector's item. Today live nuts of this species are sold to palm enthusiasts and dead nuts to tourists. The longest leaf is produced by the Central African Raffia Palm (*Raphia regalis*), individual leaves of which have been measured at 25.11 m long. Many other *Raphia* species produce leaves over 20 m long. The largest inflorescence produced by any plant occurs in the various *Corypha* species. The huge, imposing, candelabra-like inflorescence of *C. umbraculifera* is probably the largest of all and estimates suggest that a single inflorescence may carry as many as ten million flowers.

Extremes in height in the palm family are provided by comparing the small, acaulescent species with the lofty tree palms. The diminutive Lilliput Palm (*Syagrus lilliputiana*) from Paraguay is 10–15 cm tall at maturity and the Elfin Palms (*Chamaedorea stenocarpa* and *C. pygmaea*) from Central and South America are often only about 50 cm tall. At the other extreme, the Wax Palms of the Andes (*Ceroxylon quindiuense* and *C. alpinum*) are giants, both capable of producing trunks rising more than 60 m from the ground. The Wanga Palm, *Pigafetta filaris*, can also reach 50 m and may be one of the fastest-growing species. Massive palms with thick, woody trunks or huge leaves include *Copernicia fallaense, Corypha umbraculifera, Jubaea chilensis* and *Lodoicea maldivica*.

NAME CHANGES

Name changes, or more accurately corrections to names, are the inevitable result of detailed taxonomic studies carried out by botanists. Such name changes are particularly frustrating to nursery growers and amateur growers alike and may even cause confusion to other botanists who are unfamiliar with a particular group. With the plethora of names which have been applied to palms, particularly by botanists in the late eighteenth and early nineteenth centuries, it is inevitable that detailed studies of the various genera will result in changes to names.

Like all organisms, palms are classified by a binomial system which was introduced by the Swedish scientist Carolus Linnaeus in 1753. Thus the generic name and the specific epithet make up the binomial and one pair of names is correctly applied to each species—for example, *Livistona australis, Phoenix dactylifera*.

The basic unit in any biological classification system is the species. A species may be defined as a group of plants with a common set of characters which sets them apart from another group of closely related organisms. While such a definition is useful, it fails to account for the fact that species are constantly evolving to meet changes in their habitats. This factor is of particular significance for palms because so many species consist of disjunct relict populations which have been isolated from each other for long periods of time. Changes in morphology, both major and minor, which result from adaptations to each environment can cause problems in identification.

A genus is usually defined as being made up of a group of closely related species which are too distinctive to be grouped with any other. Within palms, the majority of genera are readily recognisable as distinctive natural groups and considerable refinements have been made to generic delimitation following recent studies.

Botanical nomenclature, or the naming of plants, is governed by a set of rules laid down by botanical authorities and revised every five or six years. Most changes to plant names occur as the result of detailed morphological studies, from examination of type specimens or from someone applying the botanical rules correctly.

Two of these rules are of major significance and concern priority of publication. A plant can have only one correct binomial and that is usually the oldest name available unless this name contravenes the rules. Thus if two or more correctly published names are available for one species, then the first one published (as a species under whatever generic name) is the correct name and all others are synonyms. Similarly, if the same name is used for two different plants, the second use is incorrect. As well as understanding and correctly applying the botanical rules, detailed studies of the original type specimens are of paramount importance in sorting out the nomenclature of any genus. Often, particularly with early collections, these type specimens are poor, were never retained or have been subsequently lost or destroyed (many palm types were destroyed in the Berlin herbarium during the Second World War).

PALM CONSERVATION

Along with many other groups of plants, palms have suffered at the hands of humans. Destruction of habitat is one of the greatest threats palms face and this occurs in most countries where they are found naturally. Large-scale operations such as forestry, woodchipping, mining, conversion of land to farms and shifting agriculture pose major threats, especially where they are followed by land degradation and erosion. Palm plants are often retained during these operations but can rarely reproduce in the altered conditions and exist as a forlorn reminder of their past glory.

Even localised human activity may be threatening since many species of palm are limited in number or are found in restricted habitats. Thus the construction of roads and highways, dams and drainage schemes and specific mining projects (such as limestone mining) can pose significant threats to localised populations and rare species. Unfortunately, some natural populations of palms have been destroyed before being studied by scientists and thus their diversity is lost.

Palms also suffer from other threats. The multiplicity of uses of various parts of the palm is in itself threatening. Many species have been reduced to great rarity by the collection of their cabbage—which is often considered a great delicacy and may even be served as a gourmet item in some restaurants—for example, *Veitchia montgomeryana* on Vanuatu. In New Caledonia, *Pritchardiopsis jeanneneyi* became extremely rare when its cabbage was collected and eaten by convicts and in South America, various species of *Euterpe* are on the verge of extinction as a result of excessive collection for cabbage. The continued collection of leaves for thatch, crafts (baskets, hats) or for animal fodder poses a similar threat.

More recently, the collection of seed for sale to nurseries and palm enthusiasts has assumed significance. Often collectors will concentrate on rare palms and the damaging effects are compounded because natural regeneration is reduced or eliminated. Examples are *Neodypsis decaryi* and *Ravenea xerophila*. Specialised species which have a restricted distribution may suffer adversely from even limited seed collection. Some species of *Chamaedorea* fit this category, as do certain species of *Pinanga* which have a very low fruit production—for example, *P. yassinii* and *P. rupestris*. In extreme cases, whole plants of large palms may be cut down in the seed collection process. This has occurred with *Beccariophoenix madagascariensis* in Madagascar and probably also others. Sometimes recently discovered populations or newly described species may suffer badly from poaching because excessive demand creates high prices for their seeds. This situation has occurred recently with the Foxtail Palm, *Wodyetia bifurcata*, in north-eastern Queensland.

Some species of palm may have never been particularly common in nature, occurring in small numbers either concentrated or scattered. Examples are *Pelagodoxa henryana* which occurs as scattered individuals on the Marquesas Islands; *Lavoixia macrocarpa* from New Caledonia with its total wild population of five plants which do not appear to be reproducing (collected seed has proved to be impossible to germinate); and the Bornean *Salacca dransfieldiana*, known only from about thirty plants.

Extinct Palms

About nine species of palm are thought to be extinct, although the number is somewhat controversial since the status of some of these species has been questioned. *Sabal miamiensis*, described as recently as 1985, is now extinct since its Florida habitat has been destroyed by urbanisation. *Roystonea stellata* from Cuba is apparently extinct as a result of habitat disturbance. In the Hawaiian Islands two species of *Pritchardia* (*P. macrocarpa* and *P. montis-kea*) are extinct due to alienation of their habitat. Two species of *Syagrus* from Brazil (*S. leptospatha*, *S. macrocarpa*) and one from Paraguay (*S. lilliputiana*) have become extinct for the same reason. *Thrinax ekmaniana* from Cuba is also apparently extinct. The intriguing palm *Medemia argun* from oases in Egypt and Sudan has not been seen in the living state since 1964 and is presumed to be extinct.

Endangered Palms

These palms are on the border of extinction and are unlikely to survive in the wild unless the endangering factors are removed. Probably the most endangered palm is *Hyophorbe amaricaulis*, which is known from a single individual in Pamplemousses Botanic Gardens, Mauritius. Other species of *Hyophorbe* have also been reduced to great rarity, including *H. lagenicaulis* and *H. verschaffeltii*, both from the Mascarene Islands. Also from these islands are *Latania lontaroides* and *L. loddigesii* which have been reduced to a few isolated individuals in the wild. In Madagascar, many species of palms have become endangered, including *Beccariophoenix madagascariensis*, *Halmoorea trispatha*, *Marojejya darianii*, *M. insignis*, *Ravenea madagascariensis* and *R. robustior*. In the Hawaiian Islands about thirteen species of *Pritchardia* are listed as endangered and in South America five species of *Attalea* and *Ceroxylon* are similarly categorised. In all, about 100 species of palms are regarded as being endangered, but this figure is unfortunately only preliminary. (For further information see D.V. Johnson, 'World Endangerment of Useful Palms in The Palm Tree of Life', *Advances in Economic Botany* (1986): 268–73.)

CHAPTER 2

THE STRUCTURE OF PALMS

Palms are woody monocotyledons with a distinctive appearance. While they do have basic similarities to other plants, having roots, trunk(s), leaves, flowers and fruit, these organs may be quite different in appearance and structure and some explanation is necessary to aid in an understanding of the group. Not all specialised palm structures are included in this chapter and those not mentioned may be found in the glossary or in a more detailed book such as P.B. Tomlinson, 'The Structural Biology of Palms', Clarendon Press, Oxford, 1990.

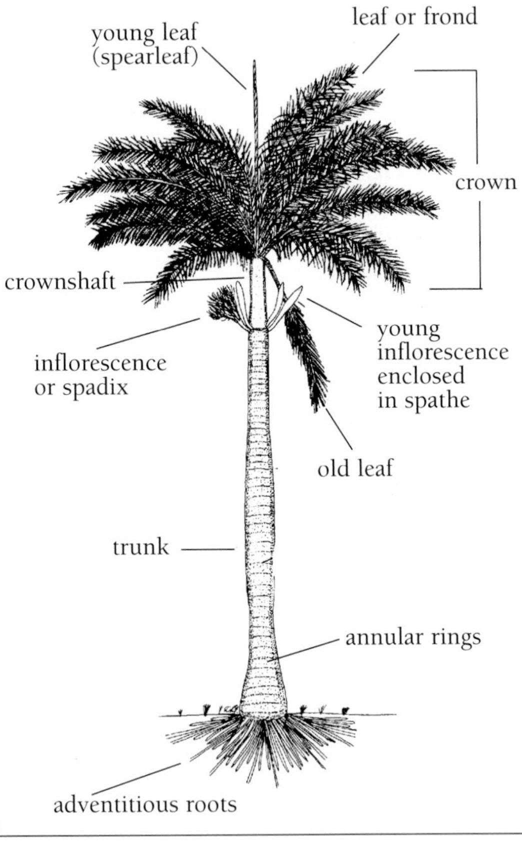

ROOTS

The roots of a palm originate from the base of the trunk and as in other plants, perform the essential functions of anchorage and the uptake of nutrients and water. Palm roots are produced from one of two different systems but essentially their functions, irrespective of their origin, are the same. The primary or seminal root system of a seedling is very important for early anchorage and initial uptake of nutrients and water, but is essentially short-lived.

A secondary root system is also produced and this is the most important one to the palm. This system arises from the base of the trunk just near the soil surface and is known as an adventitious root system (also termed accessory or supplementary root system). This is present in all palms, but in clumping palms a new adventitious root system is produced at the base of each new trunk. The roots of an adventitious root system may be crowded or sparse, short and stubby or may wander for considerable distances. In some palms—for example, *Phoenix* spp. and *Archontophoenix cunninghamiana*—these supplementary roots may be visible at the base of the trunk above the soil surface. Here they arise in a basal swelling and break through the outer bark in clusters. In some palms the roots creep over the soil surface (*Aiphanes aculeata*, *Gaussia* spp., *Manicaria saccifera*). *Raphia* spp. and a number of other palms have specialised upright roots similar to the pneumatophores of mangroves. The roots of most palms are thick and tough but some

Fig. 1
Structures of a palm

young leaf (spearleaf)

leaf or frond

crown

crownshaft

young inflorescence enclosed in spathe

inflorescence or spadix

old leaf

trunk

annular rings

adventitious roots

Stilt roots of *Areca vestiaria*, north Sulawesi.
J. Dransfield

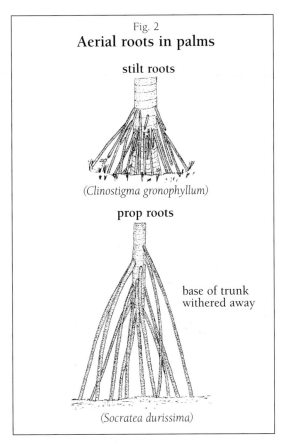

Fig. 2
Aerial roots in palms

stilt roots

(*Clinostigma gronophyllum*)

prop roots

base of trunk
withered away

(*Socratea durissima*)

are slender and wiry. The surface of the roots is generally smooth but those of *Clinostigma exorrhizum* and *Socratea exorrhiza* bear numerous spines.

A very few palms produce aerial roots—for example, *Cryosophila warscewiczii*, *Clinostigma exorrhizum*—and in even fewer species these develop into prop or stilt roots (*Areca guppyana*, *A. vestiaria*, *Eugeissona minor*, *Iriartea deltoidea*, *Pinanga aristata*, *P. tenacinervis*, *Veitchia montogomeryana*, *Verschaffeltia splendida*, *Socratea exorrhiza*, *S. durissima*). In these palms the bottom of the trunk may wither away leaving the aerial parts completely supported by the prop roots (see Fig. 2).

TRUNKS

Palms have a woody stem or trunk which gets taller as the plant gets older. The trunk is terminated by the apical meristem or growing apex. This vital organ is protected by the developing leaves and their sheathing bases and is buried well within the trunk. A young palm plant does not start developing a trunk until its apical bud has reached a certain critical size. This is the reason for the slow initial spreading development that occurs in many young palms before they form their trunk.

Palms have no bark as such, although the epidermis is hardened to form a protective layer. Thus they have no secondary thickening as in dicotyledons and once the trunk is formed it can only increase in diameter to a limited extent. Most species attain their maximum girth before the trunk grows upwards. Any thickening that does occur after the trunk develops is due to the swelling of cells with supplementary thickening in their walls. In the Bottle Palm (*Hyophorbe lagenicaulis*) the basal half of the trunk swells grotesquely. A number of palms have unusual swellings near the centre of the trunk, due to the localised swelling of cells caused by the uptake of water—for example, the Spindle Palm (*Hyophorbe verschaffeltii*), Pot-bellied Palm (*Colpothrinax wrightii*) and the Cherry Palm (*Pseudophoenix vinifera*). Variations in the thickness of individual palm trunks are believed to be caused by dry and wet years, with the terminal bud tending to swell more in the good seasons. The swelling at the base of some palm trunks is for support and is mainly

Spines forming ant galleries on stems of *Daemonorops formicaria*, Sarawak. J. Dransfield

caused by the development of accessory roots.

Palm trunks are remarkably strong and the slender ones very resilient, withstanding the tremendous bending stresses exerted by high winds, cyclones, and so on. The strength is obtained from a peripheral ring of fibre bundles which become lignified, while the flexibility is achieved because this ring surrounds a central soft cortex.

The shape of the trunk is characteristic for each species with the majority of palms having a trunk of nearly uniform width throughout, although a few, such as *Gaussia princeps,* are tapered upwards. Most trunks are straight but that of the Coconut is curved, probably as a result of a response to light. Some are very slender and may resemble bamboo (*Bactris guineensis, Chamaedorea* spp., *Dypsis* spp., *Oenocarpus panamanus*); others are slender and woody (*Gaussia maya, Lytocaryum insigne*) and then there is a tremendous range of stoutness (*Sabal* spp.) up to the massive trunks of *Jubaea chilensis* and *Copernicia fallaense.*

The surface of the palm trunk may be naked (*Jubaea chilensis, Sabal causiarum*), adorned with spines (*Aiphanes aculeata, Astrocaryum aculeatum, Bactris plumeriana, Verschaffeltia splendida*) or covered with the persistent bases of leaves which may themselves be adorned with hooks (*Washingtonia* spp.), woolly hairs (*Coccothrinax crinita*), stout interlaced fibres (*Trachycarpus fortunei*), fibres and spines (*Arenga pinnata, Zombia antillarum*) or spines (*Gastrococos crispa, Aiphanes aculeata*). In a number of palms the dead fronds persist and hang as a curtain covering much of the trunk (*Copernicia macroglossa, Washingtonia filifera*). In species with naked trunks the degree of smoothness, the colour, the presence of vertical fissures (*Dictyosperma* spp., *Sabal* spp.) or horizontal steps left by the fallen leaf-bases (*Archontophoenix alexandrae, Caryota cumingii*) are all useful diagnostic features (see Fig. 3).

The point on a trunk where the leaf arises is known as the node and the area in between as the internode. Scars representing the point of attachment of fallen leaves may be prominent on some palms showing up as a ring on the trunk. These scars occur at the nodes and are sometimes known as annular rings. Their distance apart is a reflection of the vigour of the palm and their number can be used as a guide to the palm's age. Small internodes (or closely spaced annular rings) indicate slow growth while widely spaced rings indicate rapid growth.

The basal part of the trunk of the unique, slow-growing Seychelles palm known as the Double Coconut (*Lodoicea maldivica*) is bulbous and fits into a large wooden socket produced by the plant. This socket is very durable and survives in the ground long after the palm has died. The base of the trunk of *Butia capitata* has a swollen knob from which the roots emerge. A few remarkable palms have stilt or prop roots which emerge from the base of the trunk (see *Roots*, p. 22).

Crownshaft

In some palms, such as *Archontophoenix* spp., *Rhopalostylis* spp. and *Roystonea* spp., the top of the trunk is crowned with a cylinder called the crownshaft. This is formed from the tightly packed tubular leaf-bases and is important as a protective measure for the meristem. Its presence provides a useful diagnostic feature. In most palms the crownshaft is bright green but in the colourful Sealing Wax Palm (*Cyrtostachys renda*) the crownshaft is tinged with pink or is waxy red to scarlet and is a striking ornamental feature. In some species of *Pinanga* it is yellow. The crownshafts of most species are smooth and glabrous but some are glaucous, spiny, scaly or hairy.

Fig. 3
Palm trunks

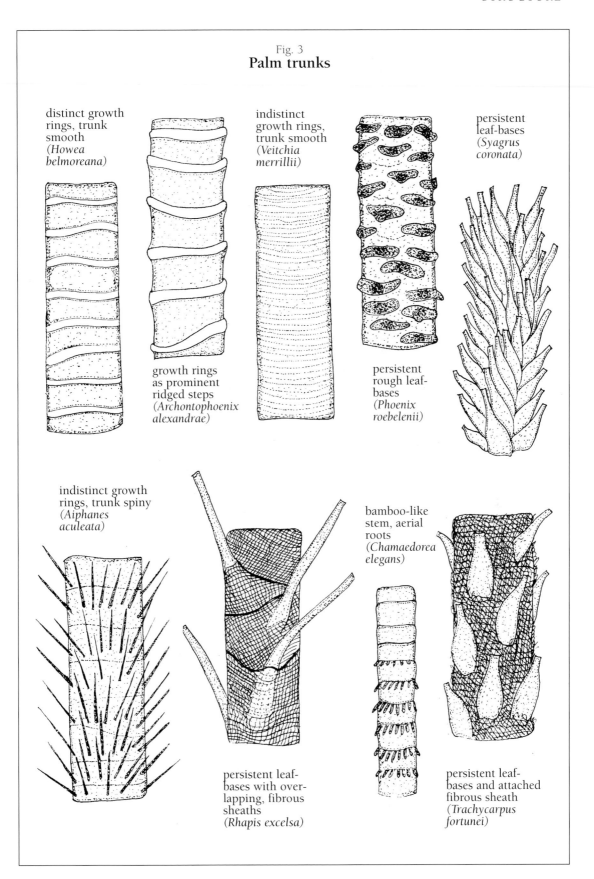

distinct growth rings, trunk smooth (*Howea belmoreana*)

indistinct growth rings, trunk smooth (*Veitchia merrillii*)

persistent leaf-bases (*Syagrus coronata*)

growth rings as prominent ridged steps (*Archontophoenix alexandrae*)

persistent rough leaf-bases (*Phoenix roebelenii*)

indistinct growth rings, trunk spiny (*Aiphanes aculeata*)

bamboo-like stem, aerial roots (*Chamaedorea elegans*)

persistent leaf-bases with over-lapping, fibrous sheaths (*Rhapis excelsa*)

persistent leaf-bases and attached fibrous sheath (*Trachycarpus fortunei*)

Diagnostic Features

The trunk is a very important feature in the identification of palms for it divides them into convenient groups. In all there are five groups of palms which can be recognised depending on their trunk—solitary, clumping, branching, trunkless and climbing palms. Palms with a solitary trunk or those which produce multiple trunks and have a clumping habit are of major significance and accomodate the majority of species. These criteria are not strict within a genus but are useful at species level. Because climbing palms have many modifications to various organs they will be dealt with separately at the end of this chapter.

Solitary Palms

As the name suggests these palms have a single trunk. They are a very common group within the palms and include many familiar, cultivated species. The trunks of these palms are variable between different species or even between forms within a species but are generally fairly constant in their principal characters within a species. Thus these characters have diagnostic significance and can be used as an aid to identification. Most have erect trunks but in some species trunks may be subterranean (*Sabal etonia*). In a few unusual species the decumbent trunk may creep over the surface of the ground before turning and growing upright (*Brahea decumbens*, *Elaeis oleifera*, *Johannesteijsmannia altifrons*, *Phytelephas macrocarpa*). In these species the trunks grow forwards, produce roots on the underside and the old parts of the trunk decay and rot.

Clumping Palms

These palms produce basal offsets or suckers at or below ground level and are therefore multiple-trunked. Occasionally some suckers are produced on the trunk—for example, *Chrysalidocarpus lutescens*, *Phoenix dactylifera*—but this is uncommon.

Within clumping palms some distinctions based on growth habit can be made. In the majority the suckers are produced on the perimeter of the clump as it spreads and they grow normally without any restriction on their development. In these types the oldest and tallest growths are found at the centre of the clump and grade down in height to the newest on the outside. Not infrequently the oldest stems appear to slow in their growth rate so that the adjacent but younger stems are not greatly different in height. Examples of this growth habit are *Chrysalidocarpus lutescens*, *Chamaedorea microspadix*, *C. costaricana* and *Phoenix reclinata*. Some species tend to form sparse clumps (*Phoenix reclinata*) while in others the stems are crowded close together (*Chamaedorea costaricana*, *C. seifrizii*, *Rhapis humilis* and *Cyrtostachys renda*).

In a few clumping palms a different growth habit is noticeable. One or two stems develop rapidly and dominate the clump while the remaining suckers develop slowly or reach a stationary phase. These do not develop any further unless a mature stem dies or is badly damaged. Thus the clump consists of the mature stem or a couple of stems surrounded by a rosette of undeveloped suckers. This growth habit can be seen in *Hydriastele wendlandiana*, *Linospadix minor*, *Laccospadix australasica*, *Phoenix dactylifera* and *Ptychosperma macarthurii*.

Branching Palms

Natural aerial branching in palms is rare and restricted to about seven genera although branching may occur in many other species following partial damage to the growing apex (see *Abnormal branching in palms*, p. 28). Branching is slightly more widespread in species with underground trunks, although sometimes it is difficult to distinguish this growth habit from the typical clumping palm which produces suckers.

Branching in palms occurs in two distinct ways, either forking at the growth apex (dichotomous branching) or lateral branching (see Fig. 4).

Dichotomous Branching Forking takes place at the actual growth apex which divides equally into two parts. These develop into branches and the dichotomy can be easily seen if the branching system is studied. This type of branching occurs aerially in the trunks of species of *Hyphaene*, *Korthalsia*, *Nannorrhops ritchiana* and *Vonitra* and in the subterranean trunks of *Nypa fruticans*. Not all species of *Hyphaene* and *Vonitra* fork and within those that do there is considerable variation in the degree of branching.

An interesting branching habit is to be found in the very attractive, verdant palm from

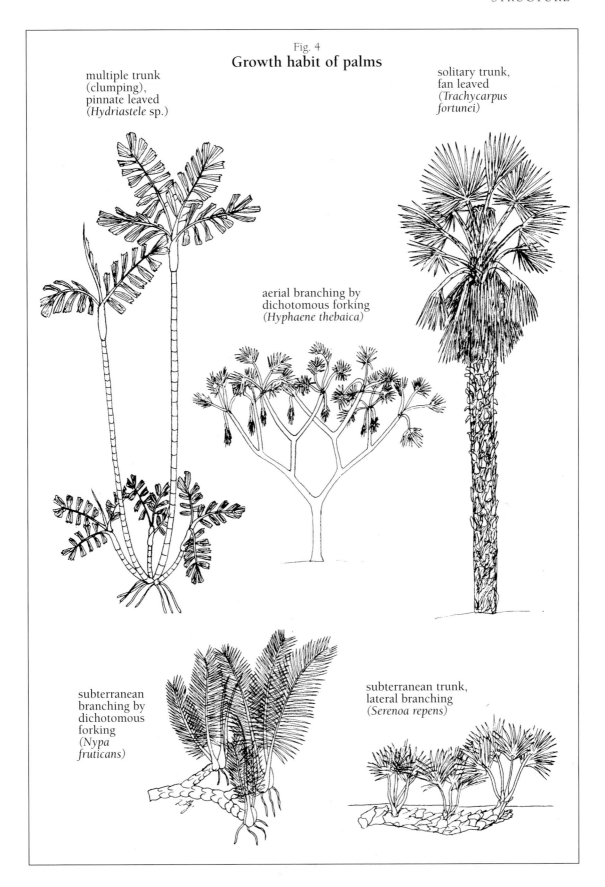

Fig. 4
Growth habit of palms

multiple trunk
(clumping),
pinnate leaved
(*Hydriastele* sp.)

solitary trunk,
fan leaved
(*Trachycarpus
fortunei*)

aerial branching by
dichotomous forking
(*Hyphaene thebaica*)

subterranean
branching by
dichotomous
forking
(*Nypa
fruticans*)

subterranean trunk,
lateral branching
(*Serenoa repens*)

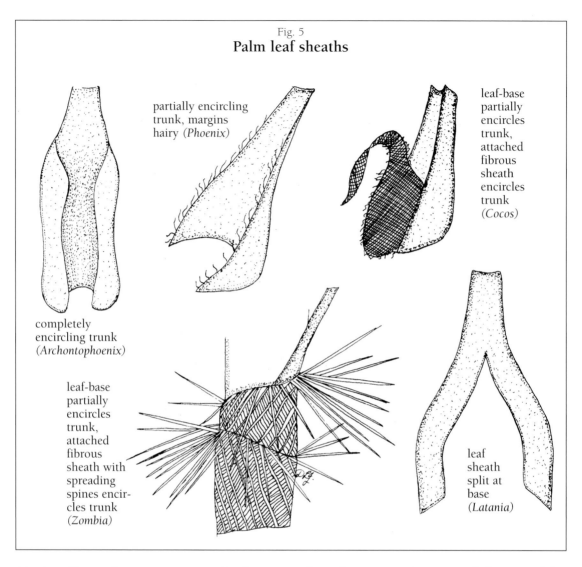

Fig. 5
Palm leaf sheaths

partially encircling trunk, margins hairy (*Phoenix*)

leaf-base partially encircles trunk, attached fibrous sheath encircles trunk (*Cocos*)

completely encircling trunk (*Archontophoenix*)

leaf-base partially encircles trunk, attached fibrous sheath with spreading spines encircles trunk (*Zombia*)

leaf sheath split at base (*Latania*)

Mexico, *Chamaedorea cataractarum*. In this species branching occurs by an equal division of the growing apex of a stem. This division takes place deep in the stem and the first sign of a change is an apparent split and pairing of the very young leaves. This split eventually becomes obvious and the two stems gradually grow away from each other until they are quite separate individuals.

Lateral Branching This method takes place in some species with subterranean trunks such as *Allagoptera arenaria*, *Serenoa repens* and *Salacca* spp. This type of branching is similar to that which occurs in trees and the branches arise from the activity of lateral meristems. Branching of this type occurs rarely in aerial palm trunks, but can be seen occasionally in *Chrysalidocarpus* spp. and *Laccosperma* spp.

Abnormal Branching in Palms Aerial branching in solitary palms is not as uncommon as is believed. It arises as the direct result of sublethal damage to the growth apex and should not be confused with the dichotomous branching that occurs naturally in some palms such as *Hyphaene* spp.

Following damage, the growing apex may proliferate in different directions; this eventually grows out and becomes branches. Two to many branches may occur and these usually survive and grow, although frequently one will dominate. As an example, several branching specimens of *Archontophoenix alexandrae* exist in gardens in Cairns, Queensland.

Trunkless palms

Trunkless palms, or those with a very much reduced trunk, are called acaulescent. Such palms usually do in fact have a trunk, although it is very greatly reduced and is subterranean. Acaulescent species are generally small palms that have become adapted to a specialised environment. In some species the trunkless condition enables them to survive adverse conditions such as drought and fire. Many of the palms which grow in the scrubby Brazilian vegetation known as Cerrado, where fires are frequent, have the acaulescent habit. Palms from Cerrado include *Acrocomia emensis*, *Allagoptera campestris*, *Butia paraguayana*, *B. purpurascens*, *Syagrus campylospatha*, *S. lilliputiana*, *S. loefgrenii* and *S. amadelpha*. Few of these palms are encountered in cultivation.

Acaulescent palms are also found growing as understorey plants in dense forests, particularly rainforests. A few species of *Chamaedorea* appear to be acaulescent (*C. stenocarpa*, *C. stricta*, *C. sullivaniorum*, *C. undulatifolia*) but in fact they may have short, buried stems. A number of species of *Aiphanes*, *Iguanura* and *Licuala* are acaulescent.

Some palms which have a solitary subterranean trunk may also be described as being acaulescent, however this is not strictly correct as the trunk is merely hidden from view. One example is *Sabal etonia*. Some palms have trunks which may be either subterranean or emergent (*Sabal minor*, *Serenoa repens*).

LEAVES

Palm leaves (often called fronds) may be scattered along the upper part of the trunk—for example, *Chamaedorea*, *Pinanga* and *Rhapis* spp.—or more commonly are borne at the top in a crown. The fronds of palms with subterranean trunks may be described as being radical or basal but they are in fact terminal on the underground trunk.

The leaves of palms are not remarkably different from those of other plants, having the same basic structure. However they are a large, spectacular feature of the plant and do possess a few unique characters. Young leaves emerge vertically or nearly so (at this stage they are often called spear leaves) and thus reduce exposure to radiation and water loss by transpiration. Some palms are quite tolerant of dry conditions and have water-conserving devices, including glossy surfaces, waxy leaf surfaces with a thick cuticle, a covering of wool or scales, corrugations and folding (to reduce the exposed leaf area), highly angled fronds, twisting of the fronds and variable angles of the leaflets (to reduce their angle of incidence to the rays of the sun).

The basal woody structure that extends from the trunk to the first leaflet or segment is known as the petiole. This is usually grooved on the upper surface and may be smooth or has the margins of the grooves adorned with teeth, hooks, scales or the whole surface covered in spines or a mixture of spines and fibres or wool.

The basal part of the petiole is expanded and clasps the trunk, often for a significant length. This part is variously known as the sheathing base, leaf base or leaf sheath. This sheathing base is entire and undivided in most species but in a few interesting ones is split at the base—for example *Thrinax* spp., *Hyphaene* spp., *Latania* spp. and *Schippia concolor*. The sheath persists on the trunk of some species long after the fronds have been shed (*Phoenix*

Net-like ochrea of *Korthalsia jata*, Sarawak.
J. Dransfield

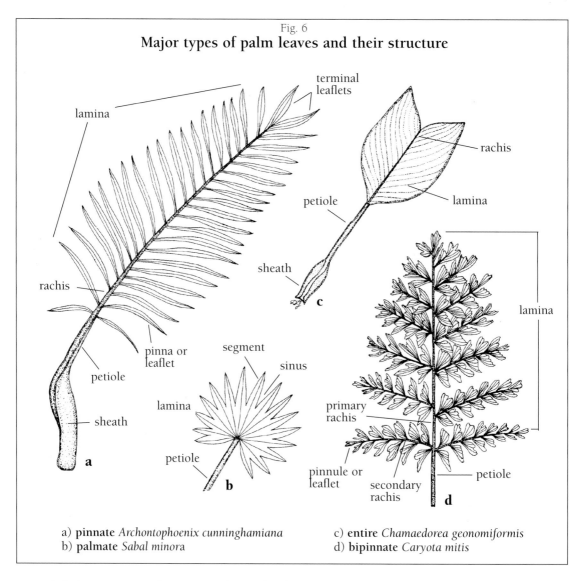

Fig. 6
Major types of palm leaves and their structure

a) **pinnate** *Archontophoenix cunninghamiana*
b) **palmate** *Sabal minora*

c) **entire** *Chamaedorea geonomiformis*
d) **bipinnate** *Caryota mitis*

spp., *Sabal mexicana, Trithrinax acanthocoma*) while in others the sheathing base is shed along with the frond, leaving a clean trunk (*Archontophoenix* spp., *Jubaea chilensis, Sabal causiarum*). In a few palms the whole leaf may persist after death (*Copernicia torreana, Washingtonia filifera*) forming a cover of dead fronds variously known as a skirt, petticoat or shag. The sheathing bases themselves may be naked or adorned with hairs, coarse fibres, scales, spines, hooks or teeth. In some palms the upper point of the leaf sheath projects forwards as a ligule (some species of *Licuala*). In a very few palms there is a specialised appendage at the upper part of the leaf sheath that is known as the ochrea. This is well developed in species of the climbing genus *Korthalsia*.

The most prominent portion of the leaf is the lamina or blade. This is the green part where photosynthesis occurs and may be entire or divided into segments or leaflets. Palms may be grouped according to frond types and this is a major diagnostic feature (see Fig. 6).

Feather-leaved or Pinnate Palms

The leaves of these palms are pinnately divided; that is, the frond is segmented by divisions which reach to the central stem. Such pinnate leaves have a distinctive appearance and are readily identified. As a further aid to identification the fronds often resemble a feather or the backbone and ribs of a fish.

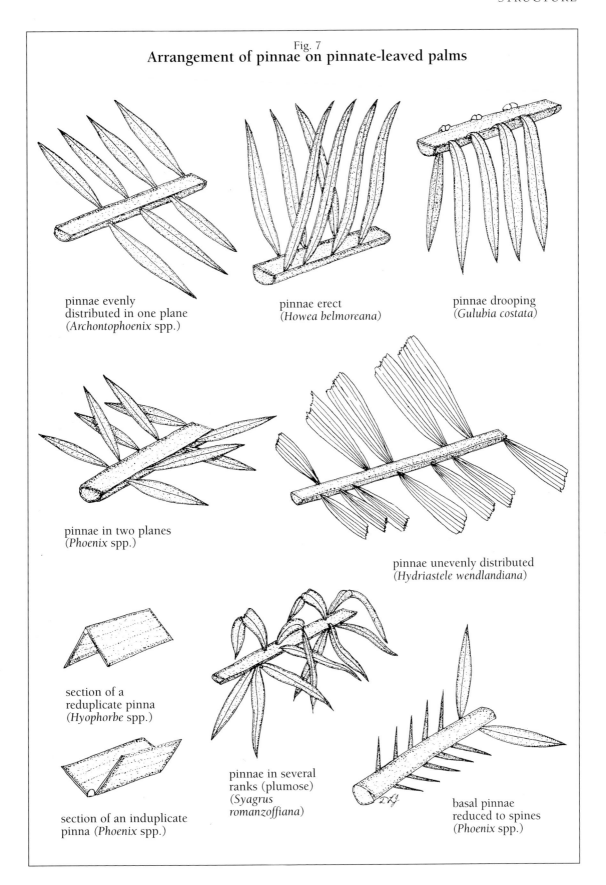

Fig. 7
Arrangement of pinnae on pinnate-leaved palms

pinnae evenly
distributed in one plane
(*Archontophoenix* spp.)

pinnae erect
(*Howea belmoreana*)

pinnae drooping
(*Gulubia costata*)

pinnae in two planes
(*Phoenix* spp.)

pinnae unevenly distributed
(*Hydriastele wendlandiana*)

section of a
reduplicate pinna
(*Hyophorbe* spp.)

pinnae in several
ranks (plumose)
(*Syagrus
romanzoffiana*)

section of an induplicate
pinna (*Phoenix* spp.)

basal pinnae
reduced to spines
(*Phoenix* spp.)

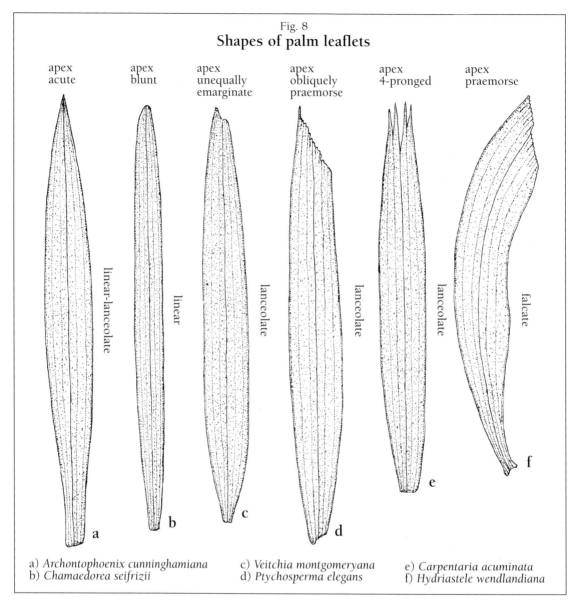

Fig. 8
Shapes of palm leaflets

apex acute — linear-lanceolate — a

apex blunt — linear — b

apex unequally emarginate — lanceolate — c

apex obliquely praemorse — lanceolate — d

apex 4-pronged — lanceolate — e

apex praemorse — falcate — f

a) *Archontophoenix cunninghamiana*
b) *Chamaedorea seifrizii*
c) *Veitchia montgomeryana*
d) *Ptychosperma elegans*
e) *Carpentaria acuminata*
f) *Hydriastele wendlandiana*

In a pinnate leaf the segments are called pinnae or leaflets and the continuation of the peti-ole to which they are attached is the rachis. The rachis is usually straight but occasionally it is curved and imparts a distinctive appearance to the crown (*Hedyscepe canterburyana, Howea bel-moreana, Hyophorbe lagenicaulis*). Leaflets vary greatly in their shape, from long and slender (*Phoenix roebelenii*) to broad and uneven (*Aiphanes aculeata*). They may be arranged oppositely or alternately along the rachis or grouped irregularly in bundles (*Hydriastele*) and may all be in one plane (*Archontophoenix* spp.) or arranged in two or more ranks to give a layered or feathery appear-ance (*Syagrus romanzoffiana, Normanbya normanbyi, Roystonea regia*)—a condition which is described as plumose.

The leaflets may be stiff (*Phoenix canariensis*) or lax and drooping (*Gulubia costata, Howea forsteriana*). In some species the basal leaflets are reduced to stiff, sharp spines (*Phoenix* spp.). In many palms the terminal leaflets are united and may resemble a fishes' tail (*Pinanga disticha, Ptychosperma elegans*). Leaflet tips may be entire and pointed or appear as if they have been bitten off or irregularly chewed (erose or praemorse as in *Ptychosperma elegans*). In most palms the edges of the leaflets turn down and are described as being reduplicate (∧ in section). In a few

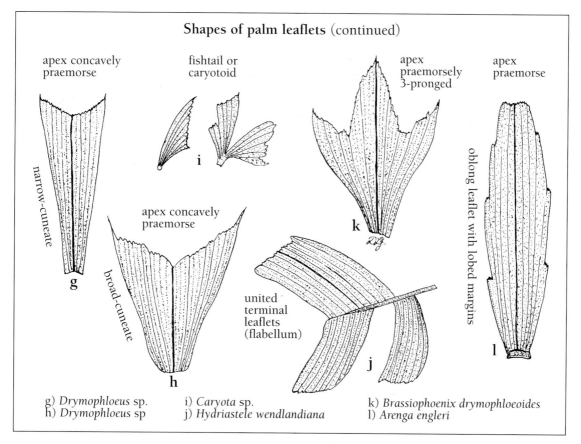

Shapes of palm leaflets (continued)

apex concavely praemorse

narrow-cuneate

g

broad-cuneate

apex concavely praemorse

h

fishtail or caryotoid

i

united terminal leaflets (flabellum)

j

apex praemorsely 3-pronged

k

apex praemorse

oblong leaflet with lobed margins

l

g) *Drymophloeus* sp.
h) *Drymophloeus* sp
i) *Caryota* sp.
j) *Hydriastele wendlandiana*
k) *Brassiophoenix drymophloeoides*
l) *Arenga engleri*

genera, of which *Phoenix* is a common example, the leaflets are folded upwards and are described as being induplicate (V in section).

In some feather-leaved palms the tips of the leaflets of new fronds are bound together by a fibrous material (*Dictyosperma*, *Neodypsis*) and this breaks away as the leaflets expand. Strips of this binding material may hang off young fronds and are termed reins.

Fan-leaved or Fan Palms

The lamina of the leaves of these palms are semi-circular, circular or paddle-shaped in outline and are divided into many segments. The whole shape resembles a partly or fully open fan and leaves with this type of division are called palmate if the segments are divided to the base (where they join the top of the petiole) or palmatifid if they are only divided part way. Palms of this group have a very distinctive appearance (see Fig. 9).

The leaf divisions of this group are known as segments and the petiole may either stop abruptly where the segments join or continue as an extension into the lamina (*Borassus flabellifer*, *Sabal palmetto*). This latter type of frond is referred to as being costapalmate and the extension from the petiole is

The distinctively shaped leaflets of *Wallichia densiflora*.

33

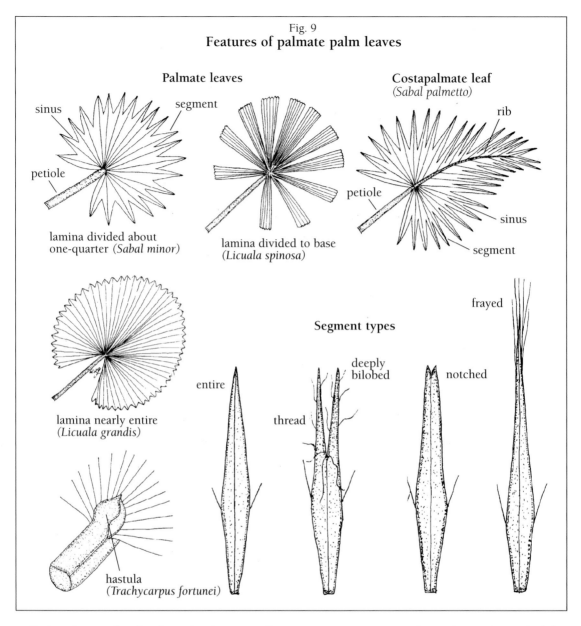

Fig. 9
Features of palmate palm leaves

Palmate leaves

sinus

segment

petiole

lamina divided about
one-quarter (*Sabal minor*)

lamina divided to base
(*Licuala spinosa*)

Costapalmate leaf
(*Sabal palmetto*)

rib

petiole

sinus

segment

lamina nearly entire
(*Licuala grandis*)

Segment types

entire

thread

deeply
bilobed

notched

frayed

hastula
(*Trachycarpus fortunei*)

called a rib. In a few fan-leaved palms a small projection juts out at right angles to the end of the petiole. This is referred to by some botanists as the ligule because of its similar position to that organ in grasses, but is more correctly named a hastula. This extension may be small and barely noticeable (*Thrinax parviflora*, *Trachycarpus fortunei*) or up to 30 cm long and prominent, as in *Copernicia rigida*.

The lamina of fan-leaved palms may be variously divided, with some species being entire and barely notched around the margins (*Licuala grandis*) and others divided for more than two-thirds of their length (*Licuala spinosa*, *Rhapis excelsa*). In the Australian Fan Palm (*Licuala ramsayi*) the juvenile fronds are deeply divided and have widely separated segments, while the mature fronds are circular with segments also deeply divided, but closely crowded together to impart an almost continuous circular or falsely peltate appearance.

The segments of fan-leaved palms may be stiff or somewhat lax, and often the segment tips may be drooping (*Livistona chinensis*, *L. decipiens*). The tips themselves may be entire, long and pointed or even thread-like (*Livistona chinensis*), emarginate (*Licuala grandis*), deeply divided

(*Brahea brandegeei, Washingtonia filifera*), or erose (*Rhapis excelsa*) (see Fig. 9).

A few fan-leaved palms produce long, cottony threads which arise at the junctions of the segments and along the margins. These palms are frequently referred to as cotton palms and the threads may be more prominent on juvenile fronds than on mature fronds. *Washingtonia filifera* is frequently called the Cotton Palm because of an abundance of these threads on the leaves.

Bipinnate or Fishtail Palms

The leaves of these palms are twice-divided or bipinnate. Though this type of frond division is prominent in ferns it is rare in palms and seems to be restricted to the fishtail palms of the genus *Caryota*. Leaves of this type have a distinctive, lacy appearance and are very easily discerned.

The ultimate leaf divisions of palms with bipinnate leaves are strictly called pinnules (but they are generally referred to as leaflets), while the primary divisions arising directly from the main rachis are called pinnae. As in feather-leaved palms the continuation of the petiole to which the pinnae are attached is called the rachis.

Entire-leaved Palms

The leaves of these palms are simple and undivided in their mature state, although they may

Cottony threads on a young, expanding frond of *Washingtonia robusta* give rise to the common name Cotton Palm.

often be damaged by the wind and then may have a pinnate appearance. Often fronds of this type are deeply notched at the apex. Examples include *Chamaedorea geonomiformis*, *Geonoma decurrens*, *Manicaria saccifera*, *Johannesteijsmannia altifrons* and *Phoenicophorium borsigianum*. Some species have similar entire fronds in their juvenile stages of development (*Verschaffeltia splendida*) while their mature fronds are pinnate.

Palms with entire leaves may in fact be primitive relics and the leaves only survive intact in very sheltered environments. It is interesting to observe that when the fronds of some of these palms shred they seem to tear at predetermined points, after which they resemble a member of the feather-leaved group. The structure of the fronds of these palms is basically similar to those of the feather-leaved group, being divided into petiole, rachis and lamina, although in this case the latter portion is entire.

THE INFLORESCENCE

Palms flower when mature but the length of time before they flower varies greatly with the species. Some dwarf palms, such as the Window Pane Palm (*Reinhardtia gracilis*) and the Parlour Palm (*Chamaedorea elegans*), may flower when three to six years old, while at the other end of the scale the Talipot Palm (*Caryota umbraculifera*) and the Double Coconut (*Lodoicea maldivica*) may not flower until they are thirty to eighty years of age.

A number of interesting palms produce one or several inflorescences from a stem which then dies. Such plants are described as being hapaxanthic. If the palm has a solitary stem, then the whole plant dies and such palms may be termed monocarpic. For example, in species of *Corypha*, vegetative growth continues for many years and culminates in the production of a massive terminal inflorescence which flowers and fruits before the whole plant dies. In the Solitary Fishtail Palm

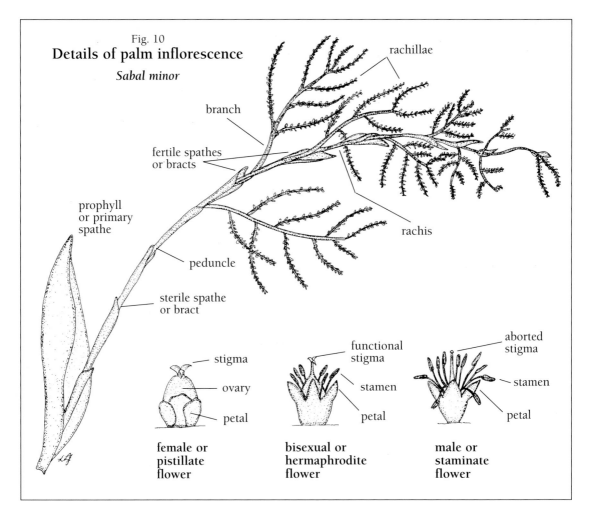

Fig. 10
Details of palm inflorescence
Sabal minor

rachillae

branch

fertile spathes
or bracts

prophyll
or primary
spathe

rachis

peduncle

sterile spathe
or bract

stigma

ovary

petal

**female or
pistillate
flower**

functional
stigma

stamen

petal

**bisexual or
hermaphrodite
flower**

aborted
stigma

stamen

petal

**male or
staminate
flower**

(*Caryota urens*) and the Sugar Palm (*Arenga pinnata*) the plant begins flowering from among the topmost leaves when it has reached maturity. Flowering continues progressively down the trunk and when the lowermost inflorescence has matured its fruit, the plant dies. If the palm is a clustering species like *Caryota mitis*, *Metroxylon sagu* and *Raphia farinifera*, then the life of each plant is prolonged by suckers even though individual stems will die. Hapaxanthic palms are a minority within the group and most palms continue to produce flowers annually (termed pleonanthic) when conditions are favourable for them.

The vast majority of palms bear an axillary inflorescence. The position of this inflorescence on the palm is useful in grouping them. Many species, especially those with a crownshaft, carry their inflorescences below the crownshaft and this condition is described as being subfoliar or infrafoliar (*Archontophoenix* spp., *Rhopalostylis* spp.). When the inflorescence arises among the leaves it is referred to as being interfoliar (*Phoenix* spp., *Sabal* spp., *Washingtonia* spp.) and in the rare cases in which it is carried above the leaves it is superior or suprafoliar (*Corypha* spp., *Metroxylon sagu*). In these latter types the inflorescence is not axillary but terminal on the main growing axis. A few small palms produce inflorescences which arise at or below ground level and are described as being radical (*Chamaedorea radicalis*, *Geonoma procumbens*). Some palms produce only a single inflorescence each year but most produce three to five. In *Gaussia maya* up to fifteen inflorescences at various stages of development may be present on the trunk at the same time. The length of the inflorescence is often of significance in taxonomy and in palms of the interfoliar type it is useful to note whether the inflorescence exceeds the length of the leaves or is encompassed by them (see Fig. 13).

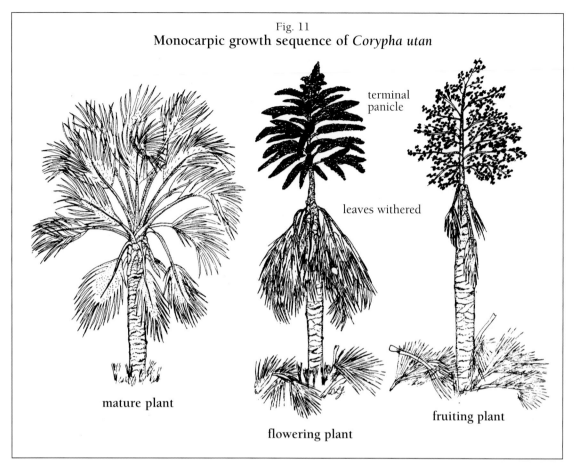

Fig. 11
Monocarpic growth sequence of *Corypha utan*

terminal panicle

leaves withered

mature plant

flowering plant

fruiting plant

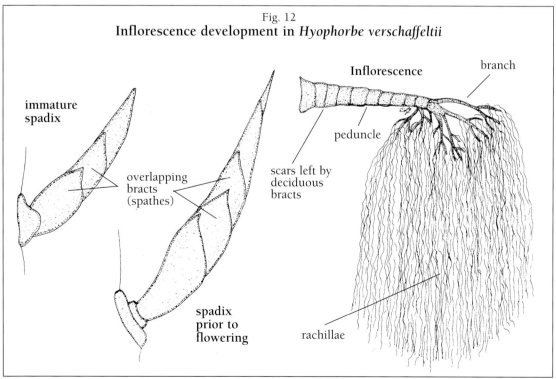

Fig. 12
Inflorescence development in *Hyophorbe verschaffeltii*

immature spadix

overlapping bracts (spathes)

spadix prior to flowering

Inflorescence

branch

peduncle

scars left by deciduous bracts

rachillae

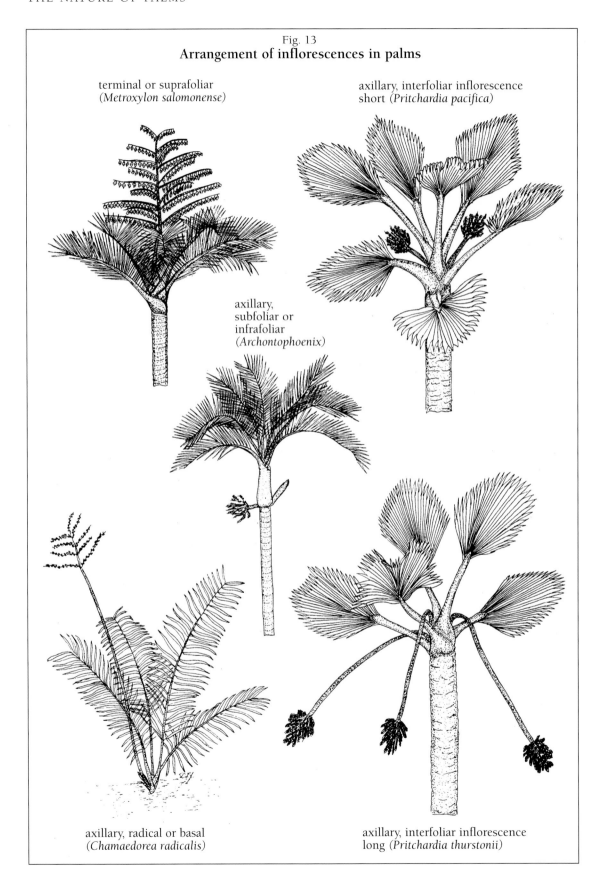

Fig. 13
Arrangement of inflorescences in palms

terminal or suprafoliar
(*Metroxylon salomonense*)

axillary, interfoliar inflorescence
short (*Pritchardia pacifica*)

axillary,
subfoliar or
infrafoliar
(*Archontophoenix*)

axillary, radical or basal
(*Chamaedorea radicalis*)

axillary, interfoliar inflorescence
long (*Pritchardia thurstonii*)

The palm inflorescence is fairly complex and contains many important diagnostic features useful for taxonomy. The whole inflorescence is called a spadix and when young is completely enclosed in one or more sheathing bracts which are called spathes. These are actually modified leaves and their number and arrangement (which can be very complex) are important diagnostic features. The spathes may be large or small and may be persistent or deciduous, falling as the inflorescence emerges. The large basal spathe which encloses all of the young inflorescence is termed a prophyll. Those spathes which cover the main stalk supporting the inflorescence are called sterile spathes, while fertile spathes are those which enclose an inflorescence branch (see Fig. 10).

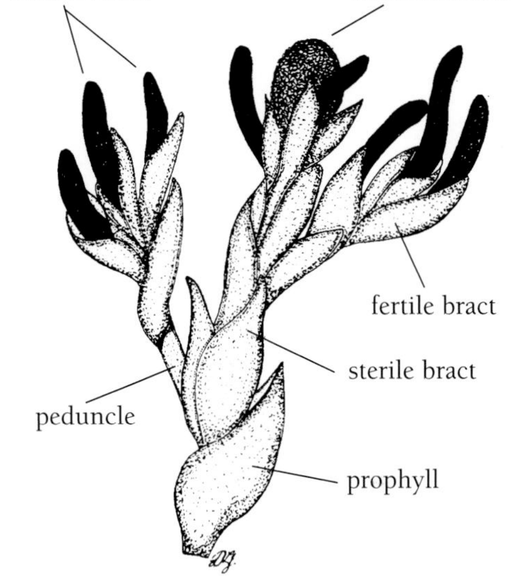

The main stalk of the inflorescence is known as the peduncle and it may be woody, thick and fleshy or thin and fairly wiry. In most palms the inflorescence is a much-branched structure called a panicle and the secondary and higher order branches are referred to as rachillae. The branching varies from sparse and simple (*Chamaedorea, Rhapis, Reinhardtia*) to complex (*Brahea, Corypha*). Rarely is the inflorescence an unbranched solitary spike (*Calyptrocalyx spicatus, Howea belmoreana*) (see Fig. 15). The curious inflorescence of *Raphia* spp. is a twice-branched structure up to 4 m long and 15 cm thick that has variously been likened to a sausage or stuffed stocking.

Fig. 14
The unusual inflorescence of
Nypa fruticans

heads of male flowers

head of female flowers

fertile bract

sterile bract

peduncle

prophyll

FLOWERS

Individual palm flowers are small and fairly inconspicuous, but are carried in profusion along the rachillae and may be stalked or sessile and often embedded in the tissue of the rachilla itself. One enterprising worker has even estimated that a single inflorescence of *Corypha* may contain several million individual flowers. Palm flowers vary in colour but those of most species are greenish to creamy-white. Those of *Archontophoenix cunninghamiana* are lilac-mauve, while some species are bright yellow (*Nypa fruticans, Polyandrococos caudescens*), orange (*Arenga engleri*) or reddish (*Acanthophoenix rubra*). The flowers are often strongly fragrant (*Areca catechu, Coccothrinax fragrans, Hyophorbe verschaffeltii, Lytocaryum insigne*) and many have an unpleasant scent (*Arenga pinnata*). Some palms are wind-pollinated (*Phoenix* spp.) while others attract insects such as beetles (*Salacca zalacca*) or flies and honey bees (*Cocos nucifera*) to carry out the necessary cross-pollination. Palm flowers are generally short-lived, seldom lasting more than one day and often less.

Individual palm flowers may be bisexual (or hermaphrodite)—having both male and female sexual parts present; or unisexual—having only one sexual

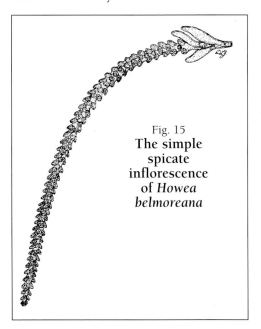

Fig. 15
The simple spicate inflorescence of *Howea belmoreana*

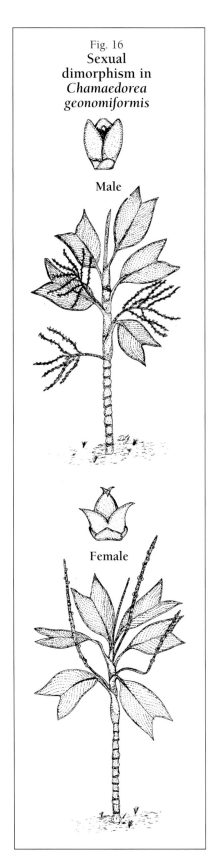

Fig. 16
Sexual dimorphism in _Chamaedorea geonomiformis_

Male

Female

part functional. Unisexual flowers will be male if they have functional stamens and aborted ovary and stigma (pistillode) and female if they have aborted stamens (staminodes) and a functional ovary and stigma. The distribution of each flower type may be uniform with the whole inflorescence consisting of one type, or there may be various combinations on the inflorescence. Unisexual flowers may be arranged separately on special parts of the inflorescence or on separate inflorescences on the same plant. Frequently, unisexual flowers are arranged on an inflorescence in groups of three (one female between two males). When the unisexual flowers are separate on the one plant, this type of plant is described as being monoecious, and if the unisexual flowers are carried on separate plants they are described as being dioecious. Some species of palm have both unisexual and bisexual flowers present on the same plant and are described as being polygamous.

Palm flowers are made up of a perianth of three sepals and three petals, either of which may be free or joined in the basal half. In unisexual flowers the petals of the females tend to overlap and remain partially closed, while in the male flowers they become widely separated with the stamens obvious. Bisexual flowers generally open widely. The stamens are borne in whorls and vary in number from six to more than fifty (_Veitchia_). The ovary is trilocular and superior and each locule contains one ovule. Unisexual male flowers retain a rudimentary but non-functional ovary (see Fig. 10).

FRUITS AND SEEDS

Palm fruits—which are botanically either a drupe, such as _Borassus_ and _Sabal,_ or single-seeded berries as in _Metroxylon_ and _Phoenix_—are often a very decorative feature of the plants. They may be conspicuous because of their size (_Borassus, Cocos_), their bright colours (orange—_Polyandrococos_; white—_Thrinax_; blue—_Trachycarpus_; black — _Livistona_; red—_Archontophoenix_) or their production in huge quantities (_Syagrus romanzoffiana, Phoenix_ spp.). The surface may be smooth or delicately patterned with symmetrical or geometrical scales (_Calamus_ spp., _Mauritiella setigera_), a roughened, compound clump (_Nypa fruticans_) or spiny (_Astrocaryum mexicanum_). The lobed fruit of _Manicaria_ spp. are covered with large tubercles.

Fruits of some maritime palms have a built-in flotation system and are distributed by the currents of the oceans. Chief among these is the Coconut, but others that may be distributed by this means are _Pseudophoenix sargentii,_ _Manicaria saccifera_ and the Mangrove Palm (_Nypa fruticans_). Infertile fruits of the Double Coconut (_Lodoicea maldivica_) float, but the fertile ones sink and hence the sea is not important for the distribution of this palm.

Many palm fruits are succulent and edible (see _Edible Palm Products_, p. 50) but a number contain toxic materials.

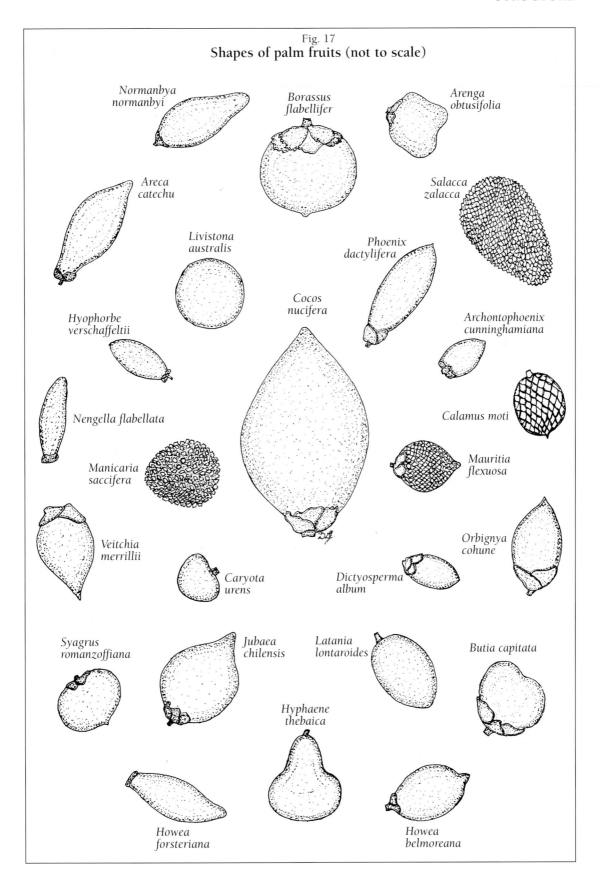

Fig. 17
Shapes of palm fruits (not to scale)

Normanbya normanbyi

Borassus flabellifer

Arenga obtusifolia

Areca catechu

Salacca zalacca

Livistona australis

Phoenix dactylifera

Hyophorbe verschaffeltii

Cocos nucifera

Archontophoenix cunninghamiana

Nengella flabellata

Calamus moti

Manicaria saccifera

Mauritia flexuosa

Veitchia merrillii

Orbignya cohune

Caryota urens

Dictyosperma album

Syagrus romanzoffiana

Jubaea chilensis

Latania lontaroides

Butia capitata

Hyphaene thebaica

Howea forsteriana

Howea belmoreana

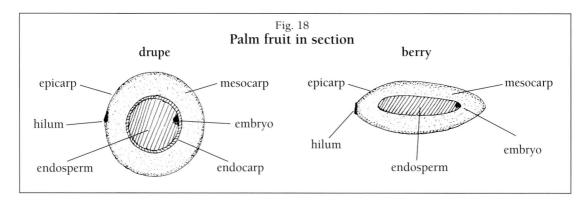

Fig. 18
Palm fruit in section

drupe — epicarp, mesocarp, hilum, embryo, endosperm, endocarp

berry — epicarp, mesocarp, hilum, endosperm, embryo

Some fruits contain caustic material (calcium oxalate) in the skin and pulp which causes severe burning or irritation if eaten or handled.

Extremes in size in the fruits of palms can be illustrated by comparing those of the Double Coconut (*Lodoicea maldivica*) which may weigh up to 20 kg with those of a *Chamaedorea* which may weigh but a few grams. Many palm fruits are rounded (*Sabal* spp.), but some are ovoid (*Archontophoenix*), pear-shaped (*Cryosophila warscewiczii*) or even sickle-shaped (*Chamaedorea falcifera*). Most contain only one seed but some species may have up to three seeds in a fruit (*Arenga, Borassus, Manicaria, Orania*), or less commonly up to seven (*Orbignya*).

Fruit and Seed Structure

Botanically, the drupe (which is the fruit of most palms) consists of a thin outer shell known as the epicarp, a fleshy layer of variable thickness known as the mesocarp and an inner hard layer which is attached to the seed, known as the endocarp. Most of the seed consists of endosperm

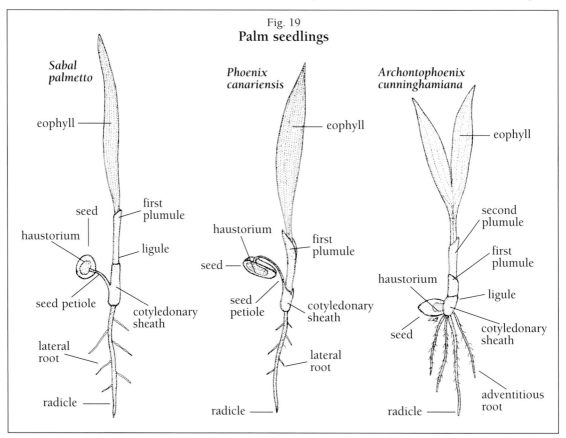

Fig. 19
Palm seedlings

Sabal palmetto — eophyll, seed, haustorium, first plumule, ligule, seed petiole, cotyledonary sheath, lateral root, radicle

Phoenix canariensis — eophyll, haustorium, seed, first plumule, seed petiole, cotyledonary sheath, lateral root, radicle

Archontophoenix cunninghamiana — eophyll, second plumule, first plumule, haustorium, ligule, seed, cotyledonary sheath, adventitious root, radicle

which is a cellulose, non-nitrogenous storage material, sometimes referred to as albumen. The embryo is embedded in the endosperm usually at one end. A berry is a similar fruit but lacks an endocarp (see Fig. 18).

Seed Germination

Germination of palm seeds occurs when the enzymes mobilise the food resources in the endosperm. Palm seedlings are characteristically slow to appear above ground but below-ground activity can occur quite rapidly. Palms have a single cotyledon but this structure and others in palm seedlings is very different from comparable structures in other commonly grown monocotyledons.

The cotyledon of palms is tubular and remains buried. Part of it closest to the endosperm expands into a structure known as a haustorium. This releases enzymes into the endosperm and absorbs materials from the endosperm which are needed for growth. At the other end the cotyledon elongates into a tubular structure known as the cotyledonary sheath. The first green leaf which expands into sight is known as the eophyll and this grows through the tubular cotyledon. Between the eophyll and the cotyledon are one to three scale-like or bract-like structures which are in fact modified plumules or primary shoots (termed scale leaves). The upper part of the cotyledonary sheath is sometimes called the ligule.

The primary root or radicle is the first root to appear on the seedling. It is usually thick, fleshy, white and very prominent. Appearances are deceptive, however, as the primary root is incapable of increasing in thickness and is therefore very short-lived in monocotyledons. A few lateral roots may grow from the primary root and these are of importance in aiding the early establishment of the seedling. Adventitious roots arise from the base of the cotyledonary sheath and quickly replace the radicle root system in importance.

The above notes on germination are basic to most palms, but some groups have variations on this theme. The basic types are illustrated in the accompanying drawing (Fig. 19). One interesting feature to note is the presence of an extension between the haustorium and the cotyledonary sheath that is known as the seed petiole.

CLIMBING PALMS

A specialised group of palms has adopted the climbing mode of growth. In so doing, they have not only occupied an important ecological niche but have also introduced some adaptive modifications to their habit of growth. As there are about 600 species of climbing palms (including about 350 in the genus *Calamus*), they are obviously a very important group and worthy of separate consideration. The fifteen genera containing clumping palms are listed in Table 3, but it should be realised that not all species in each genus are climbers, although the majority do have this habit.

Most climbing palms grow in clumps, with the new growths being produced at intervals from a basal rhizome. A few, however, are solitary and have a single climbing stem. Their slender stems, which are more frequently termed canes, are erect at first but may wander over the rainforest floor until they find some suitable support on which to climb. These canes have very long internodes, are of uniform width and are very hard, like bamboo. When young they are protected by the sheathing bases of the leaves, but these are shed on old stems to expose smooth lengths of cane (rattan).

All climbing palms have pinnate fronds scattered along the stems rather than in a dense crown. These spreading fronds may aid climbing by catching in surrounding vegetation. Often the rachis has curved hooks which further aid in climbing. In *Chamaedorea elatior* the terminal leaflets are reflexed and catch in surrounding vegetation.

Nearly all climbing palms are ferociously armed with recurved hooks or whorls of spreading spines, these being liberally distributed on the leaf sheaths, petioles, rachises and along the inflorescence. These unfriendly projections help the stems to climb into the surrounding forest and discourage grazing by herbivores. Further specialised aids are provided in the form of long filamentous extensions which are liberally armed with recurved hooks. These blow about in the wind and catch in surrounding vegetation. The extensions are of two types depending on their origin. Those projecting from the apex of the leaf rachis are called cirri, while flagella are sterile inflorescence developments that arise in the axils of the leaves. The inflorescences of climbing palms also frequently act as climbing aids, since they are long, pendulous, sparsely branched and are armed with recurved hooks.

Stems of most climbing palms seem almost capable of indeterminate growth and individual stem lengths of over 200 m have been recorded. In many countries these stems are harvested as rattan, the tough, pliant material valued for furniture construction (see *Furniture*, p. 62). Climbing palms of the genus *Plectocomia* and *Plectocomiopsis* are hapaxanthic, the whole plant or individual stems dying completely after fruiting. These climbing palms may have a synchronous gregarious flowering habit with all plants in one area flowering (and therefore dying) simultaneously.

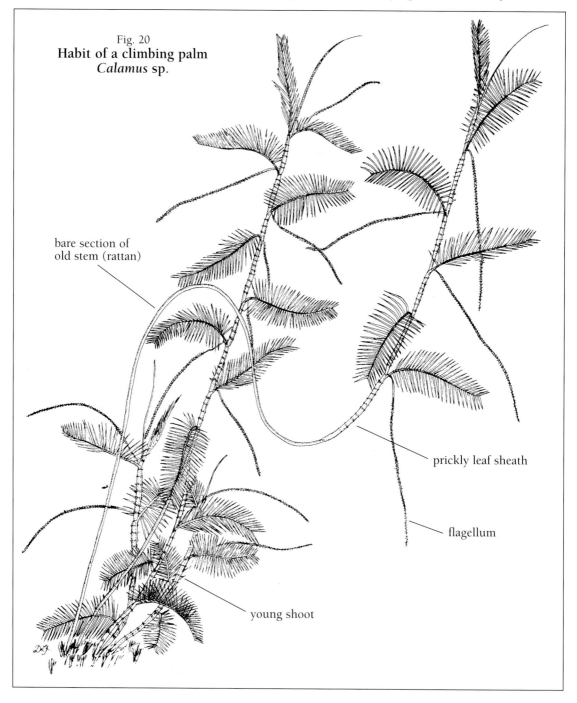

Fig. 20
Habit of a climbing palm
Calamus sp.

bare section of
old stem (rattan)

prickly leaf sheath

flagellum

young shoot

Fig. 21
Features of some climbing palms

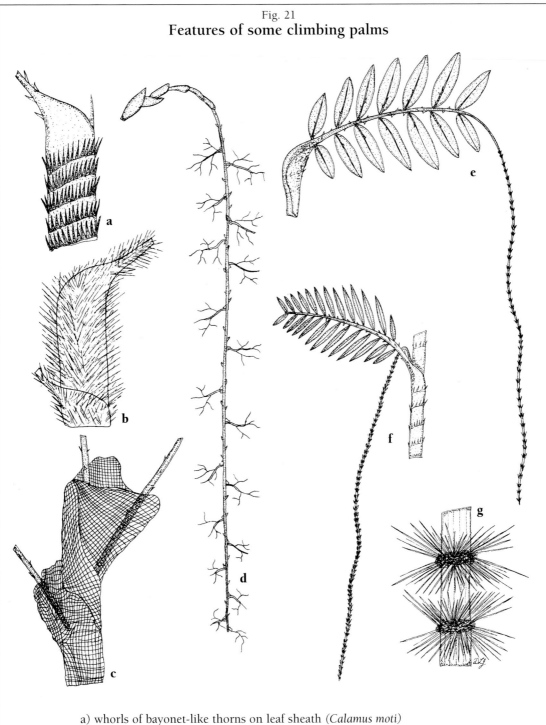

a) whorls of bayonet-like thorns on leaf sheath (*Calamus moti*)

b) numerous spines on stems, sheaths and petioles (*Calamus radicalis*)

c) net-like ochrea on leaf sheath (*Korthalsia jala*)

d) elongated, narrow, pendulous panicle armed with thorns (*Calamus* spp.)

e) whip-like rachis extension called a cirrus (*Calamus hollrungii*)

f) whip-like, sterile inflorescence called a flagellum (*Calamus australis*)

g) whorls of long spines which may be ant-inhabited (*Daemonorops formicaria*)

TABLE 3
Palm genera with a climbing growth habit

GENUS	NUMBER OF SPECIES	AREA OR COUNTRY
Calamus	370	Africa–India–Borneo–New Guinea
Calospatha	1	Malaysia
Ceratolobus	6	Malaysia–Indonesia
Chamaedorea	1	Mexico
Daemonorops	115	India–Malaysia–New Guinea
Desmoncus	65	South America
Eremospatha	12	Tropical Africa
Korthalsia	26	India–Malaysia–New Guinea
Laccosperma	7	Tropical Africa
Myrialepis	1	Thailand–Malaysia–Sumatra
Oncocalamus	6	Tropical Africa
Plectocomia	16	India–China–Malaysia–Philippines–Indonesia
Plectocomiopsis	5	Thailand–Malaysia–Borneo
Pogonotium	3	Indonesia–Malaysia
Retispatha	1	Borneo

CHAPTER 3

THE ECONOMIC IMPORTANCE OF PALMS

In the West, palms are an ornamental plant most familiar in tropical forests and gardens and elsewhere popular for indoor decoration. These same plants, however, contribute significantly to the economies of many countries and are of prime importance in the daily lives of millions of people. The contributions these plants make to the world's economy, local economies and lifestyles is quite amazing and appreciated by very few people.

A surprisingly large number of products can be obtained from palms. For instance, the various parts of the widespread Coconut Palm can be used in more than 1000 ways and the Palmyra Palm in over 800 ways. The Date Palm is not such a prolific relative by comparison and yet this species helps to keep millions of people alive by its tremendous production of energy-rich dates in a climate where little else will grow. Numerous other species of palm are also of economic significance and contribute to human survival in various ways. Thus palms provide shelter, food and drink, clothing, fuel, fibre and medicine. A sample of these uses is presented in the following pages.

The unfortunate aspect of people's use of plants such as palms is that it frequently leads to overexploitation of wild populations. Many observers have noted the absence or rarity of certain palms around native villages in various parts of the world. More significant is the destruction which results from commercial exploitation of palms for delicacies such as palm hearts. Who knows what damage the continuous collection of fruit, palm-cabbage or leaves will have on wild populations. Even if the species survives, valuable genotypes may be lost through destruction in certain climatic zones or specific habitats. Only in very few countries are cultivated plantations established to ease the pressure on the wild.

THREE FAMOUS PALMS
The Coconut Palm

The Coconut Palm (*Cocos nucifera*) has been cultivated for thousands of years and has been so widely distributed by humans that today its centre of origin is uncertain and the subject of much speculation. The nuts, which are an excellent source of food and drink, must have been prized by the early travellers since they are a self-contained unit that does not leak, dry out

Palm tapper with buckets made from the leaf of *Borassus flabellifer*, Madura, Indonesia. J. Dransfield

or become tainted with salt water. Travellers must have spread the word widely about such a useful fruit and also more than likely helped to establish the species in new areas which have a suitable climate for its growth. (For a treatise on this palm see H.C. Harries, 'The evolution, dissemination and classification of *Cocos nucifera* L.', *Botanical Review* 44 (1978): 265–319.)

Although its centre of origin is believed to have been the islands of the western Pacific, the Coconut itself is not the type of plant to remain static as it is a coloniser. The nuts float for long distances on ocean currents and if they are washed ashore on a beach with a suitable climate, plants will become established and colonise the area. The original Coconut is believed to be the tall-growing variety with a thick husk. Plants of such a wild type occur naturally in the Philippines and north-eastern Australia. Superior varieties with larger fruit, a larger quantity of water or a dwarf growth habit are the result of selection by humans.

The Coconut is regarded as the jewel of the tropics and it is undoubtedly the most economically important palm (often called the Tree of Life) and indeed is one of the ten most important tree crops. More than thirteen million people are directly or indirectly involved with Coconut products and Coconuts are now widely grown in the tropical regions of the world. World production of Coconuts is more than 3000 hectares. The major producers are the Philippines, India and Indonesia. Coconut Palms produce products which are of major economic importance to many small countries, particularly in the Pacific region. In the South Pacific, for example, it is estimated that about 400 000 people depend entirely on the Coconut Palm for their livelihood. About a quarter of the export earnings of the Philippines are derived from the Coconut and its products.

Coconut Palms are very easy to grow but need a specific regime for successful fruit production. They are essentially tropical in their requirements and need a nearly uniform temperature range throughout the year and a good, regular, annual rainfall (mean temperature of approximately 20°C and 1000 mm rainfall per annum). Coconuts will grow in colder areas but rarely produce fruit and may be killed by sudden cold snaps. They are widely planted as ornamentals in subtropical areas where their growth is slow and fruit production erratic or absent.

Coconuts need well-drained soil but plants will really thrive and grow rapidly if their roots can tap ground water. This is especially applicable in coastal districts where the soils are usually deep and sandy and dry out between periods of rain. Coconuts will grow in other well-drained soils and contrary to popular belief do not have to be planted 'within sight of the sea'. If the climate is suitable and they are not crowded by other plants they will grow and fruit some considerable distance inland, again especially if their roots can tap ground water.

Because of their importance to humans and because Coconut Palms have been grown for such a long period, numerous horticultural cultivars have been selected and are propagated by seed. A famous cultivar once widely planted in the West Indies is known as 'Jamaican Tall'. On the island of Fiji there are no less than seven different tall cultivars and a further three that are regarded as dwarf-growing cultivars. Within each group the cultivars are mainly distinguished by nut size and colour when ripe. Some produce very large individual nuts but in smaller numbers and their total yield per plant is not as heavy as that of other smaller-fruited cultivars. Some cultivars have a thin, fibrous covering (husk) around the fruit which is easy to remove; in others the husk is very thick and difficult to dislodge. Estimates of individual yields of Coconut cultivars vary somewhat from 75–350 nuts per annum.

The range of uses to which various parts of the Coconut Palm can be put is truly amazing. Every part is used in some way and even when the trees are blown over or felled for some reason, the trunk, the fibrous pith and the apical bud (cabbage) are all used.

- The meat of the nut can be cooked in more than a hundred different dishes or eaten raw or shredded, or dried and exported as desiccated coconut
- the water, especially that of a green coconut, makes a refreshing drink or it can be blended with the meat to make coconut milk, cream or jam
- the shells can be burnt, made into charcoal, or used as bowls, scoops or cups
- the fibre can be made into mats or woven into string and ropes (see *Palm Fibres*, p. 58) or shredded and made into potting mix for nursery plants

- the leaves can be used for thatching and the leaflets for weaving
- the wood for building
- the flowers for palm honey and the sap of the inflorescence for palm sugar, alcohol and the potent drink arrack (see *Palm Beverages*, p. 51)
- coconut oil can be extracted from the kernels and the dried kernels (copra) are an important component of this process as they can be stored and transported (see *Palm Oils*, p. 60)
- after oil extraction the remaining meaty material can be pressed into a cake suitable for animal feed
- coconuts are sold to tourists either whole or with designs carved into the husk
- the water of green coconuts has growth-regulatory properties and is added to the growing media used in tissue culture and for raising orchid seedlings.

Coconut Palms are very sensitive to the disease known as Lethal Yellowing. This disease is caused by a mycoplasma-like organism and results in rapid death of infected trees. The disease is prevalent in the Caribbean and is also well established in Florida and Texas. Tall Coconut cultivars are extremely sensitive and have been virtually eliminated from areas where the disease has become established. Fortunately, cultivars of Malayan Dwarf Coconuts are nearly completely resistant to this disease. Lethal Yellowing continues to spread and poses a tremendous threat to numerous Pacific nations which rely so heavily on the Coconut Palm for their survival.

The Date Palm

The fruit of the Date Palm (*Phoenix dactylifera*) is of singular importance since it is a staple part of the diet of millions of people. Dates are very low in proteins and fats but are energy rich, with some cultivars having a sugar (sucrose) content of up to sixty per cent. As well as being eaten fresh they can be successfully dried and in this state can be stored for considerable periods. Dried dates are the principal agricultural export of many Middle Eastern countries. Iraq, Iran and Egypt produce more than half of the world's crop of about two million tonnes and Dates are also grown commercially in the Canary Islands, India, Pakistan and in two states of the USA—Arizona and California. Dried dates are also used as animal food.

Date Palms can be successfully grown in many areas of the world from temperate to tropical regions, but the climatic zones where they will successfully fruit are much more restricted. They need sunny, warm, dry climates (low rainfall, low humidity) and the centres of commercial production are restricted to the areas between the latitudes $15°$ and $30°$ north. The hotter and drier the climate the higher the yield and the better the quality of the fruit. Date Palms grow well outside of these zones but may never flower, let alone produce fruit. In many subtropical regions the flowering and fruiting is upset by high humidity or rainfall.

Date Palms thrive best where there is ground water and where this is lacking they require irrigation. Propagation is mostly by removing suckers from established plants but more recently high-yielding cultivars have been successfully propagated by tissue culture. This technique has resulted in a dramatic increase in the amount of planting material available for the establishment of new plantations. The plants themselves are dioecious and so both male and female plants must be established. Pollen transference is achieved by wind with huge quantities of pollen being released early in the day. This pollen can be collected and will retain its viability for up to twelve months if stored properly. The female flowers can be hand pollinated using collected pollen (or applied via dusting machines) and this greatly increases the percentage of female trees which can be established in a plantation and thus the yield. Normally one to five per cent of the trees must be male to ensure cross pollination. The pollen variety used for pollination has a marked influence on the size, weight, shape and ripening time of the fruit and also influences seed characters. Removal of excess sucker growths, fertilising, thinning fruit to increase quality and reduce biennial bearing, and covering fruit to reduce sunburn are some beneficial cultural practices used in date production.

Dates are an extremely ancient crop, having been cultivated since about 4000 BC. The species is unknown in the wild but is believed to have originated in north Africa. Hundreds of different cultivars are known, these being suited to different situations, soil types, microclimates or the production of fruit of different size, colour and flavour. The plants themselves are very long-lived and are known to bear annually for well over 100 years. At each crop they may bear bunches of fruit weighing about 50 kg and containing 3000–5000 individual fruits. The bunches are cut when a reasonable percentage of the fruits are ripe and the remainder will continue to ripen. Quality dates to be eaten fresh are selected and picked by hand.

The Palmyra Palm

The Palmyra Palm (*Borassus flabellifer*) is widely cultivated in tropical parts of Asia for the numerous uses to which parts of it can be put. In fact it is second only to the Coconut in number of plants in the world. As many of its products are important dietary items in countries where food is scarce, this adds further to its reputation and importance. The species is native to southern India, northern Sri Lanka, Burma, Thailand, Malaysia and Indonesia. (For a comprehensive treatment of the palm, see J.F. Morton, 'Notes on Distribution, Propagation and Products of Borassus Palms (Arecaceae)', *Economic Botany* 42 (3) (1988): 420–41.)

Palmyra Palms are tall, very stately plants which usually grow in large colonies. The plants themselves have stout leaves which can be used in many different ways for paper, shelter and weaving and the petiole for the production of fibre. The trunk provides heavy, strong timber, sago and cabbage; the fruit and seeds are edible (the pulp of juvenile fruits is a delicacy) and the inflorescence can be tapped for sap which can be used for sugar, wine, alcohol or vinegar. Even the swollen roots can be eaten in times of famine. The jelly-like kernel of immature seeds has a sweet taste and is canned—whole or sliced—and exported from Thailand. The plant also has an appealing appearance and is widely grown for its ornamental features.

The fruit of the Palmyra Palm is produced in abundance (200–350 per tree each year) and is of considerable significance as a potential source of starch. While the pulpy, orange flesh of just-ripe fruit can be eaten raw or cooked or made into jam, the flesh of very ripe fruit loses its consistency and becomes dry and hard. The seeds are collected, grown on specially prepared mounds of soil and harvested when the sinker (actually the first bladeless juvenile leaf) becomes fleshy and tuberous. This sinker is regarded as a delicacy and is very nutritious, being rich in starch. Because of their prolific fruit production Palmyra Palms are obviously an underexploited source of starch. It has been estimated that a hectare of mature palms (2500 trees) could produce up to 750 000 fruit per annum. If these were converted into edible sinkers the yield of nutritious starch would be quite considerable. Palmyra Palms have numerous other minor uses. The hard, white kernel of mature seeds is made into buttons; seed shells are used for ceremonial bowls or burned as fuel; the leaves are made into fans, watertight buckets and caps; the fibres may be fashioned into rope or twine; the cabbage of old plants is harvested before they die; and the young roots have diuretic properties.

EDIBLE PALM PRODUCTS
Sago

Sago is an edible, starchy material obtained from the central pith of the trunks of many species of palm. Starch is a natural storage reserve of many plant species and is laid down in good seasons or during certain growth cycles to be drawn on by the plant in times of energy need, such as during spurts of growth, flowering and fruiting. Sago varies tremendously in its consistency, depending on the species of palm, from an almost floury substance to a coarse, granular material. It is frequently found embedded among fibres and must be scraped clean or freed by water. In all cases the collection of sago is from the trunk and involves the complete destruction of the palm or the removal of a whole trunk from a clumping species.

Sago yields may be quite substantial (more than 600 kg from a trunk) and the material is a primary source of carbohydrate for millions of people in Malesia and Asia. It is also a minor revenue earner because high-quality sago is exported to Europe by a number of countries, especially Indonesia and Malaysia. Sago can be cooked into various dishes such as meals, gruels or puddings and is also made into bread. Sago is rich in carbohydrate but low in protein and is usually supplemented with other food such as meat or fish.

Starch-rich palm trunks may also be used in a novel way to obtain protein. In Indonesia and in the Amazon region of South America, Sago Palms and other species may be deliberately chopped down and left to decay. The rotting pith makes an ideal food source for the larvae of large moths and beetles. When these grubs are sufficiently large, they are collected and eaten, providing a useful source of protein.

The most important commercial source of sago is from the various species and cultivars of Sago Palms belonging to the genus *Metroxylon*. One species (*M. sagu*) is of particular importance and it occurs in New Guinea and the Moluccas but is widely planted further afield. All species of this genus are large, handsome, pinnate palms.

Cultivars and variants of the Sago Palm (*Metroxylon sagu*) are commonly planted for the production of sago, whether it be as a local source for small villages or for the commercial production of a high-quality product. Numerous cultivars of Sago Palm have been selected by growers and their taxonomy is complex. Spineless cultivars of *M. sagu* were previously known as *M. laeve* and the spiny forms as *M. rumphii*. Plants of *M. sagu* commonly form colonies in marshy, sometimes brackish ground and have stout trunks 5–10 m tall with a few semi-dormant basal suckers. At maturity (fifteen to thirty years of age) a huge terminal panicle 4–5 m long is produced. The trunk dies after maturing the fruit and is replaced by a developing sucker. The best sago is obtained if the trunk is felled just as the inflorescence begins to emerge. The trunk is then split or cut into sections and the pith removed and pounded in water until the starch separates. This starch can be used in liquid form or dried into a flour or granular sago. Yields are somewhat variable but one palm may contain as much as 600 kg of sago in its trunk.

Two genera of monocarpic palms other than *Metroxylon* are also important in the production of sago, although these are primarily of local significance. The trunks of various species of *Caryota* (commonly called Fishtail Palms) contain edible starchy pith. That of *Caryota urens* from India and Malaysia is of very high quality, while the sago of *Caryota obtusa*, also from India, is of lesser quality and is used mainly for bread and gruel. The trunk of the Talipot Palm (*Corypha umbraculifera*) from Sri Lanka and India also yields a sago-like starch which is called Talipot. This palm is now widely cultivated in tropical countries but for uses other than sago. *Corypha utan* in Malaysia yields a kind of sago that is used locally. Species of *Eugeissona* may be locally important for sago, especially in Borneo.

Many other palms also contain starch which may be harvested as food, but often this is collected only in times of scarcity. The Mangrove Palm (*Nypa fruticans*) is one such example, as are also *Borassus flabellifer*, the Palmyra Palm, *Roystonea oleracea*, the Feathery Cabbage Palm, and *Sabal minor* of some southern states of the USA. The Gomuti or Sugar Palm of Malaysia (*Arenga pinnata*) produces sago with an unusual or even peculiar flavour that may be an acquired taste. A couple of palms produce a very floury sago which may be baked into a bread-like substance. *Phoenix pusilla* from India and Sri Lanka is sometimes called the Flour Palm for this reason. The Ita Palm, *Mauritia flexuosa*, from South America also has a very similar, floury pith. The sago of *Syagrus coronata* from Brazil is first ground to flour and then made into bread.

Palm Beverages

Drinks can be concocted from the sap obtained from various parts of some palms and, as well as being thirst-quenching, they are nutritious and energy-rich since they have a high sugar content and also contain minerals and some protein. The beverages can be drunk fresh or allowed to ferment for various periods to produce wines and even distilled further to produce spirits. Palm sap is generally termed toddy but this name may also be used for palm wine. Palm spirits may be generally referred to as arrack (or arak), although it is not uncommon for this term to be restricted to the spirit obtained from the sap of the Coconut Palm (*Cocos nucifera*).

The sap of the Coconut is tapped by either cutting off the inflorescence when it is expanding or binding and massaging the tip as detailed in *Palm Sugar*, p. 52. It is fermented to form wine

---

TABLE 4
Palms yielding sap used for wine production

SPECIES	COUNTRY	SPECIES	COUNTRY
Arenga pinnata	India, Malaysia	*Hyphaene coriacea*	Africa
Borassus aethiopum	Africa	*Mauritia flexuosa*	South America
Borassus flabellifer	India	*Nypa fruticans*	Philippines
Borassus madagascariensis	Madagascar	*Phoenix dactylifera*	North Africa
Caryota urens	India	*Phoenix reclinata*	Africa
Cocos nucifera	Polynesia	*Phoenix sylvestris*	India
Elaeis guineensis	Africa	*Raphia taedigera*	Central and S. America
Hyphaene compressa	Africa	*Raphia vinifera*	Nigeria

---

or distilled to further increase the alcohol content and form potent arrack. The Polynesians also prepare an intoxicating drink, known as kava-kava, from fermented coconut milk and the roots of *Piper methystichum*.

A number of other palms, both prominent and less well known, also yield beverages. In all cases the sap is collected as described in *Palm Sugar*, below and then either drunk fresh or fermented to form wine. The fresh sap of the Palmyra Palm (*Borassus flabellifer*) provides the pleasant drink known as sura, while saguir is the fresh drink made from the sap of *Arenga pinnata*.

Examples of palms used in various parts of the world for the production of palm wine are given in Table 4.

In certain palms the sweet sap is stored within the trunk itself and its recovery requires procedures much more drastic than the simple removal of an inflorescence. The stout trunks of the Chilean Wine Palm (*Jubaea chilensis*) yield up to 400 litres of sugary sap which can be fermented into wine, but the stately palm must first be felled. The crown is removed and the sap may flow from the upper trunk for several months. Unfortunately *Jubaea chilensis* is now a rare and endangered palm in its natural state because of overexploitation for its sugary sap. Another wine palm of South America, *Scheelea butyracea*, is also cut down and a cavity excavated in the trunk. This fills with a creamy or yellowish, sweet syrup which can be drunk fresh or fermented. The sweet sap from the Cherry Palm of Cuba, Haiti and San Domingo (*Pseudophoenix vinifera*) is removed by tapping the conspicuous bulge in the trunk and fermented to make wine. This method of collection is damaging but does not completely destroy the plant.

Fruit pulp which is fleshy and sugary can be allowed to ferment to make wine. Palms frequently produce prodigious quantities of fruit and wine-making is an ideal way to use this bounty. Those of the Bacaba Wine Palm (*Oenocarpus distichus*) from South America are very pulpy and sweet and are used in this way. Other species of *Oenocarpus* are also reported as being suitable. *Cryosophila nana* from Mexico produces abundant sweet fruit which are made into wine and the fruits of three species of Bactris (*B. guineensis*, *B. major* and *B. maraja*) from South America are made into wine. The large purple fruits of the Brazilian Assai Palm (*Euterpe oleracea*) are used to make a popular thick liquid (acai or assai) which may be either drunk fresh or fermented. The thin flesh of the fruits of *Leopoldinia piassaba* is agitated with water to make a popular drink in the area of Brazil where it grows.

Alcohol is an important fuel in this energy-conscious world and from palms it can be produced by distillation of the fermented sap. It should be noted in passing that some species of palms (*Nypa fruticans* for example) are capable of contributing very large quantities of sap and many of these occur in huge numbers in less privileged countries. They are generally easy to grow, can be closely planted (up to 2500 per ha) and will frequently grow on poor soils or those unsuitable for agriculture, such as in mangrove areas. If modern technology can prove alcohol to be an economically viable energy source, such palms could well prove to be significant in the future economies of many countries.

Palm Sugar

Palm sugar, known as gur or jaggery, is a coarse, dark sugar crystallised from the sap of many species of palms and consumed locally. This product is of tremendous importance to the people of Asia particularly India, Cambodia, Thailand and Burma. Most of the sugar used in these countries comes from palms (some 100 000 tonnes produced and consumed locally each year) and so productive are these plants that there is no need to grow the more conventional sugar cane or sugar beet. To prevent fermentation of the fresh juice, the inside of the collecting vessel is first rubbed with lime paste.

Cakes of sugar made from *Borassus Flabellifer* sap and moulded in baskets made from leaves of the same species, Madura, Indonesia. J. Dransfield

Palm sap is usually collected by cutting off a flower stalk. Sometimes the whole emerging inflorescence is severed to leave the projecting peduncle (*Arenga, Caryota*). In other cases (*Borassus, Cocos*) the tip of the inflorescence is bound tightly, beaten or vibrated (massaged) and sliced off for eight successive days, after which the sap flows for four to six months. The sugar content of the sap varies but may be up to sixteen per cent. After collection, the sap is boiled to form molasses and if boiled further, sugar crystallises and can be stored after drying.

Large quantities of sugar are made from the sap of the Palmyra Palm (*Borassus flabellifer*) which is native to India and Malaysia. This palm grows in large colonies and is dioecious. Male plants may have the inflorescence removed completely but the female inflorescence is usually bound and treated as outlined above. *Borassus* plants yield 3–4 litres of sap per day, with female trees producing a fifty per cent greater yield than the males. About 1.5 kg of sugar can be made from 10 litres of this sap. This sugar is mainly saccharose. The African species, *Borassus aethiopum*, is also used for sugar production in a similar way.

Arenga pinnata is known as the Sugar Palm because of the large quantities of sugar that are obtained from its sap (often called arenga sugar). This palm is naturally distributed from India to the Philippines and is widely cultivated elsewhere. The plants are monocarpic and die once the fruit matures on the inflorescence which arises from the lowest leaf axil. The inflorescences are removed early in their development and prodigious quantities of syrupy sap flow from the cut surface and are collected in containers. The sap is then boiled to produce sugar. Sugar is produced in a similar manner from the sap of cut flower stems of the Coconut Palm (*Cocos nucifera*). Its sap contains sixteen per cent sucrose. The sap of the stout trunk of the Chilean Wine Palm (*Jubaea chilensis*) can also be crystallised into sugar, but this is more often used to produce palm wine or the delicacy called palm honey. In this case the whole tree is cut down.

The best tasting jaggery is made from the sap of the Fishtail Palm (*Caryota urens*) which is widespread in India and Sri Lanka. The developing inflorescences are cut off and up to 50 litres of sap may be collected each day for a few months. The Mangrove Palm (*Nypa fruiticans*) is harvested in a similar way. In some areas these palms grow in huge colonies and it has been estimated that about 350 000 litres of sap could be obtained from one hectare of these palms. This sap is very sweet and syrupy.

Date sugar is the term used for the sugar produced from various species of the Date Palm genus *Phoenix*. The Date Palm (*P. dactylifera*) itself can be tapped for the syrupy sap but this is usually considered a waste since the inflorescence must be removed when the flower clusters are just expanding. Poor cultivars or excess male trees, however, may be tapped for this material. The Wild Date of India (*Phoenix sylvestris*) is tapped for its sweet juice, much of which is converted to sugar. The very widely cultivated Canary Island Palm (*Phoenix canariensis*) and *P. reclinata* from tropical Africa can also be used for sugar production.

Fruits

Edible fruits are not predominant among the palms. Of the 2800 or so species of palm it would appear that fewer than 100 are recorded as having edible fruit and, of these, most are only of significance to the people who live where they grow naturally, and some are only eaten in times of hardship. However, two of the three most important palms, namely the Date Palm and the Palmyra Palm, each have edible fruits and between them help to feed millions of people.

The fruit of the Date Palm (*Phoenix dactylifera*) is an energy-rich and important article in the diet of millions of people. The fruit can be eaten fresh, or when dried and stored properly will keep for many years. Other *Phoenix* species have edible fruit but are of much less significance than the Date. The fruit of *P. loureiri* from India is reddish-purple when ripe and has a sweet, edible pulp surrounding the hard seed. That of *P. pusilla* from South-East Asia and southern China is a shiny black berry with a dry but sweet pulp. In Africa the fruit of *P. reclinata* are relished by the local inhabitants. Those of *P. acaulis* from South-East Asia are recorded as being edible but astringent.

Both the fruits and seeds of the Palmyra Palm (*Borassus flabellifer*) can be eaten. The flesh of recently mature fruit is bright orange and quite refreshing but is detracted from somewhat by its rather fibrous texture. It can be eaten raw or roasted or made into preserves. In young fruits it is jelly-like and is then highly prized. The immature kernel of the seed is also edible. The fruit of *Hyphaene* spp. can also be used in a similar manner.

Salacca zalacca a is a handsome, clumping palm native to Indonesia and widely grown in South-East Asia. It has pinnate, erect fronds that arise from a subterranean trunk which usually grows in marshy ground. Most parts are liberally covered with long, black, sharp spines. Small clusters of fruit are carried between the leaf-bases and these fruit are highly prized for their edible flesh. They are 4–8 cm long, pear-shaped and covered with overlapping, dull orange scales. The yellowish-cream flesh is sweet and juicy. The fruit are commonly sold in Asian markets and are a popular item. Each fruit contains three seeds and these are also reportedly edible. The fruit of other species of *Salacca*, namely. *S. affinis*, *S. glabrescens* and *S. wallichiana*, together with the closely related *Eleiodoxa conferta*, are also edible.

A number of other palms produce edible fruit which are primarily used locally. *Hyphaene thebaica*, the African Doum Palm, is also frequently called the Gingerbread Palm because of the unusual flavour of its fruits which are mainly eaten in times of shortage. Some tribes of American Indians eat the fruits of the Cabbage Palmetto (*Sabal palmetto*), the Cotton Palm (*Washingtonia filifera*) and Saw Palmetto (*Serenoa repens*). Each of these palms only produces small fruits but in large quantities and the plants themselves grow in extensive colonies. The Guadalupe Palm (*Brahea edulis*), produces prodigious quantities of small, black fruit with an edible flesh that is variously described as sweet or dull and insipid. *Brahea dulcis*, the Rock Palm of Mexico, produces long hanging clusters of succulent yellow fruit each about 1 cm long. Another Mexican palm, *Cryosophila nana*, has a white, shiny fruit with a spongy, sweet pulp. It is a prolific bearer and has been planted in Trinidad specifically for its fruit. On Robinson Crusoe's island of Juan Fernandez is the palm *Juania australis* which has edible fruit. The fruit of the Fijian palm *Neoveitchia storckii* are eaten before they reach maturity and in Hawaii the immature fruits of various *Pritchardia* species are prized by the natives. These palms even have steps cut into the trunks to facilitate fruit collection. The round fruit of *Calamus rotang* have a thirst-quenching, acid flesh. These are sold in local markets and may be either eaten fresh or after pickling. Other species of *Calamus* also have edible flesh. A Borneo palm, *Daemonorops scapigera*, has a thick layer of sweet flesh in the fruit. In some areas the prickly *Aiphanes aculeata* is cultivated for its edible fruits.

Palms are prominent in some South American forests and some species which have edible fruit are of significance in the diet of the local natives. One such species is *Butia capitata*, the Jelly Palm. This handsome palm is widely grown for its arching crown of bluish-green fronds and few people realise that its fruits are quite tasty. They are rounded, about 2.5 cm across and when ripe are yellow, suffused with red. They are borne in large clusters and have a fibrous, tasty flesh somewhat reminiscent of apricots. Another species, *B. yatay*, has similar edible fruit up to 5 cm in diameter. *Allagoptera arenaria* is a trunkless maritime palm colonising the coastal sand-dunes of Brazil. Its fruit is green to yellowish when ripe and has a fibrous, acid flesh. The deep-orange fruits of *Polyandrococos caudescens*—also from Brazil—are globular, with succulent orange flesh. *Syagrus coronata,* a tall Brazilian palm with prominent, adhering leaf-bases which ascend the trunk in a spiral pattern also produces fleshy, orange fruit that are edible. The palm *Maximiliana maripa* bears a yellow fruit with a sweet, pleasant, juicy pulp, while the fruit of the Mucuja Palm (*Acrocomia aculeata*) have a thin layer of firm, orange flesh and are eaten by the people of the Guianas. The Urucuri Palm (*Scheelea martiana*) of the Amazon is a prolific bearer carrying clusters of fruit. Its pulp has the similar pleasant taste of the date. This palm is cultivated in Trinidad for its fruit. The fruit of *Orbignya spectabilis* are eaten raw by various Amazon tribes.

Three species of palm in each of the genera *Astrocaryum* and *Bactris* produce edible fruit of good quality. Those of *Astrocaryum murumuru* from Brazil have a juicy, aromatic flavour, while the yellowish fruit of *A. tucuma* are fibrous and fleshy but very rich in vitamin A. The third species is *A. acaule* also of Brazil. *Bactris gasipaes* is known as the Peach Palm because of the appearance of its fruit which are 5–8 cm across. They are yellowish when ripe, with a dry, mealy flesh and may be eaten fresh, but are more usually boiled in salt water. It is interesting to note that this palm produces two crops a year and seedless strains are known, indicating selection and domestication by humans. This palm is widely cultivated in South America for its fruit and is virtually unknown in the wild. The Maraja Palm (*Bactris maraja*) has an acid fruit, while those of *B. guineensis,* popularly known as the Tobago Cane, are dark purple and about the size of cherries. The Colombian Palm (*B. major*) also produces clusters of egg-shaped, purple fruit that are edible.

Species of *Mauritia* also produce edible and nutritious fruit. *M. flexuosa* (known as the Ita Palm) is most familiar, producing a red fruit with yellow flesh that tastes like an apple but may require cooking first. The oily pulp of the fruit of *M. flexuosa* can also be eaten after boiling. Fruit of the related *Mauritiella aculeata* can also be eaten. Even the very oily, yellow drupe of the African Oil Palm (*Elaeis guineensis*) can be eaten in times of scarcity, but may induce nausea.

Some palms have a very thin layer of edible flesh around the seed and while this may provide a tasty morsel, it is generally not worth the effort of serious collection. The scaly fruits of some species of the climbing palm genus *Calamus* have just such a layer of edible flesh. Similar remarks apply to the small, dark fruit of the Dwarf Date Palm (*Phoenix roebelenii*). The glossy blue fruits of the Himalayan palm *Trachycarpus martianus* have a thin layer of edible, but rather insipid, flesh. The Walking Stick Palm (*Linospadix monostachya*) of moist gullies in the eastern Australian bush, produces long strings of waxy, red, ovoid fruit. These are quite pleasant to chew but are merely a tasty morsel. The fruits of this little palm achieved fame when they helped sustain the survivors of a plane crash on the Lamington Plateau of south-eastern Queensland in 1927.

Seeds

Undoubtedly the most famous palm seed that is prized for its edible qualities is that of the coconut. The meat of the coconut can be eaten raw or cooked in various ways and added to numerous dishes. Its milk provides a refreshing drink and before the shell hardens the young nuts contain a jelly-like material which is both delicious and highly nutritious. A peculiar nut appears occasionally on certain varieties of Coconut Palm in the Philippines and is known locally as the macapuno. This is an abnormal nut and is considered to be a great delicacy since the entire cavity is filled with a soft, sweet, gelatinous, white curd which is chewy and can be eaten fresh or cut into strips and preserved in syrup.

The seed of Betel Nut Palm (*Areca catechu*) is used by millions of people. A concoction made from the prepared nut, rolled in a fresh Betel Pepper leaf (*Piper betel*) together with some slaked lime (or a pellet of dried lime) derived from burnt shells and a little tobacco or plant extract, is chewed or sucked for about thirty minutes at a time. This preparation is a mild stimulant which promotes salivation. It is also claimed to have other beneficial side effects, such as aiding digestion and controlling dysentery and internal parasites such as worms. A deleterious effect is the intensive red staining on the gums, lips and tongues of habitual betel chewers. The practice of betel chewing is very popular and widely spread throughout Asia, Polynesia and parts of Africa. The palm—*Areca catechu*—itself is native to either Malaysia or the Philippines but is now widely grown throughout Asia, often in groves or plantations.

The preparation and sale of betel nut is a very big industry. The egg-shaped fruits are harvested just before they are ripe (they are colouring but not yet a deep orange or scarlet). The outer fibrous husk is cut off and the reddish-yellow, mottled seed is removed. This is then softened by boiling, sliced and dried in the sun until it is a dark brown colour and is then ready for use. Sometimes the seeds may be roasted before drying. Betel Nut Palms live for about thirty years and produce fruit for about twenty-five years.

Seeds of many other palm species are chewed in a similar manner to betel nut and may be used as a substitute for it. In Sri Lanka, the seeds of *Loxococcus rupicola* are prepared and chewed like betel nut and sometimes with it. In the Andaman Islands, the seeds of *Areca laxa* are chewed in this way and similarly in New Guinea and the Solomon Islands, *Areca macrocalyx* and *A. guppyana* are used as betel nut substitutes. In the Philippines, betel substitutes include *Areca caliso*, *A. ipot*, *Heterospathe elata*, *Oncosperma* spp., *Pinanga* spp., and *Veitchia merrillii*.

A few other palms have edible seeds but these are rather insignificant compared with the very large industry surrounding coconut and betel nut production. The kernel of the Mexican and Central American palm *Astrocaryum mexicanum* is edible and is reported to have a flavour similar to the coconut. Its fruits, which are about 3.5 cm across, are spiny and borne in large clusters. Another with a coconut-like flavour (but more oily) is that of *Orbignya cohune* from Honduras. The seeds as well as the fruits of *Salacca zalacca* are edible, each fruit containing two to three yellow kernels. A similar situation exists for *Borassus flabellifer*. The South American Ivory Nut Palm (*Phytelephas macrocarpa*) have strangely shaped aggregated fruits, each of which contain up to four large seeds about the size of a hen's egg. When immature and soft, these can be eaten and are reported to have a nutty flavour. The very cold-tolerant *Nannorrhops ritchiana* from the mountains of Afghanistan has an edible seed which is harvested locally. The seeds of the West Indian palm *Bactris major* are sold in local markets. The kernel of the Mucuja Palm, *Arocomia aculeata*, from South America is reported to be sweet, while that of *Jubaea chilensis*, commonly known as the Chilean Wine Palm, is nutty. Those of *Veitchia joannis* and *V. vitiensis* are somewhat astringent but are eaten in Fiji, especially by children. Seeds of *Balaka* spp. are also eaten in Fiji. In Central and South America, two species of *Parajubaea* have a sweet, edible seed which is prized by the local people. Other palm genera which have edible seeds include *Hyophorbe* spp. *Jubaeopsis caffra*, *Latania* spp., *Pelagodoxa henryana*, *Ptychococcus* spp., and *Sclerosperma* spp.

Some kernels may require preparation before they are edible. In India and South-East Asia, the immature seeds of the Sugar Palm (*Arenga pinnata*) are preserved in sugar and sold as a delicacy known as palm nuts. Young seeds of the Mangrove Palm (*Nypa fruticans*) may be treated in a similar way. Those of the African Oil Palm (*Elaeis guineensis*) are roasted and are reported to taste like mutton. A coffee substitute can be made from the roasted seeds of *Phoenix reclinata*, the Senegal Date Palm.

Palm-cabbage

The apical bud of a palm is central in the upper part of the trunk and is surrounded by the undeveloped leaf-bases and leaves. In a large number of palms, this apical bud, together with the very young leaf sheaths and leaves that immediately surround it, are edible and highly prized by native tribes and western gourmets alike. This cabbage may be cooked or eaten raw in salads and usually

TABLE 5
Some palms with edible cabbage

SPECIES	ORIGIN	SPECIES	ORIGIN
Acanthophoenix rubra	Mascarene Islands	*Hyophorbe verschaffeltii*	Mascarene Islands
Archontophoenix alexandrae	Australia	*Juania australis*	Juan Fernandez
		Licuala paludosa	Malaysia–Thailand
Archontophoenix cunninghamiana	Australia	*Linospadix monostachya*	Australia
		Livistona australis	Australia
Areca catechu	Asia	*Livistona benthamii*	Australia
Arenga listeri	Christmas Island	*Livistona humilis*	Australia
Arenga pinnata	Indonesia	*Livistona rotundifolia*	Malaysia–Philippines
Arenga undulatifolia	Philippines–Borneo	*Livistona speciosa*	India–Burma–Malaysia
Astrocaryum murumuru	Brazil	*Lodoicea maldivica*	Maldive Islands
Astrocaryum tucuma	Brazil	*Maximiliana maripa*	Brazil
Bactris gasipaes	South America (cult.)	*Metroxylon sagu*	Melanesia
Borassodendron machadonis	Malaysia	*Normanbya normanbyi*	Australia
Brahea brandegeei	Mexico	*Oenocarpus bacaba*	Brazil
Carpentaria acuminata	Australia	*Oenocarpus bataua*	Brazil
Caryota urens	India–Malaysia	*Oncosperma horridum*	Malaysia–Indonesia
Cocos nucifera	Pacific region	*Oncosperma tigillarium*	Malaysia
Corypha utan	Asia	*Pritchardiopsis jeanneneyi*	New Caledonia
Deckenia nobilis	Seychelles	*Ravenea madagascariensis*	Madagascar
Euterpe edulis	Brazil	*Rhopalostylis sapida*	New Zealand
Euterpe macrospadix	Costa Rica–Belize –Panama	*Roystonea borinquena*	Puerto Rico
		Roystonea elata	Florida, USA
Euterpe oleracea	Brazil	*Roystonea oleracea*	Cuba–West Indies
Gronophyllum ramsayi	Australia	*Sabal palmetto*	USA
Gulubia palauensis	Palau Island–Carolinas	*Satakentia liukiuensis*	Ryukyu Islands
Heterospathe elata	Amboina	*Scheelea martiana*	Brazil
Hyophorbe lagenicaulis	Mascarene Islands	*Syagrus romanzoffiana*	South America

has an appealing nutty flavour. In some countries such as Brazil, Paraguay and Venezuela, these apical buds are canned and exported as 'palm hearts' or 'palmito', although they may be more generally referred to as 'palm-cabbage'.

The collection of palm-cabbage relies on the destruction of the palm, for the tree must be felled and all of the surrounding tissue removed in order to reach the delicacy. For this reason, dishes containing this material may be referred to as 'millionaire's salads' although this term is generally confined to the cabbage obtained from the versatile and important Coconut Palm (*Cocos nucifera*). As well as being of a pleasant taste, palm-cabbage is very nutritious and has been (and still is) an important item in the diet of many tribes and races.

Nutritional and culinary aspects aside, the harvesting of palm-cabbage is a tremendously wasteful process. The vast majority of palms harvested are cut from wild populations and in certain areas this exploitation has reached such a degree that the forest ecology is being changed. It has been estimated that each year in Brazil and Paraguay more than six million palms are destroyed just for the export trade. In the Dominican Republic (a former exporter) the situation became so bad that a permit system was introduced. The over-harvesting of palm-cabbage has directly threatened a number of species and caused them to become rare or endangered in their natural habitat, perhaps already with the loss of certain genotypes. Examples include *Acanthophoenix rubra*, *Euterpe edulis*, *E. oleracea*, *E. precatoria*, *Hyophorbe* spp., and *Pritchardiopsis jeanneneyi*. Not only are mature palms harvested, but when young plants reach about 4 cm in diameter they are susceptible to destruction for a mere handful of cabbage and without the opportunity to reproduce.

Palm hearts may well be referred to as millionaire's salad if their production is to result in extinction of species and cause a change in the forest ecology. Either western gourmets should stop creating the demand for this product or plantations should be established to grow palms for harvesting. In western Costa Rica an experimental station is examining the possibilities of close planting the very fast-growing Peach Palm (*Bactris gasipaes*) for cabbage. Results show that these palms yield about 1.5 kg of cabbage after four years and are faster growing and higher yielding than local palms. Peach Palms also sucker and thus the harvested stem is automatically replaced.

Palms which are commonly called Cabbage Palms more than likely derived this vernacular name from their edible apical bud. Perhaps the best known of these is the Feathery Cabbage Palm or Iraiba Palm (*Roystonea oleracea*) from South America. About 5 kg of the white delicacy called cabbage can be obtained from a mature tree and it can be eaten either raw, cooked or pickled and is sometimes canned for export. The related species *R. elata* from Florida, *R. oleracea* from the West Indies and *R. boriquena* from Puerto Rico, also yield a similar cabbage. The Cabbage Palmetto from Florida (*Sabal palmetto*) similarly yields an excellent cabbage and was widely eaten by the Indians of the area. In Australia a number of species of the genus *Livistona* are known as Cabbage Palms. Many of these contain edible cabbage and were probably harvested by the Aborigines. For example, *L. benthamii* and *L. humilis* were eaten in the Northern Territory. The most widely spread species is *L. australis* and its cabbage became a popular article of diet with the early settlers.

The apical buds of other palms can be eaten. That of the Inaja Palm (*Maximiliana maripa*) from Brazil and the Nibung Palm (*Oncosperma tigillarium*) from Malaysia are described as being excellent with a very nutty flavour. The major species harvested and exported from South American countries appear to be *Euterpe edulis* and *E. oleracea*. A list—although by no means comprehensive—of palms with edible cabbage is presented in Table 5. It should be noted that not all palms have edible cabbage. Some are bitter (*Bismarkia nobilis*, *Hyophorbe amaricaulis*, *Plectocomiopsis geminiflora* and *Ravenea amara*) and others are highly poisonous. The most toxic appear to be species of *Orania* but others include some *Hyophorbe* spp. and *Plectocomiopsis* spp.

Other Edible Parts

Young developing parts (other than the cabbage) of some palms are eaten, usually after cooking. The very young leaves of *Acrocomia aculeata*, *Coccothrinax argentea* and *Nannorrhops* make an excellent vegetable either raw or cooked. In Italy the young suckers of *Chamaerops humilis* are eaten cooked and in parts of Asia the young suckers of various *Calamus* are used in a similar way.

In some countries young unopened palm inflorescences may be collected and eaten raw or after cooking—for example, *Chamaedorea elegans* and *C. tepejilote* from Mexico and South America, *Rhopalostylis sapida* from New Zealand and *Trachycarpus fortunei* from China. A substantial industry has developed around *Chamaedorea tepejilote* in South America. Plantations of selectively propagated plants are grown for the young male inflorescences (called pacaya) which are sold locally in markets and used either as a salad vegetable or eaten after cooking.

Palm Honey

Relatively few palm flowers are attractive to bees but those species which bees do visit produce a honey that is highly esteemed. Coconut honey is perhaps the best known in this regard. That from the Saw Palmetto (*Serenoa repens*) and *Borassus flabellifer* are also prized.

In Chile 'palm honey' refers to an entirely different product, although it also is highly esteemed and very popular. The sap obtained from the trunk of *Jubaea chilensis*, popularly termed the Chilean Wine Palm, is boiled to a treacle-like consistency and is sold as palm honey.

OTHER MATERIALS
Palm Fibres

A number of important fibres are obtained from palms and even in this modern age of synthetics, many of these retain their place because of certain unique properties. Fibres are obtained from various parts of the palm and are prepared in some way before use.

Perhaps one of the best known fibres comes from one of the most familiar palms, the Coconut (*Cocos nucifera*). This is the fibre known as coir and it is obtained from the fibrous husk which surrounds the nut. Coconut fibres may be up to 30 cm long and are used to produce various types of mats, brushes and brooms, carpets and filling, or are twisted to form string and rope. The individual fibres are light because they are hollow. Coir fibre rope is still of major importance today because it is not only strong, but also durable and resistant to sea-water and bacterial action.

Coir is commonly a by-product of the copra industry, but another type of fibre is taken from nuts that are not fully ripe, and these nuts may be harvested for the fibre alone. This fibre is difficult to obtain and may require soaking for up to six months before it can be separated. The long fibres obtained from these green coconuts are spun into yarn which is then used for carpets, mats and rope.

Coir is stripped from the copra husks after they have been soaked for up to six weeks. After soaking, the fibres are teased, beaten to remove rubbish and dried in the sun. This may be carried out by hand, but today it is more commonly done by machine. Long-bristle fibres suitable for brushes and brooms are separated from short, fine fibres which can be used for filters, fillings in mattresses or compressed into panels. It is not often realised that long fibres are also found on the petioles of coconut leaves and these are twisted into strong string by the Polynesians.

Piassaba (or piassava) fibre is a group name for a series of important palm fibres produced in South America and Africa. The name is mainly associated with the Brazilian palm, *Leopoldinia piassaba*, but has also been attached to fibres from other palms. In South America two species of palm are used, namely Pará Piassaba (*Leopoldinia piassaba*) from the Brazilian state of Pará and Bahia Piassaba (*Attalea funifera*) from the state of Bahia. West African piassaba is obtained from the petioles and rachises of the long pinnate leaves of *Raphia* species (*R. hookeri*, *R. palmapinus* and *R. vinifera*).

The fibres from all of the palms mentioned above are collected from natural stands and either exported or used in local industries. Those of the Brazilian palms arise from the leaf bases and cover the trunk. They are untangled and straightened on the trees before cutting. Only flexible fibres less than five years old are cut and these may be up to 1.5 m long. After cutting, these fibres are tied together in bunches and a number of bunches are securely tied to form a cone-shaped bundle which is exported. The fibres are used to produce heavy brooms and industrial brushes as well as cords and ropes. The ropes are excellent for marine purposes (ropes, hawsers) since besides being resistant to salt water they will not sink.

The West African piassaba fibre is rather pale and brittle by comparison with the fibre from Brazil. It is also more difficult to obtain since the leaves must be cut and the petioles and rachises soaked in water for about two months before the fibre can be separated by pounding and beating. It is principally produced in and exported from Sierra Leone. It is generally used in the production of brooms and brushes and may be strengthened by mixing with Brazilian piassaba fibre.

Other species of *Raphia* produce a different fibre known as raffia. This is harvested from the Madagascan palm *Raphia farinifera* and *R. taedigera* from the Amazon basin. These are short-stemmed palms with exceptionally long leaves. The fibre is obtained by stripping the surface from young leaflets which may be up to 2 m long. Raffia fibre was widely used in plaiting, basket work and as tying material but has now largely been replaced by synthetic materials.

Palmyra fibre from the Palmyra Palm (*Borassus flabellifer*) is used in brooms, scrubbing brushes and in the manufacture of carpets, twine and rope. It is obtained from the petioles and leaf sheaths which are beaten to separate the fibres. The fibres themselves are wiry and inelastic. They are widely used locally and are also exported from India. Tucuma fibre is harvested from the pinnate leaves of the Brazilian palm *Astrocaryum tucuma* and is useful for marine ropes and fishing nets.

Valuable fibres are obtained from the petioles and leaf-bases of the Fishtail Palm (*Caryota urens*) of India and Malaysia. These black, bristly fibres are known as 'kittul' and are used to produce rope, brooms, paint brushes and other specialised brushes for raising the pile on fabrics such as velvet. Fibres from the Nipa Palm (*Nypa fruticans*) are also known as kittul.

The Sugar Palm (*Arenga pinnata*) is also known as the Gomuti Palm because of the production of fibres of that name from the leaves. These coarse, black fibres are very resistant to dampness and are used in various types of filters, for caulking boats and in the production of rope for use in salt water.

Miscellaneous fibres of small scale or local interest are produced by a number of palms. Buriti and muriti are two fibres obtained from the petioles of the Brazilian palm *Mauritia flexuosa*. Vegetable horse-hair or African hair, used as a substitute for horse-hair in upholstery, is obtained from the black leaf fibres of the European Fan Palm (*Chamaerops humilis*) which is widespread from the Mediterranean coast of Europe to Africa. Fibres from the unopened leaves of a species of *Corypha* are used in the Philippines to make fine hats, while those from the grossly hairy trunks of the Cuban palms *Coccothrinax crinita* and *C. miraguama* are collected and woven for various purposes. The lattice fibre from the trunk of the Chinese Windmill Palm (*Trachycarpus fortunei*) and that of the Chinese Fan Palm (*Livistona chinensis*) is very useful in nurseries for lining hanging baskets. Fibres from the petioles of the Ita Palm (*Mauritia flexuosa*) of South America are woven into threads.

Palm Oils

The fruit and seeds of a number of palms contain oils which have been found to be very useful for cooking, for the manufacture of margarine and lubricants and in the production of many other materials.

The most famous of these palms is again undoubtedly the Coconut (*Cocos nucifera*), the oil of which makes up about twenty per cent of all vegetable oils used in the world. The oil is extracted from the copra which is the flesh of the coconut dried in the sun. The expressed oil is used in the production of dressings and margarines, soap, shampoos and so on. The waste material is pressed and used as animal feed.

The most important palm grown solely for oil is the African Oil Palm (*Elaeis guineensis*) which is native to West and Central Africa. This handsome and very hardy palm is grown in plantations in various parts of the world, including Indonesia, Malaysia, Nigeria, New Guinea and Central and South America. Oil Palm plantings were also tried in parts of northern Australia but their production was found to be uneconomic. One problem was the slow growth of the plants in their early years of establishment.

Mature plants of the African Oil Palm produce three to six dense clusters of fruits annually and each contains about 4000 individual fruit. The fruit are 3–6 cm long and are black when ripe. The flesh contains thirty to seventy per cent oil which is extracted by a fermentation process and is known as palm pulp oil. It is very similar in its make-up to olive oil and is used in the manufacture of cosmetics, margarines and dressings and in fuels and lubricants. The seed kernels also contain about fifty per cent useful oil and this is extracted after crushing and is known as palm kernel oil. It is less valuable than the pulp oil and is used for the manufacture of soap and glycerine. A selection program has produced horticultural cultivars which have small seeds and therefore give a greater yield of the more valuable pulp oil. These superior forms are now propagated by tissue culture.

Other species of palms produce useful oils but these are of less importance than the African Oil Palm and may only be of local interest. The American Oil Palm (*Elaeis oleifera*) from South America has an oily flesh and produces large clusters of fruit similar to that of the African Oil Palm. The oil has been extracted experimentally and was used by the local tribes and early settlers. A valuable oil is also extracted from the seeds of the Cohune Palm (*Orbignya cohune*) which is native to Honduras. Two other species of *Orbignya* from Brazil produce fruits that are rich in oils. These are harvested from the wild and are known as Barbassu Palms (*O. martiana* and *O. oleifera*). The oil is used in margarine and soap, for cooking and in lubricants. The fruits of two species (*Oenocarpus bacaba* and *O. bataua*) from the Amazon region give a colourless, sweet oil which can be used for cooking. Two other South American palms also provide useful oils: *Acrocomia sclerocarpa* provides oil from the husks of the fruit and *Syagrus coronata* from the seeds.

Palm Waxes

Many palms have waxy coatings which act as a water-conserving mechanism for the plant. These waxes may be found on leaves, petioles and trunks and have some properties useful to humans. The most important wax is obtained from the Wax Palm (*Copernicia prunifera*) and is known as 'carnauba wax'. This material covers the upper surface of the leaves and is the hardest of all commercial waxes, with a very high melting point. Mature leaves are collected from both wild and cultivated palms, allowed to dry and the wax is then collected by flailing these leaves mechanically. The wax flakes off freely and is thus termed caducous. It is used for a variety of purposes including candles, lipsticks and other cosmetics, records, shoe polish, car polish and floor wax. This wax is exported from Brazil where the palms are grown in plantations. More than 20 000 tonnes a year is produced. Breeding and selection programs are being carried out to produce high-yielding cultivars.

Other, less important waxes are present on various palms. The trunks and underside of the leaflets of *Ceroxylon quindiuense* from the northern Andes of Colombia bear a wax which has been harvested. A wax is also harvested from *Syagrus coronata* in Brazil but it is not caducous and must be scraped from the leaves. Several other palms such as *Serenoa repens* have waxy layers but these are not harvested commercially.

PALMS IN CONSTRUCTION
Housing

Various palms contribute to the housing and shelter of people in many countries of the world. In Polynesia, huts may be built entirely from parts of the Coconut Palm, while in Panama, huts also made entirely from palm products may be built around the still-living trunks of a local species of *Astrocaryum*. In Thailand, huts may be composed entirely of local palms (*Salacca wallichiana* and *Nypa fruticans*). The Nipa huts of the Philippines are also constructed largely or entirely of palm products (particularly *Nypa fruticans*) as are huts in the lowlands of New Guinea, Indonesia and the Amazon region of South America.

House thatched with leaves of *Johannesteijsmannia altifrons*, Selengui, Western Malaysia. J. Dransfield

A house constructed from leaves of *Salacca wallichiana*, south Thailand. The walls are made from the petioles (top left) and the woven panels from leaflets (top right). J. Dransfield

Lumber for the supporting posts and main structural timbers can be supplied from the trunks, these either being sawn into sections or split length-wise (*Borassus flabellifer, Cocos nucifera, Neoveitchia storckii, Socratea exhorrhiza* and *Astrocaryum* spp.). Palm trunks are not readily sawn into planks. Minor structural support can be obtained from the petioles (*Cocos nucifera, Salacca wallichiana*) and the floor can be formed from split and polished trunks (*Astrocaryum* spp., *Cocos nucifera*).

The leaves thatched on the roof form an excellent waterproof covering and internal partitions can be woven from the leaflets. Internal decorations and utensils can also originate from palms and a notched trunk can form a ladder if one is needed. Such houses are excellent for tropical climates, maintaining coolness, dryness and air movement. Magnificent examples can be seen in Fiji.

Thatching of roofs is quite an art and around the world the leaves of numerous species of palms are used for this purpose. Many pinnate-leaved palms are used (*Astrocaryum* spp., *Metroxylon vitiense, Nypa fruticans, Geonoma baculifera*) but some species of fan palms are also suitable (*Cocothrinax argentea, Corypha umbraculifera, Borassus flabellifer, Lodoicea maldivica* and *Livistona benthamii*). Some palms are commonly known as Thatch Palms (*Thrinax* spp.) because of this use. In coastal districts of the tropics the leaf of the Coconut Palm is widely used. Leaves of the South American Monkey Cap Palm (*Manicaria saccifera*) have proved a very long-lasting thatch.

Furniture

The harvesting of rattan cane is an important industry in South-East Asian countries such as Indonesia, the Philippines and Malaysia. The cane is actually the stems of several species of climbing palms in the genus *Calamus* and a few other genera and it may be exported in bundles or made into items of furniture. *C. rotang* is strictly known as the Rattan Cane but many other species are also harvested, including *C. caesius, C. rudentum, C. scipionum, C. trachycoleus, C. manan* and *C. tenuis*. The vast majority of rattan is harvested from natural stands, but there is considerable interest in its cultivation and successful plantations have been established in Malaysia and Borneo, where *C. trachycoleus* has proved superior.

Rattan stems are usually smooth and naked in the older sections, but in the young regions are covered by spiny leaf sheaths which are an aid to climbing. When being harvested the stems are cut near ground level and dragged out of the forest canopy. They are commonly 50–100 m long but may occasionally reach as much as 160 m. The leaves, cirri and inflorescences are trimmed off and the leaf sheaths removed by dragging through closely spaced pairs of pegs. The supple stems are left smooth and are then cut into manageable lengths and left to dry in the sun for at least a month. They may be used whole or split and can be manufactured into an amazing range of furniture items, baskets, bins and so on (see photos p. 155).

OTHER USES
Horticulture

Palms are a very important component of the nursery industry and each year in many countries, hundreds of thousands are grown and sold solely for their ornamental appeal. This results in the direct employment of many thousands of people engaged in their propagation and growth. It also creates demand for seeds and thus provides employment in this area. As an example, palm seed collecting was a major industry on Lord Howe Island and in 1975 produced a gross return to the island of A$100 000. Germinated seed is now exported and in 1992, 3 million seedlings were exported earning $1 million.

Fans, Hats and Weaving

Not surprisingly, the leaves of various species of fan palms are used to produce fans and sun shelters. The fans may be crudely cut from the leaves or carefully processed into decorative, as well as useful, implements. Fans may be made from a range of palms but species of *Sabal, Livistona, Pritchardia* and *Corypha* are prominent in this use. Primitive but effective shelters from the sun and rain are made from various species. It is claimed that the large leaves of *Corypha umbraculifera* may shelter up to fifteen people at once. The undivided leaves of various species of *Johannesteijsmannia* provide a refuge from sudden storms in the jungles of Malaysia and Borneo.

Many species of fan palms are used to produce light, fibre hats that provide excellent protection in the tropics. The most famous, the Panama Hat, is mainly produced from a palm relative, the Panama Hat Plant (*Carludovica palmata*). Similar-looking hats are made from palms in the same region—for example, *Sabal causiarum* which may be known as the Puerto Rican Hat Palm. The leaves of this (and other Sabal species) are collected at a certain stage of development, soaked in boiling water and the fibrous leaf segments separated and dried before plaiting into hats. The segments of many other palms are used in a similar way, including *Corypha, Coccothrinax,*

Borassus, Livistona, Lodoicea and *Pritchardia*. Those of *Livistona jenkinsiana* are used to make a curiously shaped hat. In Australia, several species of *Livistona* were used to make Cabbage Tree Hats which were very popular with the early settlers. Their mode of preparation was similar to that for the *Sabal*.

Buntal is a fine fibre removed from the unexpanded leaves of *Corypha umbraculifera*. It is used in the Philippines to make quality hats. A curious cap known as the Temiche Cap is made from the pouch-like spathes of the Monkey Cap Palm (*Manicaria saccifera*) which is native to parts of South America. The cap is cut by the local Indians before the spathe is ruptured and is worn as protection from the sun.

Leaflets lend themselves well to various forms of plaiting and thatching and can be used for a large range of articles, including internal house partitions, hammocks, mats, baskets and a range of containers used for carrying.

Walking Sticks

Palms have traditionally produced the best walking sticks, probably because the trunks of some of the smaller species are about the right thickness, are usually very straight and are tough enough for the job. Also they generally polish very well and present a good appearance.

Walking sticks may be made from a number of palms which have thick, straight trunks. These are split and suitable pieces may be turned or sanded and polished. The hard, mottled porcupine wood of the Coconut is very suitable, as is the very straight trunk of the Fijian palm *Balaka seemannii*. The stems or canes of many of the smaller palms such as *Rhopaloblaste singaporensis* may also be used.

In Australia, *Linospadix monostachya* is commonly known as the Walking Stick Palm because of its suitability for this use. This little palm is widespread throughout the moist forests of southern Queensland and north-eastern New South Wales and has a solitary, tough, slender stem which has a curious swelling just below ground level. This swelling provides a convenient hand grip and thousands of these palms were harvested and turned into walking sticks for use by wounded servicemen returning from the First World War. This little palm is also known by the curious vernacular of 'Midginball'. Walking sticks have also been manufactured in Australia from the hard, dark trunks of the Black Palm (*Normanbya normanbyi*).

Malacca canes are the traditional cane used for officer's batons and once also for punishment in schools. These are straight, stout sections of rattan obtained from the climbing palms of the genus *Calamus* (in particular *C. scipionum*). The only drawback with these palms is the pale colouration of their stems and these are generally stained or darkened with smoke before sale.

Vegetable Ivory and Other Carvings

Many fine ornaments are carved from the large seeds of some species of palm, the endosperm of which is hard and similar in texture to animal horn. This material has a remarkable similarity to ivory and for this reason is called vegetable ivory. In the past the endosperm was widely used for the production of buttons, billiard balls, dice and such but has been largely replaced in this use by synthetic plastics and is now mainly used for decorative carving. *Phytelephas macrocarpa* from Columbia and Ecuador is known as the Ivory Nut Palm, the name arising from its Puseful seeds, which are about the size of a hen's egg. These are collected and exported in large quantities. *Metroxylon amicarum* from the Caroline Islands also produces vegetable ivory and is perhaps not surprisingly known as the Carolina Ivory Nut Palm. The seeds of a number of other *Metroxylon* species, including the Sago Palm (*M. sagu*) can be used as vegetable ivory.

The hard shells of the fruits of certain palms can be used for carvings. Probably the best known of these is the coconut which may be carved while still in the husk or the polished shell may be used. Both are widely sold to tourists in various parts of Polynesia and the Pacific region. The shells of species of *Hyphaene* from Africa are carved to produce animal replicas which are sold locally.

Minor Uses

The number of other uses to which palms can be put is almost beyond recording and only a few will be presented here.

Writing Material The material known as olla is used as a substitute for writing paper in India. It is made from the dried and flattened sections of the leaves of *Corypha utan* and *C. umbraculifera* and when finished is parchment-like and will take ink. This writing material has been used for centuries and samples more than 1000 years old are preserved in museums. In Java and parts of South-East Asia *Borassus flabellifer* was a very important source of writing material.

Sails An interesting resourcefulness is shown by the Warao Indians of the Orinoco Delta in Venezuela who use the large, entire leaves of *Manicaria saccifera* as sails for their canoes.

Potting Media Coconut fibre or coir can be used to produce a peat substitute which can be used in nursery potting mixtures to grow plants. The dust and rubbish left over from coir production can be used for this purpose or coir itself can be hammer-milled to produce a higher-quality product.

Charcoal Coconut shells can be used for such purposes as a fuel for fire, for producing a very high-quality charcoal and for eating and drinking vessels.

Water Drinking water can be obtained from the stems of some climbing palms in the genera *Calamus* and *Daemonorops*.

Resins and Dyes A red resin is obtained from the fruit of the Indonesian climbing palm *Daemonorops draco* and related species. Dyes can also be obtained from fruit. *Oenocarpus batua* which gives a dark blue dye is one example.

Warfare and Hunting The strong, straight trunks of the Fijian palm, *Balaka seemannii*, were much in demand for spear shafts. In the Philippines the various species of *Livistona*, *Oncosperma* and *Pinanga* were used. In Brazil the trunks of *Iriartella setigera*, and in Colombia those of *Chamaedorea linearis*, are hollowed and polished to make blowpipes; and the 30–90 cm long black spines from the trunk of *Oenocarpus bataua* are used for the darts. Similar spines from the trunk of the Malaysian palm *Arenga pinnata* were used by native peoples to make arrows. In the Amazon region, bows may be fashioned from the trunks of *Oenocarpus bataua* and *Oenocarpus bacaba* while brittle arrow tips are made from many other palm species. In New Guinea, the stems of *Ptychococcus lepidotus* are used to make bows and arrowheads while in the Philippines the same weapons were manufactured from *Livistona rotundifolia*. Arrow poisons have been concocted in Malaysia and the Philippines from the very poisonous fruit of some species of *Orania*. Sturdy clubs can be made from the very heavy, dense wood of the trunks of some palms such as *Oenocarpus bataua* from the Amazon region and *Normanbya normanbyi* from Australia.

CHAPTER 4

THE BIOLOGY OF PALMS

Palms have evolved some interesting specialised features. Aspects of the biology of these features are examined here.

PALM POLLINATION

Palms are commonly believed to be wind-pollinated but that myth has been dispelled by recent studies. In fact many early naturalists realised that the pollination system of palms was far more complex than merely involving wind. These observations were made as early as 1823, but still the erroneous notion that palms are basically wind-pollinated has persisted and is even mentioned in some modern texts. (For a comprehensive review of pollination in palms see A. Henderson, 'A Review of Pollination Studies in the Palmae', *The Botanical Review* 52 (3) (1986): 221-259.)

The original idea of palms being wind-pollinated probably arose because they often produce masses of relatively simple flowers and some species (such as the male flowers of *Phoenix*) produce large quantities of pollen. Today it is known that some palms are definitely wind-pollinated but these are relatively few and most species have a much more complex pollination syndrome, using insects and other animals as vectors.

Biological features found in the flowers of many palm species which show that they are not wind-pollinated include:

- the production of heat in the buds prior to flowering
- floral odours of various types
- the production of nectar in flowers
- the production of sticky pollen which can only be transported by animals.

In addition, it is common for many small, understorey palms to grow in closed forests where wind currents are much reduced and unpredictable.

Heat Production in Palm Buds

The production of heat in the inflorescences of some aroid species of the family Araceae is well known and recently the mechanism of heat production in mature cones of cycads has been investigated. It is less well known that the buds of some species of palms produce heat just prior to opening. This emission is probably associated with insect pollination and indeed was recognised as such by Carl F.P. von Martius as early as 1823. The heat emissions of the buds may act as an attractor to insects, enhance odour release, facilitate the production of nectar or release of pollen; or perhaps play a part in all of these processes.

Various species of palms have been reported to produce heat from their buds, but careful studies are needed to determine the extent of the process in the family and to measure details such as length of heating period, its coincidence with anthesis and the temperatures reached by the organs. The buds of *Nypa fruticans* become sufficiently hot to be noticed when touched and may smell like hot rubber. At certain stages in its development the inflorescence of *Bactris gasipaes* also becomes noticeably warm to the touch. This temperature of a mature inflorescence has been measured and found to be 4–5°C above ambient throughout the day. Other palms in which temperature elevation has been recorded include *Astrocaryum* spp., *Nenga pumila*, *Phytelephas macrocarpa* and *Pinanga patula*.

Floral Fragrances of Palm Flowers

Flowers release odours to attract insects for the purposes of pollination. These odours may be pleasantly fragrant (and so enhance the horticultural appeal of the plant) or unpleasant and unattractive. The nature of the odour can be of significance in pollination by attracting different groups of insects. Thus flies are attracted to unusual odours whereas bees and wasps respond to sweet fragrances. These insects may specialise even further with fruit flies and vinegar flies being attracted to fruity odours and carrion flies preferring decaying smells.

Palm flowers produce various fragrances, some of which are pleasant, others much less so. In some species the fragrance may be noticeable only in the flowers of one sex—for example, male flowers of *Leopoldinia piassaba*—whereas in others both male and female flowers are scented (*Cocos nucifera*).

Pleasant scents are noticeable in some palms. The colourful flowers of *Chamaedorea fragrans* and *Hyophorbe verschaffeltii* are described as being intensely and sweetly fragrant and those of *Cryosophila albida* resemble the delightful fragrance of lilacs. The flowers of *Gronophyllum ramsayi* have a strong lemon fragrance. Others with a pleasant fragrance include *Asterogyne martiana*, *Barcella odora*, *Chamaedorea angustisecta*, *C. pauciflora*, *Cocos nucifera*, *Copernicia* spp., *Johannesteijsmannia perakensis*, *Mauritia flexuosa*, *Sabal palmetto* and the male flowers of *Leopoldinia piassaba* and species of *Phoenix*. The flowers of *Chamaedorea deckeriana* are reported to release a spicy perfume.

The flowers of many palms produce musty or offensive odours. These include *Bactris gasipaes*, *B. guineensis*, *Chamaedorea elatior*, *C. graminifolia*, *C. pochutlensis*, *C. seifrizii* and *Hydriastele microspadix*. The odour of *Johannesteijsmannia altifrons* resembles that of sour milk and sewage; *Nenga gajah* has been described as having a penetrating, musty, sickly-sweet smell; that of *Ceratolobus glaucescens* is also penetrating. The flowers of *Corypha* species produce a powerful, offensive odour, whereas the odour of *Calyptrogyne sarapiquensis* is like garlic. Some species have distinctive chemical smells, thus *Pinanga coronata* resembles the smell of ethyl acetate; *Eugeissona* flowers have an odour like alcohol and *Johannesteijsmannia lanceolata* of coumarin.

Nectar Production in Palm Flowers

The flowers of many palms produce nectar which is both attractive to insects and acts as a reward for their visits. Nectar production usually coincides with floral maturity and the droplets of nectar are often visible in the flowers soon after opening. Nectar is produced from specialised glands known as nectaries. These are often found in the carpel walls and the droplets of nectar collect in the base of the flower. In male flowers the sterile pistil (pistillode) may be important in nectar production.

Pollination Syndromes in Palms

Various species of palms have adopted different systems to ensure pollination. Flower colour and presentation, floral fragrance and the presence or absence of nectar all combine to attract pollinating agents. These floral combinations vary between genera (and sometimes within a genus) and result in the attraction of a particular group of organisms to the flowers. The aggregation of the various characteristics into a syndrome allows species which have opted for a particular pollination system to be grouped together. Examples of pollination syndromes in palms follow. It should be noted however that the details provided are merely a guide. Living organisms do not adhere to rigid rules and it is not uncommon for the flowers of any palm species to be visited by a whole suite of insects. Not all of these, however, will be pollinators.

Wind Pollination (Anemophily)

A few species of palms are known to be wind-pollinated. These include some—but not all—species of *Thrinax* (*T. parviflora*); some such as *T. excelsa* have colourful, scented flowers which are insect-pollinated. *Zombia antillarum* is also believed to be wind-pollinated as are some species of

Coccothrinax and the commercially important genus *Phoenix*. The large genus *Chamaedorea* is mostly adapted for insect-pollination, but a number of species have light, powdery pollen and appear to be wind-pollinated. These include *C. allenii, C. alternans, C. arenbergiana, C. cataractarum, C. crucensis, C. deckeriana, C. nationsiana, C. oblongata, C. oreophila, C. pochutlensis, C. seifrizii, C. tepejilote* and *C. zamorae*. Some palms which are basically insect-pollinated, may also have a proportion of their flowers pollinated by the wind (*Cocos nucifera, Orbignya martiana*).

Beetle Pollination (Cantharophily)

Various reports suggest that many species of tropical palms are pollinated by beetles. Beetles were one of the earliest groups of insects to evolve and those species which have been found to be involved with palms are often members of primitive beetle families, in particular Curculionidae (weevils) and Nitidulidae (nitidulid beetles). The flowers of beetle-pollinated plants are commonly simple, open and dull-coloured. The floral fragrance is often musty and chemical or fruity and nectar is present. Pollination may occur at night. Beetle pollination is somewhat crude, for the insects are clumsy and they commonly chew parts of the flowers, including pollen.

The syndrome whereby pollinating agents congregate on the inflorescence of a plant, mate and lay eggs is believed to be the most primitive pollination system known which involves insects. This syndrome could be interpreted as occurring in palms such as *Butia leiospatha, Rhapidophyllum hystrix* and *Sabal palmetto* where the weevil adults feed in the flowers (and achieve pollination) mate and lay eggs and the larvae feed on the carpels and developing seeds of the palm. Pollen-eating beetles have been found in species of *Thrinax* and *Chamaerops excelsa*. Other palms which are pollinated by weevils or nitidulid beetles include *Cryosophila albida, Socratea exorrhiza, Wettinia hirsuta, Hydriastele microspadix, Nenga gajah, Orbignya martiana, Elaeis guineensis, Bactris guineensis, Phytelephas aequitorialis, Ammandra decasperma, Phytelephas macrocarpa, Salacca* spp. and *Serenoa repens*. Scarab beetles may also act as pollinators in *Astrocaryum alatum, Bactris gasipaes* and *B. porschiana*.

Bee Pollination (Melittophily)

Bees are common and widespread and involved in the pollination of most flowering plants, including many palms. Bees are a complex insect group which include primitive families—for example, Trigonidae—as well as highly advanced groups like the honey bee (Apidae). Bee-pollinated flowers are commonly brightly coloured, sweetly fragrant and produce nectar, which is often hidden. It should be realised that not all bee visits to flowers coincide with pollination; trigonid bees, for example, are avid pollen collectors and often steal pollen from flowers without transferring it to the stigma. Similarly, honey bees often visit the male flowers of palms to collect pollen and ignore female flowers.

Palms which may be pollinated by trigonid bees include *Archontophoenix* spp., *Plectocomia dransfieldiana, Iriartea gigantea, I. deltoidea* and *Eugeissona* spp. Other palms which are generally thought to be pollinated by bees include *Hyphaene* spp., *Pritchardia* spp., *Cocos nucifera, Welfia georgii, Arenga tremula, Maximiliana martiana, Areca* spp., *Calamus* spp., *Asterogyne spicata, Euterpe oleracea, Roystonea* spp. and *Ptychosperma macarthurii*.

Fly Pollination (Myophily)

Typical fly-pollinated flowers open widely, are often white or yellow, are fragrant and produce drops of nectar. Flies are a very large group of insects and specialised fly families may be attracted to flowers by specific odours, such as those resembling the smell of fermenting fruit (vinegar flies and fruit flies) or decaying matter (carrion flies and blowflies).

Palms of the genus *Corypha* produce a terminal inflorescence which bears millions of flowers. These give off a strong, offensive odour and are pollinated by blowflies of the family Calliphoridae.

The Mangrove Palm (*Nypa fruticans*) is pollinated by vinegar flies (Drosophilidae) which gather in thousands and breed in the inflorescence. The sticky pollen is transported from the male flowers to the female flowers by these insects. Drosophilid flies are associated with the flowering of other palms, including *Synechanthus warscewicianus, Hydriastele microspadix* and *Geonoma* spp.

Syrphid flies (also known as hoverflies) are the main pollinators of *Asterogyne martiana*, an understorey palm from the rainforests of Central America.

Ant Pollination (Formicophily)

Ants are frequent visitors to palm blossoms but they rarely act as pollinators, more usually just stealing nectar from the flowers. Ants are, however, the pollinating agents for *Iguanura wallichiana*, a common undergrowth palm of Malaysia, Sumatra and Borneo. They may also be involved in the pollination of *Iriartella setigera* and species of *Hyospathe*.

Bat Pollination (Chiropterophily)

A number of small bats have adapted to feed on nectar and pollen and have become important pollinating agents of certain plants. Some of these bats are capable of hovering near the flowers while probing with an extensile tongue; others cling to the inflorescence or flower while feeding.

In Costa Rica it has been demonstrated that the flowers of the palm *Calyptrogyne sarapiquensis* are pollinated by bats. It is also possible that the flowers of *Corypha* species are visited by bats at night. Some species of *Arenga* from Malaysia may also be bat-pollinated.

SEED DISPERSAL

Plants put a considerable amount of energy into the production of seeds and it is important that these seeds be dispersed to a new site if their potential is to be realised. Palms are no exception to this rule. Palms produce a great range of fruit, from those which are quite small to large types such as the coconut. The smaller fruit are often produced in large quantities and may be brightly coloured whereas the larger fruit are fewer in number and are usually dull coloured. There is also a tremendous range of shape within palm fruit and seeds (see Fig. 17).

Palms seem to be dispersed by three main methods—by gravity, by water and by animal vectors. Those palms which are dispersed by animals attract a wide variety of species and it is rare for close mutualistic relationships to be involved. (For a comprehensive review of animal dispersal see S. Zona, 'A Review of Animal Mediated Seed Dispersal of Palms', *Selbyana* 11 (1989): 6–21.)

Gravity

Palm fruits fall when ripe and aggregate beneath the crown of the parent tree. These commonly germinate resulting in a mass of seedlings, but very few, if any, of these plants survive to reach maturity. Seeds of palms which are growing on ridges and slopes may be dispersed for short distances on the downhill side of the parent plant. This movement may also be enhanced by the feeding activities of animals, especially rodents. Dispersal of this type may be of significance in building up a colony of a palm species, but is of limited importance in its wider distribution.

Sea-water

The fruit of at least two species of palm are commonly dispersed by sea-water. These are the Coconut (*Cocos nucifera*) and the Mangrove Palm (*Nypa fruticans*). The flotation qualities of the Coconut fruit are well known and its dispersal onto seashores is well documented, with nuts retaining the ability to germinate after 200 days' immersion in salt water. The fruit of the Mangrove Palm is dispersed by currents around estuaries, deltas and mudflats, but in this case the seed has already germinated before it falls from the palm and it merely takes root after lodging in a suitable site. There is some evidence that the fruit of *Pseudophoenix sargentii* may also be dispersed by the sea.

Fresh Water

Many palms grow beside streams and the fruit which fall into the water may be dispersed by the currents. However, only one palm *Ravenea musicalis* from Madagascar, is a true aquatic. Its seedlings have an unusual hook-like structure which catches on obstructions and helps hold the seeds in place on the bottom of the river while they establish. The seeds of many streamside palms including rheophytic species, are probably also dispersed by currents but the germination takes place on the bank of the stream (see p. 17). In North America a proportion of the seeds of *Sabal minor* are distributed by floods. Species of *Leopoldinia* from South America have fibrous fruit which float and are dispersed by streams.

Animals

Fish

The fruit of some palms are a source of food for fish and it has been observed that the feeding fish may congregate under a fruiting tree waiting for the ripe fruit to fall. In the Amazon region of South America, a large range of fish species are known to feed on palm fruit (among the fruit of other plant species), including the electric eel. Some seeds are destroyed during digestion, however others are voided intact and may germinate if they lodge in a suitable site on the bank.

Birds

Birds are the most common dispersal agents of palm seeds. Mature palm fruit advertise their availability by colour (commonly red to scarlet or orange) and are often also glossy. Back-up display features include massed fruiting on the infructescences and contrasting colours between the fruit and swollen rachillae of the infructescence (*Chamaedorea* spp., *Pinanga* spp.). Some palm fruit are often dull-coloured (green, purple, brown to black) but are still attractive to fruit-eating birds as these fruit often have an oil-rich, fleshy mesocarp. The oilbird of Central America and the Caribbean region, is an important dispersal agent of seeds of this type. Birds may void seeds after passage through their gut or regurgitate the seed after digestion if it is too large to be passed.

Many birds are known to feed on the fruit of palms. These include crows, mockingbirds, hornbills, parrots, cotingas, pigeons, emus, rheas and cassowaries. Fruit-eating pigeons are particularly significant in some areas and because they are nomadic, may be responsible for the island distribution of some palms. In South America, cotingas are heavily dependent on the fruit of *Euterpe* species and the distribution of the birds closely overlaps that of the palms. A similar close relationship occurs in South Africa between the palm nut vulture and *Raphia australis*, with the bird feeding on the thin layer of fleshy mesocarp which surrounds the seed.

Even flightless birds may be involved in palm seed dispersal. Cassowaries are important dispersal agents in Queensland and New Guinea, eating fallen palm fruit of a range of species and plucking the ripe fruit from the infructescence of small palms such as *Calyptrocalyx*, *Licuala* and *Linospadix*. In Australia, emus have been observed to feed on the fallen fruit of *Livistona*. In South America, rheas feed on the fruit of a range of palms, including *Acrocomia* spp., *Butia* spp. and *Syagrus* spp.

Mammals

Many mammals feed actively on palm fruit. Fallen fruit are often targeted by rodents such as rats, mice and squirrels. These animals gnaw the outer flesh and may eat the fruit on the spot or carry it away to a secure site which often results in the aggregation of seeds, some of which may germinate later. Large mammals such as bears, baboons, civets, coyotes, monkeys, gibbons, apes, peccaries, wild dogs, opossums, agoutis, raccoons and hyraxes feed on ripe palm fruits. In Florida, the bear is an important dispersal agent for a number of palms, including *Rhapidophyllum hystrix*, *Sabal etonia*, *S. palmetto* and *Serenoa repens*. Monkeys and baboons may carry away sections of infructescences for their own consumption. In Indonesia, gibbons feed on the fruit of *Arenga obtusifolia* and the orangutan is partial to the fruit of *Borassodendron borneense*. The palm civet is an important dispersal agent in Malesia feeding on a wide range of palm fruit. It is an excellent climber, eating the fruit while still attached to the infructescence and capable of climbing even lianes (*Calamus*, *Daemonorops*, etc.). In America and Mexico, coyotes have been shown to be important dispersal agents of *Washingtonia* species. Being opportunistic feeders they probably also eat the fruit of other palms such as *Brahea* and *Sabal* species. A number of animals found in rainforests and other heavy forests are opportunistic feeders of fallen fruit, including that of palms. The agouti is known to feed on the fruit of many palms in South America (such as *Acrocomia* spp., *Chamaedorea tepejilote*, *Orbignya* spp., *Phytelephas* spp., *Syagrus* spp. and *Welfia georgii*). Peccaries also feed on similar palm species. In Malesia tapirs may feed on fallen fruit and wild dogs are known to eat fallen palm fruit. In Africa, elephants are important dispersal agents for many fruits including a number of palm fruits (*Phoenix* spp., *Hyphaene* spp. and *Borassus aethiopum*). Asian elephants also defecate palm seeds in their dung.

Bats Fruit-eating bats are adept fliers and clamber over fruit clusters when feeding, often knocking many fruit to the ground. They gnaw the outer flesh of palm fruit and discard the seed. They are probably unimportant as long-distance dispersal agents of palm fruit but nevertheless many different species of bat feed on many different palms in different countries and they must make some contribution to seed dispersal.

Fruit which attract bats often have a distinctive odour reminiscent of butyric acid. It is uncertain whether this is the case in palm fruit or whether other factors apply. Palms on which bats feed include *Arenga* spp., *Borassus aethiopum*, *Coccothrinax* spp., *Corypha* spp., *Hyophorbe* spp., *Hyphaene thebaica*, *Livistona* spp., *Phoenix* spp., *Raphia hookeri* and *Roystonea* spp. It is interesting to note that bats are not attracted to the fruit of various climbing palms, presumably because of the danger of becoming entrapped by the thorny, entangling climbing devices employed by these palms.

Humans

The influence of humans on the dispersal of palms cannot be underestimated. Species such as the Peach Palm *Bactris gasipaes*, the Date Palm *Phoenix dactylifera* and *Parajubaea cocoides* are unknown in the wild and must have been tamed by humans over thousands of years. Similarly, distribution of the Coconut (*Cocos nucifera*) has been greatly influenced by humans as has, more recently, that of the African Oil Palm (*Elaeis guineensis*). The process of human dispersal is continuing with many commercial species being grown in local plantations for a wide range of products. Further, the amazing dispersal of palms to various countries by palm enthusiasts interested in them solely for their ornamental qualities should not be ignored.

THE EFFECTS OF FIRE

Few species of palm grow in habitats that are affected by fire. Thus the majority of palms—which are found in rainforests and other moist forests—have no mechanism to cope with fire.

However, palms growing in some open habitats which are subject to frequent fires cope well with these events. The common result of such fires is the above-ground destruction of living plant parts. Palms with subterranean trunks are well insulated by the soil and the growth apex is protected. The overlapping leaf bases may also offer protection.

A number of palms found growing in vegetation known as Cerrado in Brazil and Paraguay are well adapted to coping with regular fires. This open, scrubby vegetation burns on a regular basis and the palms which grow in it have deep subterranean trunks and small crowns which are readily replaced. When the seeds germinate, a sinker grows deep into the ground before the first true leaf is produced from its apex, thus conferring early protection on the seedlings. Palms which grow in Cerrado vegetation include *Acanthococos emensis*, *Allagoptera brevicalyx*, *A. campestris*, *A. leucocalyx*, *Attalea geraensis*, *Butia paraguayensis*, *B. purpurascens*, *Syagrus flexuosa*, *S. lilliputiana* and *S. loefgrenii*.

In Australia, many species of *Livistona* grow in open forests and woodlands which are subject to regular burning. Recovery in these palms after a fire is generally very rapid. Species include *Livistona benthamii*, *L. eastonii*, *L. humilis*, *L. inermis*, *L. loriphylla* and *L. muelleri*.

Palms which grow in savanna vegetation in parts of South America and Africa are also subject to fire on a regular basis.

LITTER-COLLECTING PALMS

The erect leaves of some palms channel fallen litter towards the centre of the palm, which eventually becomes covered with detritus in various stages of decay. These palms may derive some benefit from this litter collection, including protection of the crown from herbivores and the release of nutrients from the material as it decays. Litter-collecting palms include *Areca tunku*, *Asterogyne martiana*, *Bactris militaris*, *Daemonorops verticillaris*, *Eugeissona minor*, *Johannesteijsmannia* spp., *Pinanga ridleyana* and *Salacca* spp. One of the most spectacular litter-collecting palms is *Marojejya darianii* from Madagascar. This palm has semi-erect, nearly entire leaves which overlap at the base and form an almost watertight container which accumulates litter and debris. This litter apparently rots rapidly and the nutrients may be important for the growth of the palm.

ANIMAL INTERACTIONS

Ants of certain species live within some structures on various climbing palms, and they may offer some protection to the plants. Organs inhabited by ants include net-like ochreas, overlapping whorls of thorns and leaf sheath auricles. The ants build nests with covered walkways between them and commonly farm scale insects, milking them for their honey dew. Disturbed ants swarm and bite ferociously, thus discouraging herbivores or other intruders. When disturbed the ants which inhabit the ochreas of *Korthalsia hispida* in Malaysia make a distinct rattling noise by banging their abdomens against the dry surface.

Various species of birds and bats may be associated with palms for shelter and roosting purposes, as well as feeding. In South Africa, the Palm Nut Vulture (*Gypohierax angolensis*) has very close links with *Raphia australis*, so much so that the distribution of the bird and the palm closely coincide. The vulture feeds nearly exclusively on the thin, sweet, fleshy layer around the seed. Also in South Africa, palm swifts favour palms for nocturnal roosting and also make their cup-shaped nests in the leaves. In Asia, the Baya Weaverbird (*Ploceus philippinus*) favours palms for nesting, with surveys showing that about eleven species of palm are used. In South America, the distribution of cotingas and species of *Euterpe* closely overlap, with the birds being heavily dependent on the fruit of the palm for their food.

PNEUMATOPHORES

Pneumatophores are specialised roots which grow upwards (negatively geotropic) and facilitate gas exchange—particularly oxygen. They are commonly found on the roots of mangroves and they enable these plants to survive in soils which are deficient in oxygen. These roots also occur in palms which grow in swampy habitats, thus enabling them to survive in the oxygen-deficient soils of swamps and bogs. Pneumatophores extend well above the soil surface and are an obvious feature of such communities. Internally, these roots contain an abundance of air spaces that facilitate gas exchange. They are often prominent close to the trunk of the palm but in *Mauritia flexuosa* may appear more than 30 m away. Palms which produce pneumatophores include *Euterpe oleracea*, *Mauritia* spp., *Metroxylon* spp., *Phoenix paludosa*, *Prestoea montana*, *Raphia* spp. and *Zombia antillarum*. Strangely, the Mangrove Palm, *Nypa fruticans*, does not produce pneumatophores on its roots.

STINGING CRYSTALS

The flesh of the fruits (and other parts) of some palms contain needle-like crystals (raphides) of calcium oxalate which are highly irritating to sensitive membranes. These crystals can produce an intense, painful reaction if eaten and even if the fruit are handled excessively—such as during fruit collection or seed sowing—can produce a range of sensations from warm and tingly to an intense, hot, itching reaction which can last for up to an hour. Fruits of this type are best handled with gloves. Palms which have stinging crystals in their fruit include species of *Arenga*, *Caryota*, *Chamaedorea*, *Drymophloeus*, *Gaussia*, *Pseudophoenix* and *Wallichia*.

It is interesting to note that the fruit of these same palms, which can produce such a bad reaction in humans, are eaten avidly by animals such as civets, gibbons and wild dogs and many birds.

CHAPTER 5

THE CULTIVATION
OF PALMS

Palms are very popular subjects for cultivation and have an enthusiastic following in many countries. Although the majority of palms are of tropical origin there is quite a range that can be grown in temperate regions and even a few that are very cold-tolerant and will still survive in areas with a cold winter climate. A number of species thrive in inland areas and a few extremely tough palms will even tolerate semi-arid conditions. In fact, given water, a good range of palms can be grown in inland towns and can significantly add to the atmosphere of such places.

Palms are very easy plants to grow given the basic requirements of good soil, plenty of water and protection from hot sun at least while they are young. When mature, most palms are sun-loving plants and must have full sunshine to achieve maximum growth. This is of special significance in temperate regions where the winter sun is weak and is often obscured by clouds. Within the group there are however, shade-loving plants which will not tolerate hot sun without bleaching or burning.

Any garden can support a palm or a selection of palms. The more hardy sun-lovers need an open position while the smaller types generally need more protection and can be successfully mixed with shrubs. Houses with large areas of protected garden can grow a good selection of small to moderate-sized palms and these can be complemented with plants such as cordylines, dracaenas, cycads and large ferns. With the exception of large or very bulky palms, such as *Phoenix canariensis*, most species do not take up a great deal of room in a garden. This is especially true once a trunk is formed and the crown is elevated. Grass, shrubs or annuals can be grown right up to the base of the trunk if so desired. Palms like company and look good when planted in groups. A collection of mixed species in a suitable situation in the garden will always be a source of interest and conversation.

LANDSCAPING WITH PALMS

The advantage palms have over other plants used in landscaping is that their growth habit and dimensions are entirely predictable. Thus a palm can be chosen to fill a particular niche and the landscaper can be certain, given suitable conditions of soil and growth, that the palm will grow as predicted. Few other plants can be chosen accordingly.

Palms are well suited to planting as specimens in lawns or in groups, provided that they are spaced so a mower can be used between them. The palms impart an interesting appearance to a lawn planted in this way and do not interfere with the growth of the grass. Grass, provided that it is regularly watered and fertilised, will grow right up to the base of a palm trunk. Mowing in these circumstances is no major problem, since a palm tree cannot be ringbarked and the mower or 'whipper snipper' can be run close to the trunk without damage. If long grass continually grows too close to the trunk and is unsightly a small garden bed can be dug around the trunk.

Palms that produce suckers may be a problem in lawns because the suckers appear at various distances from the trunk. These can be controlled by regular close mowing or else digging out the lawn and creating a circular garden bed around the palm.

Tall, noble palms such as *Pigafetta filaris* and *Roystonea regia* make excellent specimen plants and a well-grown subject is always a source of admiration. Careful placement of such plants will ensure that the maximum benefit is received from their stateliness. They are very effective as lawn specimens which can be viewed from windows or terraces, or framing stairways and balconies.

Tall palms can also be planted in pairs or groups of one species or mixed with other palms or other plants. Tall palms look especially effective when planted close together in pairs. If planted side by side and at a slight angle away from each other their trunks tend to curve apart and add an interesting aspect to the garden. Tall palms also look good beside driveways and paths and some species may be spaced regularly to form impressive avenues.

Palms, especially the smaller- to medium-growing types, prefer company and are best planted in groups or mingled with other plants. The grouping helps create a congenial atmosphere and produces a better overall effect. The small shade-loving palms such as *Chamaedorea*, *Laccospadix*, *Licuala*, *Linospadix* and *Reinhardtia* mix very well with ferns. As with ferns, these plants help to create an atmosphere of coolness and are a welcome retreat on hot summer days.

Palms provide excellent shade despite the deceptive open appearance of their crowns. A palm grove provides a shady retreat on a hot day and if combined with water features, such as a pond, stream or a pool, the atmosphere is further enhanced. An oasis, after all, is a shady retreat in the desert where water comes to the surface and the shade is, of course, provided by palms. Because of their umbrella-like structure and predictable growth, palms can be strategically sited to provide shade for outdoor living areas, barbecues and swimming pools. In hot, dry, inland areas the shade provided by hardy palms such as *Phoenix canariensis* and *P. dactylifera* is greatly appreciated. With their generally slender trunks, palms have the added advantage of not hindering air movement or breezes while still providing shade.

When planning to landscape with palms numerous factors should be considered in selecting a suitable species. One factor which is often neglected is the debris created by the palm. Fruits of some species are messy and are generally borne in profusion. Coconuts are heavy objects and these palms should not be planted where their fruit can create a hazard. Some palms retain their fronds when dead and this may be regarded as an untidy habit.

Palms with spiny trunks should not be planted beside paths or where children run or play as the long spines can easily damage sensitive areas, especially the eyes. The basal pinnae of *Phoenix* species are modified into stiff, sharp spines which can inflict painful wounds. Although commonly planted in parks and along streets they are usually severely trimmed to keep the spines out of reach.

ASPECT AND CLIMATIC CONSIDERATIONS

Many factors are involved in the choice of a site suitable for palms. These include aspect, slope, soil type and drainage, exposure to frost, wind and sun and competition from neighbouring trees. The species of palm and its tolerance of climatic conditions also significantly influences the decision.

The aspect selected for palms will be influenced by the conditions existing in the area. In temperate regions of the Southern Hemisphere, a southerly aspect is generally unsuitable since such an area receives little or no winter sunshine; but in the Northern Hemisphere such a position will be suitable. Aspect is generally unimportant in inland regions since only hardy palms can be grown in such areas and these tolerate hot sunshine even when quite small. In tropical and subtropical areas, many species of palm thrive in full sun, especially if watered heavily during dry periods. Most palms are of tropical origin and have generally adapted to the humid atmospheres that exist in the hot seasons. Exposure to full sun, while tolerated by the palm, may be to the detriment of the plant's appearance and the same plant may look lusher and less tattered if grown in partial shade.

Shade can be of special significance in subtropical and tropical regions, since sensitive species burn very quickly when exposed to full sun. Sun filtered through the canopy of established trees is excellent for shade-loving palms provided that the plants are not exposed to long periods of hot sun through breaks in the canopy. Many tropical trees shed their leaves rapidly during dry periods in the summer, often just prior to flowering. This must be considered when planting shade-loving palms since irreparable damage can be done during a comparatively short period of exposure.

Palms, especially some of the smaller types, can be grown under deciduous trees in temperate regions. Here the exposure to winter sun is greatly beneficial, as is the annual mulch of leaves. Shade-loving palms will tolerate exposure to full sun for short periods during the day, provided the sun is not too hot at the time. Morning sun or late afternoon sun is most suitable.

Palms grown under established trees suffer from root competition for nutrients and water, and must be fertilised, watered and mulched at regular intervals if they are to flourish. If the palms were planted at the same time or soon after other trees, then the competition is of less significance since the palm's root system can become established before competition is severe. Once established, small-growing palms appear to be able to cope very well with root competition.

Palms are generally quite tolerant of wind and some species such as the Coconut are renowned for their resistance. Physical damage such as shredding of the leaves of fan palms or entire-leaved palms, or the tips of leaflets of feather palms occurs with mild to strong winds, while gales or cyclones may tear the plants out of the ground or break the trunk. Such damage is obvious, but insidious effects such as reduced vigour and stunting

Fig. 22
Effect of growing conditions on a palm's trunk

stunted growth caused by poor growing conditions as indicated by close growth rings and pinched trunk

vigorous growth resulting from good growing conditions as indicated by widely spaced growth rings and fat trunk

may accompany cold winds and draughts as often occur around large buildings. Hot, drying winds, especially on hot summer days, may cause desiccation of small plants and sensitive species. Salt-laden onshore winds may similarly damage species in coastal districts. Some wind protection can be achieved by establishing windbreaks or planting in the lee of buildings or existing shrubs and trees.

Frost greatly limits the range of palms that can be grown and most tropical species collapse after exposure to even light frosts. Even moderately hardy palms may suffer damage to young and developing leaves, especially if the frosts are heavy or prolonged or the plants are in active growth. Some frost protection can be obtained by planting on slopes where air circulation is unimpeded, near buildings or under established trees.

SOIL

Palms are not exacting in their soil requirements and will grow in a tremendous range of soil types. In fact, some palm or other will probably grow in all but the very worst types of soil. Most garden soils are reasonably well structured and fairly well drained and will support a range of palms. The better the soil type, the better the growth of individual palms and the greater the variety that can be grown.

A good guide to the palms that will succeed in an area can be obtained by looking around the region—especially in municipal parks and gardens and old established gardens—for palms were a very popular subject early this century. When looking at existing palms, however, and the soils and conditions where they grow, be aware that an established palm is tremendously tolerant of neglect and the present conditions may not reflect the nurturing required in their establishment phases.

Most palms seem to prefer an acid soil with a pH of between 6 and 6.5. Where soils are more acid than these levels, some correction with dolomite or ground limestone may be necessary. Palms may grow equally well in light, sandy soils and heavier types, but the techniques of soil preparation, planting, watering and mulching may differ. Red soils, whether they be well-structured mountain soils, heavier basalt soils or laterites, have proved to be ideal for the growth of a range of palms. Those derived from granitic and other grey soils can also be good if plenty of organic material is dug in before planting.

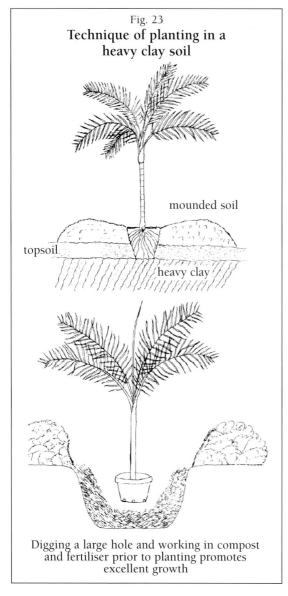

Fig. 23
Technique of planting in a heavy clay soil

mounded soil

topsoil

heavy clay

Digging a large hole and working in compost and fertiliser prior to planting promotes excellent growth

Clay Soils

Heavy subsoil or clay will support hardy palms, but their establishment will be fraught with difficulty and better results will be obtained if the clay is first worked to the stage where it is friable. Clay is difficult to cultivate whether it is wet or dry but can be improved by adding a soil conditioner such as gypsum, organic matter and/or topsoil. This mixture should be gradually worked into the clay until the friability is improved. The more effort that is put into working up such heavy clay soils, the better will be the results. Regular applications of organic matter will not only improve the friability, but will prevent surface compaction, cracking and drying out. Also, as palms like soils rich in organic matter, it will greatly aid the establishment of young plants. If the clay soils are excessively wet, surface or underground drains may be needed to remove the water. For a planting technique in clay soils (see Fig. 23).

Light, Sandy Soils

Palms thrive in light, sandy soils because they are usually warmer than heavy soils (and this may be important in winter), are very well drained and aerated and are often quite deep. Such soils are often poor nutritionally however, and are usually deficient in organic matter. When excessively deficient in the latter and when very dry (especially fine sands where the granules pack together) they may become water-repellent and difficult to wet. Heavy mulching with organic materials will help prevent this and will greatly aid the establishment and growth of young palms. As sandy soils tend to dry out rather rapidly, regular watering will be necessary, especially in the establishment phases.

Alkaline Soils

Alkaline or calcareous soils have an excess of calcium salts in the soil and this creates nutritional problems with elements such as zinc, iron, magnesium and manganese. Such soil types are not uncommon in inland areas. Hardy palms such as *Syagrus romanzoffiana*, *Brahea armata*, *Butia capitata*, *Chamaerops humilis*, *Coccothrinax* spp., *Gaussia* spp., *Livistona mariae*, *Phoenix canariensis*, *Phoenix dactylifera*, *Pseudophoenix sargentii*, *Sabal* spp. and *Washingtonia* spp., can be established quite readily in such soils, however the range of suitable species is not as great as in an acid soil (see also *Calciphiles*, p. 17). Sometimes the palms may grow while suffering a nutritional problem and then their appearance is not good.

The range of palms and their vigour can be increased by adding large quantities of organic matter to the soil surface and acidifying with sulphur.

Waterlogged Soils

Many palms grow naturally in wet or swampy conditions and most of these can be established in well-drained garden soils if they are watered regularly. Palms are rather difficult to establish in waterlogged conditions and the small plants tend to linger and appear unthrifty or die. Such soils can be improved by drainage. *Salacca zalacca* will grow in wet conditions but is tropical in its requirements. *Livistona australis* can be established around the margins of wet or boggy areas and is adaptable to a range of climates. While most palms do not like stagnant water many will happily tolerate wet spots where water is moving, such as soaks or springs in a hillside. Palms grown in such sites will be much more tolerant of exposure to hot sunshine.

Saline Soils

Salted or saline soils have generally lost their structure and the excess of minerals in their profile makes the establishment of palms very difficult. Some hardy palms such as *Phoenix dactylifera* and *P. canariensis* will continue to grow in salted soils after they have established themselves, but seedlings will not grow in such situations. Saline soils must be drained so that the salt can be leached through the profile and removed in the drainage water.

PLANTING

The technique of planting palms is not significantly different from that of any other type of plant, although the correct procedure will greatly aid the establishment of the young palm.

The palms should be thoroughly soaked in their containers before planting and this is best achieved by immersion of the container in water. Pre-soaking is especially important if planting in dry times or in dry areas. The planting hole should then be dug and this should be sufficiently wide and deep to accommodate the root system. Palm enthusiasts advocate the digging of a hole much bigger than the root system, a concept described by one professional horticulturist as a $5 hole for a $1 plant. Not only is the soil broken and loosened to aid root penetration, but compost, organic matter and fertilisers may also be mixed into the soil. This work certainly results in rapid establishment and healthy growth, but it is a question of labour and time.

If the soil in the planting hole is dry it should be filled with water and allowed to drain before planting. Some fertiliser (either inorganic, organic or slow-release) should be thoroughly worked into the bottom of the hole before planting. About a handful is usually adequate. Some growers advocate working rotted manure and compost into the soil surrounding the hole and this technique definitely aids establishment and is of particular value in sandy soils. If planting into clay soil avoid forming a sump by digging too far into the clay (see Fig. 23).

Once the planting hole is ready, the palm should be removed from its container and dead or badly coiled roots trimmed back or straightened out. The plant should be placed in position with the top of the container soil just below that of the garden soil and the soil firmed around the roots and watered thoroughly (5–10 litres per plant). The area around the plant should then be mulched (see *Mulching*, p. 79).

WATERING

Many palms are extremely tolerant of dryness, but all look and grow better with supplementary watering during dry periods. Generally the type of area palms come from is a good guide to their dryness tolerance. Shade-loving palms are generally rather sensitive to dryness and quickly wilt and look tatty if allowed to become very dry. Similarly, dwarf or small palms have few reserves to cope with dryness and are best if watered regularly. It is important to recognise the water needs of a palm before planting and also it is most convenient to grow species with similar moisture requirements together. The shade- and moisture-loving species prefer plenty of water and it is very difficult to over-water them—especially in dry times.

Established palms have a large root system which ramifies a considerable distance through the surrounding soil. This is true of even small-growing palms and it means that the plants are able to tap a vast reserve of soil for their nutrient and water requirements. It also means that water should not be applied only to the soil immediately near the base of the plant but rather should be more widely distributed. If the soil where the roots are growing is well soaked at each watering, the plants can last for longer intervals without water.

Water is best applied during cool periods or in the evening or early morning. The best technique is to thoroughly soak the root area so that the maximum amount of water becomes available to the plant. The method of water application (that is, whether it be by hose, sprinkler or drip irrigation) depends on availability and is not of major significance to palms. Sometimes with dwarf palms it is handy to have overhead sprinklers installed as these can be used for both watering and cooling and humidifying the area on hot, dry days. Mulches greatly improve the efficiency of watering and their application is strongly recommended (see *Mulching*, p. 79).

FERTILISERS

Palms are strong-growing plants and once established are very responsive to fertilisers and manures. Large palms such as *Syagrus romanzoffiana*, *Phoenix canariensis* and *Roystonea regia* are gross feeders and their growth rate can be increased very significantly by heavy applications of fertilisers. Such rates would be about 3–5 kg applied each time per tree, in two to three dressings per year. For the medium and smaller species, light dressings of fertiliser at regular intervals are beneficial to their growth and appearance. Fertilisers are best applied during the warm months when the plants are in active growth. This is particularly true for temperate regions where the plants must become hardened in the autumn in order to endure the cold winter. Fertilisers are best applied to moist soil and if watering is not possible they should be applied just before, during or after rain. They should be scattered over the surface of the ground within the drip line of the canopy.

Many growers advocate the use of organic fertilisers such as blood and bone, bone meal or hoof and horn. Animal manures are also very valuable but should not be applied when too fresh. These materials release a steady supply of nutrients for growth and are also very beneficial to the soil. They are excellent for small-growing palms, although they are quite expensive and may be difficult to obtain. Inorganic fertilisers are cheaper and more readily available and despite some claims to the contrary, are an excellent means of promoting growth. Most inorganic fertilisers are readily soluble and supply nutrients quickly to the plants. A balanced, complete fertiliser should be applied annually. The benefit can be maximised if any fertilisers are applied in conjunction with organic mulches.

Palms seem to require high levels of nitrogen and two supplementary dressings with nitrogenous fertilisers each year promote growth and improve their general appearance. Suitable materials include urea (46 per cent nitrogen), ammonium nitrate (35 per cent nitrogen), ammonium sulphate (20 per cent nitrogen) and calcium nitrate (15 per cent nitrogen). Nitrogenous fertilisers are best watered-in soon after application. Coconut Palms are particularly responsive to applications of nitrogenous fertilisers.

Slow-release fertilisers applied at planting will aid in the establishment and early development of a palm. A handful of such fertiliser, well dispersed through the soil in the planting hole, is all that is necessary.

PRUNING

Pruning in palms is generally limited to the removal of unwanted or unsightly material such as suckers, clusters of fruit or dead fronds. In a number of species the dead fronds are retained for many years and hang as a brown skirt against the trunk. Although regarded as unsightly by some people, this feature can add interest to a palm. Good examples are *Washingtonia filifera*, *W. robusta* and *Sabal* species. This feature, despite being natural, does have some drawbacks such as providing shelter for vermin like rats, sparrows, starlings and mynahs, as well as leaving the plants susceptible to fire and vandalism.

Dead fronds can be cut off regularly as they age and shredded for garden mulch. Maintenance of this type becomes a problem as the palms grow taller. *Archontophoenix* spp. and *Roystonea* spp. are self-shedding while in some others the fronds may hang for a considerable time before falling. This is one of the more significant disadvantages of the popular Queen Palm (*Syagrus romanzoffiana*). The Royal Palm (*Roystonea regia*) is basically very similar in general appearance to the Queen Palm and would be a better choice if maintenance is seen to be a major problem.

Palms growing in public areas, particularly those planted near paths, must be observed regularly for potential hazards. Many species have large thorns on their petioles and in species of *Phoenix* the basal pinnae are modified into stiff, sharp spines. In these circumstances fronds must be removed when hazardous. Similarly, in public areas the spiny trunks of some palms such as *Aiphanes aculeata* can be kept smooth to a height of about 2 m to avoid injury. Climbing palms, such as species of *Calamus* with hooked cirri or flagella are best planted away from public areas or else the trailing cirri/flagella should be trimmed out of reach.

Suckers can be a nuisance in gardens, especially if the palms are planted in lawns. This is requently the case with *Phoenix dactylifera* and *P. reclinata*. Large suckers can be removed by digging and severing the rhizome below ground. Small suckers can be kept under control by regular close mowing.

Fruits are quite a decorative feature of many palms, especially when carried *en masse* on the tree. On the ground, however, they are often regarded as being unsightly and in some species may even be soft and messy. If this is a problem the inflorescences are best cut off early in their development prior to fruit production.

Some people believe that the trunks of tall palms can be cut off at a manageable height and the trunk will resprout below the cut. This is entirely erroneous and any such attempt will result in the death of the plant or, if it is a suckering palm, the death of the stem which is cut. Palms have no capacity to survive once the growing apex is damaged beyond repair and while the fronds can be trimmed back or removed entirely, on no account should the crown be touched.

MULCHING

Mulching around palms, especially those which have been recently planted, will aid significantly in their successful establishment. Palms are by their nature shallow-rooting plants, with a large percentage of the roots being found near the soil surface. Mulching helps to keep the roots cool and the soil sufficiently moist to encourage new root growth. Its presence on the soil surface greatly reduces the stresses which plants experience during hot, dry weather. As a bonus, the mulch also reduces the development of weeds which compete with the plant for nutrients and moisture. It also aids in water penetration.

Mulches should be applied thickly as soon as possible after planting to minimise drying of the soil surface and weed germination. After application the mulch should be watered heavily to compact the surface and reduce dispersal by wind. Although palms are fairly tough once established, the mulch should be maintained at least in the early years after planting.

A range of materials can be used as mulches for palms, but some are available only on a localised basis. Mulches can include inorganic materials such as gravels, screenings and water-worn pebbles; and organic by-products such as bark, woodchips, shavings, sawdust, peanut shells, grass hay, seaweed and so on. A layer of black polythene sheeting covered with water-worn pebbles is frequently used in tropical regions. Extra fertiliser must be applied to palms mulched with organic materials because these substances will use plant nutrients from the soil as they are broken down, thus depriving the plants of these materials for growth.

TRANSPLANTING PALMS

Advanced specimens of many species of palms can be transplanted quite successfully, provided a significant portion of the root system is removed intact and that correct after-care is provided.

The amount of root system to remove varies with the species, its age and size and the soil type where it is growing. In some cases the position also influences the amount of root that can be excavated—for example, palms that are growing against walls. Some palms such as *Syagrus romanzoffiana* can be removed with very little of the root system left, while others such as *Sabal* spp. need a considerable proportion for success. As a general rule a good solid root ball 1–1.5 m across should be removed with any palm 3 m tall or more. Naturally the larger the plant the bigger the root ball that will be necessary for success. For very large palms, mechanical equipment such as backhoes, cranes and low-loaders are necessary.

In light, sandy soils the soil tends to fall away from the roots in the transplanting process, particularly if it is dry. In such situations the palms may need to be pre-wet or watered regularly during the move. Palms in heavy soils do not have this problem and tend to be slightly more amenable to shifting.

Transplanting Time

The time of transplanting can be a very significant factor in the success or failure of the project. In temperate regions, palms make very slow growth over winter and should not be transplanted during late summer, autumn or winter. The best time in these areas is spring or early summer. In subtropical regions the timing is similar, except that spring is often dry in such regions and if irrigation is not available then it is best to wait until November or December when regular rains are usual. In tropical regions, temperature is not a limiting factor but the plants are best left in the ground during the dry winter months and transplanted during the wet season when conditions are ideal for growth.

Palms transplanted from the wild are generally difficult compared with garden-grown subjects whose roots were probably moulded by the nursery container in which they were originally grown. It has also been noticed that palms grown in large containers transplant much more readily than those which were planted-out while quite small. In tree farms where the palms are grown in the ground until they are of a size suitable for transplanting, the root system of the palm may be cut regularly every six months to keep it compact and lessen the shock when the plant is finally transplanted.

Technique

The transplanting technique for a palm is basically similar to that of any other plant. The roots are severed in a circle around the trunk by cutting with a sharp instrument and a trench is gradually excavated to reach the deep roots. Once the roots are cut right through, the root ball can be surrounded by hessian or similar material to hold in the soil and reduce drying of the root system until the plant is repositioned. This material is tied and the plant can then be removed and transported to its new position where it is placed in a well-prepared hole of similar size.

After-care

It is essential to realise that the care following transplanting is probably more important to ensure the success of the project than is the actual transplanting process. Following planting, the soil should be firmed thoroughly around the root ball and a temporary reservoir for the retention of water should be made on the soil surface. The plant should then be thoroughly watered by filling the reservoir several times until the entire profile and root ball are wet. This watering should be undertaken as soon as practicable after transplanting. Watering should be maintained and the reservoir filled at regular intervals during dry periods for at least two to three months.

During transplanting any damaged fronds should be removed and a few of the others shortened to reduce both transpiration and wind resistance. It is extremely important to stake the palm to prevent movement of the roots in the new soil and this should be done immediately after watering. The most effective system of staking is to use three stakes and guy wires to support the trunk. Further care consists of regular watering, mulching and the control of any pests which may attack the palm in its weakened condition.

As already mentioned, some species of palms transplant more readily than others. A generalisation that seems to hold is that those palms with a compact root system transplant better than those with fewer but long-spreading roots.

EFFECTS OF RENOVATIONS

Major renovations are not uncommon in gardens especially if a house changes hands or is redeveloped into flats, units etc. It is difficult to start new plants among established ones and for this reason garden renovations often consist of building up soil levels to give the new plants a start.

Palms will withstand having a garden bed built up around them provided that the process is fairly gradual. An existing palm will quite happily tolerate an increase in soil height around its trunk of 15–30 cm but if a metre or more of soil is suddenly applied, the results will probably be catastrophic. Palms have the ability to form adventitious roots in the lower part of their trunk and if the filling is applied gradually the new roots can grow into it without harmful effects. If suddenly covered with a deep layer of soil the root system can rot and the whole plant can die quickly.

RECLAIMING UNTHRIFTY PALMS

While palms generally tolerate neglect well, if the growing conditions are unsuitable or severe in the extreme, plants can become unthrifty and may linger like this for many years. Nitrogen deficiency is a very common cause of this malady but other causes include poor drainage, lack of organic matter and regular periods of dryness. Unthrifty palms frequently become the target for pests such as mealy bug or scale, but these attacks are usually because of the weakened condition of the plant, not the cause of it.

Unthrifty palms can be invigorated by identifying the cause and correcting it. Nutrition is the most frequent cause and a good dose of a balanced fertiliser fortified with extra nitrogen will usually promote a dramatic response. A heavy mulch of some organic material should also be applied and the plants should be regularly watered during dry periods. Dead fronds and leaf sheaths should be removed and burnt. This will expose any pests and these can be killed by spraying or natural predators.

Following these steps will usually be sufficient to promote strong new growth. If strong growth is to be maintained, however, it is essential that the fertilisers, mulches and water be applied at regular intervals.

CHAPTER 6

PESTS, DISEASES AND AILMENTS OF PALMS

PESTS

Although a large range of insects feed on palms, very few cause serious damage and most of the attacks are of a minor nature or are of nuisance value. Some pests may be very damaging at certain times of the year or during seasons that favour their build up. Some may be closely linked with the life cycle of the palm and may be very restricted in their distribution. The following notes may help in their identification and control.

Palm Weevil *(Rhynchophorus cruentatus)*

Palm Weevil is a devastating pest of palms that is particularly injurious to *Phoenix* species and Coconut Palms. The adults are large (about 6 cm long), shiny weevils with a long, curved snout. The larvae bore into the leaf sheaths of the upper crown and feed actively in the soft central tissue of the upper trunk. They pupate in a case made of fibres from the palm. Severe infestations of this pest can kill healthy palms. Frequently, the affected leaves sag and the tops of damaged palms may blow out in strong winds. Control is by careful observation and drenching with a contact insecticide at the first sign of insect activity.

Palm Leaf Beetle *(Brontispa longissima)*

The Palm Leaf Beetle is a serious threat to some palms in tropical countries. A native of Indonesia, it has become distributed through many islands of the Pacific, east as far as French Polynesia. It has been relatively recently introduced into Australia (1979) where it is now known around Darwin in the Northern Territory, the Torres Strait Islands and the mainland of north-eastern Queensland, south as far as Cairns. Details of its life cycle and habits have been studied by research workers from the Northern Territory Department of Primary Production and have been published as an *Agnote* written by T.L. Fenner (Ref. 81/13, May 1981). Details of the pest are reprinted here with their permission.

The adult beetle is about 1 cm long, narrow, flat, with short legs and is orange and black. The larva is plump and cream with a series of spines down each side and a pair of curved hooks at its rear which resemble those of an earwig. The adult beetles are sluggish during the day and move at dusk and during the night. They do not fly long distances but have the ability to increase in numbers very quickly and to become entrenched rapidly once in a new area. It is quite conceivable that the pest will spread further in tropical Australia and other countries. Under no circumstances should palms be taken from infested areas into new areas.

Palm Leaf Beetle attacks only the very young palm leaves (spear leaves) before they have unfolded. The eggs are laid in small groups and are covered with excreta. Both adults and larvae shelter within the folds of the young leaf and chew large areas of the surface of the developing leaflets. The eaten area turns brown and as the leaf expands it becomes disfigured and takes on a scorched appearance. Once the leaflets expand, the pests move to a new unexpanded leaf. The life cycle of this beetle from egg to adult takes about forty-five days.

Attacks by this pest are not only unsightly, but if they persist can reduce the palm's vigour

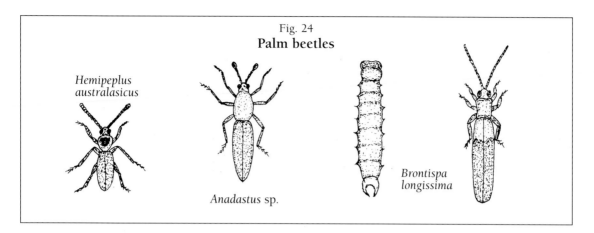

Fig. 24
Palm beetles

Hemipeplus australasicus

Anadastus sp.

Brontispa longissima

and in extreme cases result in its death. Seedlings as well as adults are susceptible to attack. Palm Leaf Beetle favours as its main host the Coconut (particularly the cultivar Malay Dwarf) but attacks have also been recorded on about twenty-six other species, including Royal Palms (*Roystonea* spp.), *Carpentaria acuminata*, *Ptychosperma macarthurii*, *Archontophoenix alexandrae*, *Caryota* spp., *Phoenix* spp., *Washingtonia* spp. and *Syagrus romanzoffiana*.

Control of Palm Leaf Beetle is by spraying the unfolded leaves with a contact and residual insecticide such as carbaryl. A thorough penetration of the spray is essential and a second spray about seven days later is recommended. Studies into biological control of the pest using a tiny parasitic wasp (*Tetrastichus brontispae*) are being carried out in Australia.

Another similar beetle (*Brontispa mariana*) is known to attack palms in the Mariana Islands and the Marshall Islands. This species bores into the terminal bud and has been known to kill Coconut Palms on the islands of Saipan, Tinian and Rota.

Small Palm Beetles

A couple of Australian beetles feed on the leaves of palms, attacking both natural populations and garden-grown plants.

In Darwin, palms may be damaged by a species of beetle in the genus *Anadastus*. The adults are slender, about 1 cm long and are orange and black in colour. They can be distinguished from the introduced Palm Leaf Beetle by the conspicuous knobs on the ends of the antennae. They feed on developing fronds and also recently expanded leaves, damaging the leaf surface.

Another beetle, *Hemipeplus australasicus*, feeds on the leaves of some palms in tropical areas of Australia. It is a small, somewhat elongated beetle about 4 mm long, which is pale yellow with a reddish-brown spot on the thorax. It chews the surface of the leaf, leaving brown, sunken areas.

Both of these beetles are mainly minor pests but if outbreaks become severe they can be controlled by spraying with endosulfon.

Palm Dart Butterflies

Palm dart butterflies are destructive pests which severely damage the leaves of a number of palms including *Archontophoenix alexandrae*, *A. cunninghamiana*, *Chrysalidocarpus lutescens*, *Syagrus romanzoffiana* and *Cocos nucifera*. The caterpillars are slender, translucent and greenish-grey or brownish in colour depending on the species. They have a prominent, striped head and wriggle actively when disturbed. They feed within the protection of a shelter which they make by sewing the edges of leaflets together. Usually the damage is confined to the outer half of a frond and the distal half of the leaflet is left projecting from the eaten part.

The adults are small, fairly colourful skipper or dart butterflies that are active on sunny days. Two species are known to damage palms. The Yellow Palm Dart (*Cephrenes trichopepla*) is widespread in New Guinea and tropical Australia and the Torres Strait Islands extending as far down the east coast as Rockhampton, north-eastern Queensland, Australia. Its larvae feed on a variety of palms and the species is commercially significant because of its damage to young leaves of sprouting Coconuts. The Orange Palm Dart (*Cephrenes augiades*) is distributed from the Torres Strait Islands down the east coast of Australia to Sydney. Its larvae feed on a wide range of native and exotic palms.

Control is readily achieved by the application of a stomach poison such as carbaryl or the spores of *Bacillus thuringiensis* applied as recommended by the manufacturer. Regular spraying is necessary throughout the summer months.

Palm Moth (*Agonoxena phoenicia*)

Palm Moth is a relatively minor pest of Alexandra Palm. The larvae feed on the underside of the leaves, while hidden beneath a flimsy silken web. The adults are a small, grey moth with hairy, fringed, slender wings. The upper wing has a longitudinal dark band. It seems to be confined to north-eastern Queensland, Australia, and can be readily controlled by spraying with carbaryl.

Palm Butterfly (*Elymnias agondas*)

Palms on Cape York Peninsula, Australia are eaten by the larvae of the Palm Butterfly which is especially fond of species of *Calamus*. The caterpillars are green with paler stripes and an unusual forked tail. The butterflies are quite handsome, having dull brown and white wings with two or three prominent eyes on the hindwing. This insect is of interest but as a pest is insignificant.

Palm Case Moth

The Palm Case Moth is a very minor pest of palms, the caterpillars of which feed on the leaflets of such species as *Archontophoenix cunninghamiana* and *Dictyosperma album*. The caterpillar itself grows to about 2.5 cm long and is pale coloured with orange or dark red bands. It constructs a silken bag in which it lives and this is decorated on the outside with pieces of palm fronds. The pieces used increase in size as the caterpillar grows and they provide an effective camouflage. Apart from chopping off pieces of leaflets to adorn its case, the caterpillar feeds by grazing the surface tissue of the palm fronds. Damaged sections eventually turn brown. Damage to a palm is usually minimal and populations of the pest can be controlled by hand picking and squashing. Should spraying be necessary carbaryl or pyrethrum are effective.

Other Caterpillars

A few caterpillars can be troublesome on palms, especially in the late summer and autumn months. Sections are usually eaten from the leaflets or fronds while they are young and these symptoms are quite obvious. Control of these pests is the same as outlined for palm dart butterflies.

Light Brown Apple Moth (*Epiphyas postvittana*) This is a particularly insidious pest because the larvae shelter in the developing frond and join adjacent leaflets together with silk strands. They feed on very young fronds and the damage is obvious only when they expand.

Painted Apple Moth (*Orgyia anartoides*) This pest has a conspicuous hairy caterpillar which can be recognised by the tufts of hairs on its back. It feeds on palm fronds while the tissues are soft. Other hairy caterpillars, which coil when disturbed, are also occasionally troublesome.

Borers

These pests bore holes into the trunks and crownshafts of some palms. Their incidence is generally minor and the scattered holes do not appear to cause any major damage or reduce the structural strength of the trunks. The holes are frequently visible in the older parts of the trunk but these are the vestiges of borers long gone.

The borer that most commonly attacks palms seems to be a species of longicorn beetle. Its larvae bore shallow tunnels, usually in the softer upper part of the trunk (where it exudes sawdust) or sometimes in the crown and crownshaft (where it exudes a white waxy sap that resembles toothpaste).

Healthy palms are normally able to tolerate borer attack without any setback but unthrifty palms may be weakened further. Control can be achieved by probing the active holes with soft, pliable wire or by injecting them with solutions of contact insecticide.

Cane Weevil Borer (*Rhabdoscelus obscurus*)

Cane Weevil Borer is a pest associated with sugarcane which can also cause severe damage to palms, even resulting in their death. It is widespread from northern Queensland, Australia, through Malesia and Polynesia to Hawaii. The adult is a weevil, the larvae of which tunnel into the lower parts of the stem and leaf-bases, feeding actively on the inner tissues. In young palms, this feeding activity frequently leads to collapse and death of the plant. The leaf-bases may also be riddled in older palms resulting in premature leaf death. In green tissues, whitish jelly exudes from the holes and on the trunk sap exudates may cause local staining. Severe infestations may result in trunk splitting. This pest attacks a wide range of palms and may be severe in nursery situations. Control is by spraying regularly with a suitable insecticide.

Palm Seed Borers and Feeders

Various species of insect feed on the fleshy endosperm of palm seeds rendering them incapable of germination. Indicators which can be used to identify affected seeds include a marked reduction in weight and a telltale exit hole (although sometimes it may be covered with a flap of tissue). Seeds of species of *Calamus* are frequently subject to attack with often a high percentage being eaten.

Most damage is done by small beetles of the families Bruchidae, Platypodinae and Scolytinae. The presence of a fat, white, legless larva in the seed is usually their hallmark. Adults of bruchid beetles are unusual, curved beetles about 4 mm long with a deflexed head, while the others are small cylindrical beetles. Seed-boring beetles and their larvae will attack palm seed while it is hanging on the tree. Bruchid beetles of the genus *Caryobruchis* feed on the seeds of a range of *Sabal* species in the southern states of the USA, Mexico and islands of the Caribbean region. The degree of predation is variable and in some years up to ninety per cent of the seeds may be destroyed. In Central America, bruchid beetle larvae destroy a high proportion of the seeds of *Scheelea rostrata* that are produced annually.

The larvae of the Australian moth *Blastobasis sarcophaga* feed on fallen palm fruits. The adults of this species are small, dull-grey moths with hairy, fringed wings.

On Cape York Peninsula, Australia, the fruit of *Caryota rumphiana* are frequently eaten-out by the larvae of a butterfly known as the Cornelian (*Deudorix epijarbas dido*). The butterflies are fairly small; the males have dull orange markings on the wings and the females are brown.

Control of seed-eating pests is usually impractical, although if it is suspected that a batch of seed is contaminated it can be soaked in a solution of maldison before sowing.

Aphids or Plant Lice

Aphids are small, soft-bodied insects which are also called greenfly or blackfly. They congregate in colonies on young, developing leaves and feed by sucking the sap. They range in colour from yellowish to green, black or reddish. They feed in mixed colonies consisting of adults and young in various stages of development. Winged and non-winged adults may be present depending on the time of the year. Young are mostly born alive and these pests can build up in numbers very rapidly in favourable conditions.

Aphids are very messy feeders, excreting considerable amounts of honeydew which may stain the area where they are feeding. This material also attracts ants and is a good substrate for sooty mould which renders infected plants unsightly. Aphids are also vectors of plant virus diseases.

Aphids are not a major pest of palms but they sometimes feed on small foliage palms. Small infestations can be squashed or dislodged with a jet of water. Severe infestations may require spraying with a contact insecticide.

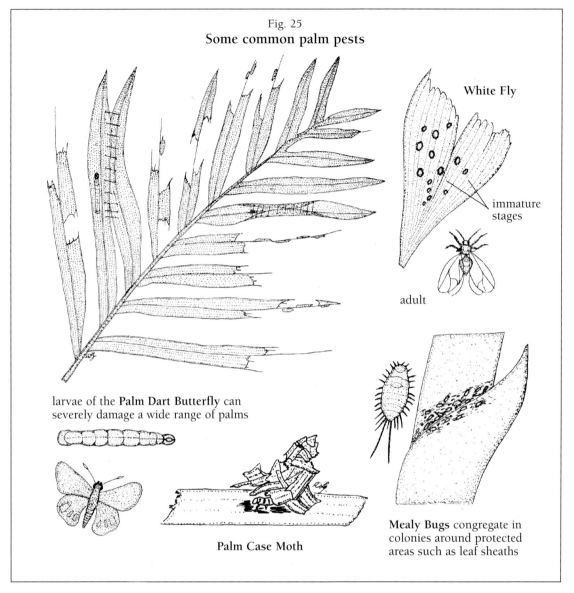

Fig. 25
Some common palm pests

White Fly

immature stages

adult

larvae of the **Palm Dart Butterfly** can severely damage a wide range of palms

Palm Case Moth

Mealy Bugs congregate in colonies around protected areas such as leaf sheaths

White Fly (Family Aleyrodidae)

A few species of white fly attack palms, sometimes quite severely. The adults are small moth-like flies (1–1.5 mm long) and hardly noticeable (except when they fly in clouds), but the juvenile stages may be quite conspicuous and are often mistaken for scale insects. In one common Australian species the juveniles are 2–3 mm long, oval in shape and black with a conspicuous white fringe of waxy segments. Another widespread species has greenish, waxy juveniles which are easily mistaken for scales. Like scales, they cluster under leaf sheaths, etc. and suck the palm's sap. Their feeding activities are followed by sooty mould which develops on their exudates. White flies are a sporadic problem, severe in some years and absent in others. They can be somewhat difficult to control and systemic sprays such as dimethoate may be needed to clean up persistent attacks. Small outbreaks can be dispersed by hosing.

Thrips

Thrips are a minor problem in palms but in some years they may congregate in considerable numbers and feed on flowers. The resulting damage causes premature browning of the flowers and reduces fruit set. Some foliage palms such as species of *Chamaedorea* may suffer leaf damage from feeding thrips. These pests favour years with mild seasons, can build up in numbers quickly and produce a dry, granular appearance on the damaged areas. The symptoms are similar to those of spider mites but no webbing is produced and the elongated shape of adult thrips is obvious when viewed under a hand lens. Control is rarely necessary but can be achieved with a contact spray such as pyrethrum.

Mealy Bugs (Family Pseudococcidae)

Mealy bugs are sucking insects which are readily distinguished by the covering of waxy, white secretions and filaments which exude from their body and impart a mealy or powdery appearance. The insects are plump and soft and feed by sucking the plant's sap. They tend to congregate in dry, protected areas such as under leaf sheaths, in spathes and in the developing crown. Discarded white, waxy threads and debris litter the areas where they feed. As their numbers build up, the damage they cause can be quite significant and may weaken the plant. Developing fronds in particular may be distorted and open misshapen because of attacks which occurred while they were still folded in the crown. Weakened plants such as those grown in very dry situations or those held in pots for too long may be subject to very severe infestations. Neglected indoor palms are often a favoured host for this debilitating pest.

Mealy bugs are not easy to control. Small infestations can be eradicated by dabbing with methylated spirits but this is impractical on large plants. Systemic sprays such as dimethoate provide the answer but they must be used carefully. A better starting point is to keep the plants healthy and hope that the mealy bugs will be controlled naturally by their predators. Cleaning, including the removal of leaf sheaths and other materials that cover infestations of the pest, aids in their control by exposing them to the elements and their predators.

Red Spider Mite (*Tetranychus urticae*)

Red Spider Mite, also called Two Spotted Mite, is a major pest of palms and other plants. It is a tiny eight-legged animal that clusters on the underside of leaves and feeds by sucking the sap. The females are dull yellow with two conspicuous dark blotches on the back. Winter populations are red all over. The animals usually live in colonies and spin a protective webbing above where they feed. They relish dry conditions and can build up in numbers extremely rapidly. Affected leaves lose their lustre and take on a dry, often granular appearance as if gently sandblasted. If feeding continues the leaves yellow and may shed prematurely. Mites may be particularly severe on weakened palms planted in a dry situation or on neglected indoor palms in a dry atmosphere. Regular syringing or hosing of their leaves will help reduce build up of numbers but spraying is usually necessary for their control. Sprays such as difocol or dimethoate are usually suitable and may need to be applied in tandem to reduce the development of spray resistance in the mites.

Scale Insects (Families Coccidae and Diaspididae)

Scale insects shelter beneath readily recognisable shells which may be leathery, waxy or cottony. The insect feeds by sucking the plant's sap. Scale insects may be solitary (causing minimal damage) or cluster in colonies and may cause localised yellowing and distortion of young growth. They release sugary exudates as they feed and sooty mould may grow on these exudates. Ants commonly attend scales to collect this sugary material.

Significant infestations of scale insects should be controlled by spraying with successive applications of white oil. This material may damage plants, particularly at high temperatures. White oil should not be applied at temperatures above 25°C. The addition of the contact insecticide maldison to the white oil often enhances control.

Pink Wax Scale (*Ceraplastes rubens*)

Pink Wax Scale is very common in tropical and subtropical regions and attacks a wide variety of plants including some palms. The adults are usually 3–4 mm across, round in shape with irregular margins and a dull, waxy, red colour. They are gregarious and congregate around the crown of the palm. The young scales attack the developing leaves but are not obvious until they are adults and the leaves are mature. They are frequently guarded by ants which feed on their exudates.

Healthy palms are not worried by the attacks of this scale but unthrifty palms may be weakened further. Some yellowing of the leaflets is associated with these attacks and black smut or sooty mould may develop on the sugary exudates of the scale.

Palm Scale (*Parlatoria proteus*)

Palm Scale is a common pest of palms, belonging to the group of armoured scales, so called because of the very hard, conical shells which cover each feeding insect. Adult Palm Scales are white, about 3 mm across and have a couple of concentric rings near the top. They usually cluster on protected green parts of the palm such as under leaf sheaths, in rachis grooves or in folded leaflets. The scales are generally well hidden in early stages of infestations but as the attacks become more severe the scales spread to open areas and become obvious. If leaf sheaths or the protective coverings are removed, a mass of scales in all stages of development will requently be exposed. Heavy infestations of Palm Scale can be quite debilitating and the pest should be controlled as soon as noticed.

Premature yellowing of leaves and the presence of sooty mould around the leaves are sure indications of infestation. When infestations spread onto the undersides of leaves they cause small, yellow patches in the area of the leaf around which they feed. Small infestations may be cleared up by dabbing with methylated spirits (one part to four parts water), or sponging with soapy water, but severe outbreaks must be sprayed with a material such as maldison, methidathion or dimethoate. Plants should also be fertilised to help restore their vigour.

Circular Black Scale (*Chrysomphalus aonidum*)

Circular Black Scale is a pest of palms in the tropics and subtropics. The covering is deep, purple-black and circular and grows to about 2 mm across. It is one of the armoured scales and the shell has a raised central point which is of a lighter brownish colour than the rest of the covering. This scale is found in clusters on the upper and lower surfaces of the leaflets of palm fronds and along the petiole and around the leaf sheaths. Tissue where the pests feed yellows prematurely. Severe infestations of this scale can badly debilitate palms and comments and control as outlined for Palm Scale apply equally to this species.

Cottony Cushion Scale (*Icerya purchasi*)

Cottony cushion Scale is frequently found on palms, especially in subtropical areas. It is a very distinctive, large scale, readily recognised by its plump, white, soft body which can be easily squashed between the fingers. This scale normally feeds in small, crowded clusters and is often to be found on the smaller palms. In *Chamaedorea* species it feeds in places such as the rachis of the inflorescences and under the upper leaf sheaths. Cottony Cushion Scale is not a major problem and infestations can frequently be removed by hand. It is generally kept under control by the activities of a predatory ladybird beetle and its larvae.

Flat Brown Scale (*Eucalymnatus tessellatus*)

The young stages of Flat Brown Scale are rather difficult to discern, producing an almost transparent covering which is pressed flat to the leaf. The adult is of similar shape but is easily seen because it is dark brown and about 5 mm long. This scale is a relatively minor pest and usually appears in small numbers. It attacks cycads and some palms such as *Howea forsteriana* and *Chamaedorea elegans* and produces localised leaf yellowing. Control is as for Palm Scale.

Soft Brown Scale (*Coccus hesperidum*)

Soft Brown Scale is mainly a pest of young fruit trees but it also attacks palms. It is a messy species and its presence is usually advertised by the masses of secretions and associated Sooty Mould found wherever it feeds. The adults can be readily identified by their soft, waxy coverings, which are brown or of a mottled colouration. They are flat, oval and about 5 mm long. They congregate in dense colonies on the fronds, petioles and leaf sheaths of a variety of palms in temperate and subtropical regions. Infested palms are very unsightly because of the sticky nature of the secretions of the scale and the heavy growth of Sooty Mould. The appearance of the palms can be improved by regular hosing. Control of the scale is by spraying with white oil.

Fern Scale or Coconut Scale (*Pinnaspis* sp.)

Fern Scale attacks palms in subtropical and tropical regions and is frequently associated with infestations of the Palm Scale. Colonies of this scale resemble a sprinkling of shredded coconut on the surface where they are feeding. These are the coverings of the male scales and a close inspection shows the dull, inconspicuous females scattered among the males. On palms, this scale favours sheltered situations such as the folds of leaflets, the junction of stems, under bracts, sheaths, etc. It causes yellowing of tissue where it feeds and generally debilitates the plant. Control can be difficult, requiring the use of chemicals such as dimethoate.

Nigra Scale (*Saissetia nigra*)

Nigra Scale is not a major pest of palms but occasional outbreaks occur. The species can be readily identified by the adults which have a large (4–6 mm long), shiny-black, raised covering which is usually oval to oblong in shape. Nigra Scale are very fast growing and it is not

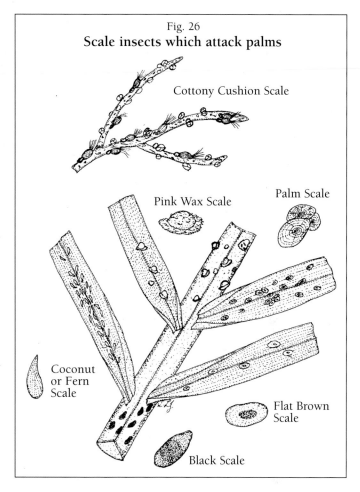

Fig. 26
Scale insects which attack palms

Cottony Cushion Scale

Pink Wax Scale

Palm Scale

Coconut or Fern Scale

Flat Brown Scale

Black Scale

uncommon to find mature specimens on a recently expanded palm inflorescence. Individuals are easily dislodged from the plant and it is a wise precaution to remove them before they build up in numbers.

Grasshoppers and Locusts

Palms, particularly those grown in the tropics, may be literally decimated by large grasshoppers or locusts. Such ravages not only impart an unsightly appearance to the plants, but they may also retard or stunt their growth, particularly if the attacks are maintained. Severe, continued attacks may even result in the death of young plants.

The symptoms of damage are obvious from the tattered leaves. Huge chunks of the leaflets are eaten at random and the midribs are left to impart a skeletonised appearance. In severe attacks all of the leaflets, often including the midribs, are eaten and the palms look very sad indeed.

Control of grasshoppers and locusts is a major problem since they are generally nomadic and are such ravenous feeders that much damage is done before insecticides can work. Squashing is effective on small palms but requires regular observation. Hand-picking by torchlight at night is a useful procedure. A continual cover of stomach poison such as carbaryl may be necessary during times of severe infestation. Attacks on large palms are difficult to remedy and all that can really be done is to ensure that the plant is sufficiently healthy to recover from them.

Nematodes

Nematodes or eelworms are tiny, microscopic round worms that are common in soils and decaying organic matter. A few species are parasitic and may cause economic losses. Most damaging nematodes attack the root systems but some infest the leaves and other above-ground parts of plants.

About twenty-two species of nematode have been found infesting *Chamaedorea* in Florida. These include the serious pest, Burrowing Nematode (*Radopholus similis*), which is commonly found on *Chamaedorea elegans*. This pest induces root stunting and rotting, lesions on the stem and yellowing and wilting of the fronds. In severe cases thousands of palms have been infected. Nematodes are difficult to control and sanitisation is the best option; sowing clean seed in sterilised propagation mixes and avoiding any later contamination with unclean containers or plants.

Red Ring Disease (*Rhadinaphelenchus cocophilus*)

Red Ring Disease is a devastating disease of Coconuts and Oil Palms in the Caribbean region, which is caused by a nematode. Leaflets develop reddish spots which have a darker perimeter, sometimes appearing as rings. As the disease progresses into the growth apex the leaves droop conspicuously and the plant collapses and dies. This disease can spread quickly into adjacent plants and the only control is to destroy plants in the early stages of infection to prevent its spread.

Termites

Termites are major destroyers of palms in tropical areas. They are communal insects that feed on wood, both living and dead, and the most destructive species attack from below ground level and work their way up the trunk. The attacks are usually rapid and affected plants quickly decline in health and the leaves lose their lustre, become pale green or yellow and wilt badly even if watered. Once plants exhibit these symptoms, there is little that can be done for them and they usually collapse and die.

Termites are extremely difficult to control and regular inspections of palms are necessary in tropical areas. Their feeding lines, which are subterranean, can range a long way and once they establish contact with a tree invasion is rapid. Any obvious colonies in the area should be destroyed along with dead or dying trees, stumps, etc. Colonies can be destroyed with strong contact insecticides applied in solution or as dusts.

Rats and Mice

At certain times of the year, these rodents may become very destructive to young palms by gnawing through the stems. Significant damage may be caused within a short time to seedlings in nurseries. Control is by baiting.

DISEASES

Few diseases affect palms, although there are some which are lethal and highly contagious. Most, however, are of a minor or sporadic nature and may become worse during adverse conditions. The best control is avoidance by good cultural conditions. A healthy plant is better able to resist disease than an unhealthy plant. In nursery situations disease problems often arise as a result of overcrowding, poor potting mixes and lack of air movement.

Lethal Yellowing

Lethal Yellowing is a devastating disease of palms which is caused by a mycoplasma-like organism (an organism closely related to a virus). It was first noticed late last century (1891) in Jamaica and has more recently spread to Florida (first being noticed on the mainland in 1971), Texas and the Yucatan Peninsula in Mexico. It is also apparently more widespread in the Caribbean region than was previously thought and very recently has been identified in West Africa and Tanzania.

Lethal Yellowing has killed millions of palms wherever it has appeared. In particular, the disease has had a major impact on Coconut Palms and in Jamaica it decimated the copra industry, since the tall Jamaican varieties were extremely sensitive to the disease. Recently these have been somewhat replaced by the Malayan Dwarf variety which is resistant to the disease. Lethal Yellowing also caused havoc in Florida by decimating populations of ornamental palms, particularly Coconut Palms, but also *Veitchia merrillii* and *Pritchardia pacifica*. In Texas, it has decimated populations of *Phoenix* palms. The resulting plant losses have changed forever the landscape of many urban areas in Florida and elsewhere. To date, at least thirty species of palm are known to be affected by Lethal Yellowing.

Young palms (eighteen months or younger) can become infected as well as mature specimens. After infection and an incubation period of about six to twelve months, symptoms begin to appear, followed by rapid death of the palm. Symptoms are: yellowing and death of the young fronds in the crown; death of the roots; flower-blackening; premature nut-shed; a change in colour of mature fronds which then droop and become necrotic, beginning with the basal ones and moving upwards in the crown; collapse of the spear leaf and plant death.

This disease has been the subject of intensive study, and while much has been learned, no cure has come to light. It is believed to be transmitted by a leafhopper—*Myndus crudis* is strongly suspected to be the culprit. Lethal Yellowing is not present in the Pacific region and its spread to island communities which rely so heavily on the Coconut for survival would be disastrous.

Hartrot, Fatal Wilt or Sudden Wilt

Hartrot is a lethal disease which affects Coconuts and Oil Palms in Central America, South America and the Caribbean region. It was first reported from Surinam in 1908 and produces similar symptoms to Lethal Yellowing, however the phloem sap of infected palms has been found to contain trypanosome-like flagellates of the genus *Phytomonas*. Symptoms start as a yellowing of the older leaves. Other symptoms include root death, leaf browning, premature nut-shed, blackening of inflorescences and rotting of the spear leaf associated with a foul odour. After about two months the infected plant collapses and dies. No known cultivars of the Coconut are resistant to this disease.

Cadang-Cadang

Cadang-Cadang is a disease which has caused severe losses of mature Coconut Palms in the Philippines and Guam since it was first noticed in the early 1900s. It is caused by a virus-like organism and also affects palms other than Coconuts, including the Oil Palm (*Elaeis guineensis*) and *Corypha utan*. In the Philippines, it has been estimated that more than thirty million Coconut Palms have been killed by this disease. Symptoms include loss of vigour, leaf mottling and yellowing, a reduction and eventual inhibition of fruiting followed by death of the palm.

Cinnamon Fungus (*Phytophthora cinnamomii*)

Cinnamon Fungus is a vigorous root pathogen that attacks a wide range of plants with devastating effects. Most palms seem to be resistant to its effects but some species are sensitive, particularly those originating in dry climates. Examples include *Syagrus romanzoffiana* and other *Syagrus* spp., *Caryota mitis*, *C. urens*, *Coccothrinax* spp., *Cocos nucifera* and *Trithrinax* spp. Most damage from this fungus occurs when such palms are planted in heavy soils where drainage is poor.

Symptoms of this disease in palms are an unthrifty appearance with a crown mainly of pale or yellowed leaves. In times of severe stress the plants may wilt or even collapse suddenly and die.

Control of affected plants is very difficult and prevention is the best procedure. Sensitive palms should only be planted in well-drained soils. Mulching is beneficial.

Red Top

Red Top is a disease caused by a fungus about which little appears to be known. It attacks the growing apex of the palm and new leaves which develop at the time of the attack open out yellow. The centre of the plant eventually collapses and dies and when inspected the tissue is covered with rusty red sporing bodies. Palms known to be attacked include species of *Archontophoenix* and *Howea*. Bordeaux mixture controls this disease but it must be applied thoroughly on all surfaces when symptoms are first noticed.

Core Rot

Some palms, such as *Livistona australis* and *L. chinensis*, are subject to a blackening of the upper part of the trunk in the crown. In severe cases the whole crown rots and becomes a black, soggy mess. This problem is believed to be caused by a fungus but it is not commonly encountered and there are no known control measures.

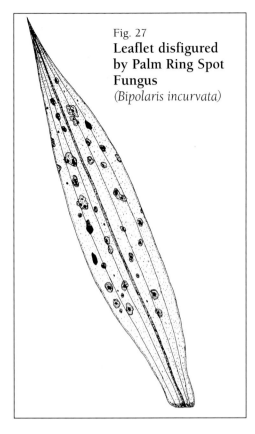

Fig. 27
Leaflet disfigured by Palm Ring Spot Fungus
(*Bipolaris incurvata*)

Palm Rust

The leaves of certain species of palms, such as *Howea forsteriana*, *Livistona chinensis*, *Phoenix dactylifera* and *Washingtonia filifera*, are attacked and disfigured by a rust fungus. These attacks are generally minor and the disfigurement is mostly limited to a peppering of rusty red to orange spots on the leaves. Young or old leaves may be attacked and there is no known means of control, although sprays of copper materials such as copper oxychloride may be beneficial.

Palm Leaf Blight (*Gleosporium palmarum*)

The leaves of some feather-leaved palms may be infested with Palm Leaf Blight fungus which causes brown spots and blotches on the pinnae. Sometimes these spots are surrounded by a pale or yellowish area. Attacks are usually of a minor nature and can be controlled with copper sprays. This disease has been noticed on *Archontophoenix cunninghamiana*, *A. alexandrae* and *Ptychosperma elegans*.

Palm Leaf Spot (*Pestalotiopsis sp.*)

Palm Leaf Spot causes small spots on the leaves of some palms, such as *Caryota* spp., *Roystonea* spp. and *Syagrus romanzoffiana*. The fungus is not very aggressive and attacks are usually of a minor nature. The fungus is mainly found on palms in shady positions and is hardly known on plants in the sun. Control is rarely necessary but persistent attacks can be cleaned up with copper sprays such as cuprox.

Palm Halo Disease (*Stigmina palmivora*)

Palm Halo Disease is a common pest of palms in humid, tropical climates. It starts as tiny brown spots which enlarge and darken, eventually developing a distinct yellow halo around the lesion. Lesions may coalesce if the disease worsens. This is a common disease of unthrifty palms, especially those in shady conditions. Control is by improving the palm's health and also spraying with cuprox at an early stage of infection.

Stem Bleeding Disease (*Endoconidiophora paradoxa*)

Stem Bleeding Disease enters through wounds in the trunk and shows up as dark, shiny stains which on closer examination are seen to be from exuding sap. If unchecked, the sap flow will continue and the stained area will enlarge. Control is by cutting out the infected tissue until healthy tissue is exposed and treating with a copper fungicide. This disease can be common on Coconut Palms.

Palm Ring Spot (*Bipolaris incurvata*)

Palm Ring Spot is a serious fungal disease which causes leaf spotting, often with associated yellowing. The lesions are sunken, brown to black and circular to irregularly shaped. It may cause the death of the spear leaf, resulting in a rot which may spread into the crown. This disease is a significant problem in tropical areas when humidity is high for long periods and is worse in crowded situations (such as nurseries) or where air movement is restricted. Affected species include *Archontophoenix* spp., *Howea* spp. and *Chrysalidocarpus lutescens*. Copper oxychloride should be used against severe infestations while Mancozeb will give control of lesser outbreaks.

Golden Cane Palm Disease (*Phytophthora nicotianae*)

This disease infects the Golden Cane Palm (*Chrysalidocarpus lutescens*) and in severe cases may cause plant death. Early symptoms often show as brown flecks or spots which may spread or coalesce and severely damage the infected leaves. Sometimes the disease attacks the terminal bud resulting in death of that stem; occasionally whole plants succumb. This disease has been found in Hawaii. Control is by using copper sprays at an early stage of infection. Flecks on the leaflets of this palm may also be caused by boron deficiency.

Coconut Bud Rot (*Phytophthora palmivora*)

Coconut Bud Rot is a common disease of Coconuts which infects developing leaves, eventually spreading to the apical bud where it may cause rotting. Severe infestations can result in death of the infected plant. Often the first symptoms show up in an emerging spear leaf. This disease is widespread in countries where Coconuts are grown.

Date Palm Stem Rot (*Fusarium oxysporum*)

This disease, which affects Date Palms in California, USA, has potential as a serious pest. It attacks the trunks of palms causing brown to blackish necrotic lesions which may exude dark gum. Leaves yellow and fall prematurely giving the plants a spindly appearance. The virulence of this disease is enhanced when the fungus Chamaedorea Palm Blight (below) is also present. Copper sprays may offer some control.

Chamaedorea Palm Blight (*Gliocladium vermoeseni*)

Chamaedorea Palm Blight is a serious disease of clumping species of *Chamaedorea* in Florida. The disease often begins in the lower leaf sheaths and then frequently enters adjacent stem tissue, often resulting in stem death. Lower leaves develop necrotic patches, yellow and die prematurely. Damaged stems exude sap and characteristic pinkish-brown spore masses are produced from damaged areas. Affected plants have a sparse appearance and sucker growth may be suppressed. Control of this debilitating disease is by hygienic practices, allowing free air movement around the plants and spraying with copper-based fungicides.

Betel Nut Rot (*Colletotrichum gloeosporoides*)

This disease affects the Betel Nut (*Areca catechu*) and is of commercial significance since it can proliferate in plantations. It affects developing leaves causing the rachis to wither, as well as the inflorescences and young fruits which drop prematurely. Mature fruit can also shrivel and rot. The same disease can also cause severe leaf damage to young seedlings of a range of palms. Control is by spraying with copper sprays at an early stage of infection.

Sooty Mould

Sooty Mould or Black Smut is a widespread and common fungus disease that attacks a large variety of plants. It causes very little damage but does render the plants very unsightly. Since it grows on the excretions of sucking insects such as scales, white flies and mealy bugs, its appearance should be used as an indication of the presence of these pests. Control of Sooty Mould involves the control of the sucking pests and it usually disappears within a couple of weeks of their removal.

NUTRITIONAL PROBLEMS

Palms, like other plants, extract nutrients from the soil so that they can grow, flower and fruit. Most elements are generally available from soils in sufficient quantities for plant growth. Soil reserves may, however, become depleted, as after heavy, leaching rains, or some elements may become unavailable in soils which are excessively acid or alkaline. Palms growing in such soils will then suffer and may show deficiency symptoms. Palms grow better if a balance of nutrients is available to them in the soil, rather than if there is an excess or deficiency of one particular element. The same applies to potted plants, especially those growing in soil-less potting media.

Boron Deficiency

Boron is required in minute quantities for protein synthesis in actively growing areas, such as stem and root tips. The availability of boron to plants diminishes with increasing pH. Some leaf spotting in *Chrysalidocarpus lutescens* has been linked with boron deficiency. Nutrient trials with this palm also showed transverse yellow streaks on the leaflets, reduced growth and eventual death of the terminal bud. Correction is by a foliar spray of boric acid at 1 g per litre of water. Excessive quantities of boron are highly toxic to plants.

Iron Deficiency

Iron is needed in the plant for the proteins used in chlorophyll synthesis. The earliest symptoms of deficiency are pale green new growth, followed by yellowing of the areas between the veins (which remain green). The older leaves on the plant remain green. Iron deficiency is common in plants growing in calcareous soils since the iron is frequently held in a form unavailable to the plants. Iron chelates applied to the soil are useful for correcting iron deficiency in such circumstances.

Magnesium Deficiency

Magnesium is an important constituent of chlorophyll and is used in photosynthesis. Deficiency symptoms show up in the older leaves which develop yellow areas between the veins, with the midrib and veins themselves remaining green. In severe cases the whole leaf becomes pale yellow and eventually necrotic. Correction of magnesium deficiency is by applying magnesium sulphate to the soil (50 g per 10 square metres) or as a foliar spray (15–20 g per litre of water).

Nitrogen Deficiency

Nitrogen is important in the formation of chlorophyll and in the synthesis of amino acids. Palms suffering nitrogen deficiency have a yellowish appearance in the crowns with the older leaves becoming quite yellow or even bleached, with necrotic patches on the leaflets. Younger leaves may remain green. Such afflicted plants stand out quite dramatically, especially if they can be compared with a healthy green specimen. Nitrogen deficiency in palms is particularly common in coastal districts with deep sandy soil. It can be readily corrected by applying fertilisers rich in nitrogen (see *Fertilisers*, p. 78).

Potassium Deficiency

This element is essential for the production and transport of carbohydrates (especially starch) in plants, as well as strengthening tissues and being involved in chlorophyll synthesis and respiration. Deficiency symptoms first appear on older leaves as necrotic spots and blotches, often with a marginal necrosis. Control is by soil applications of potassium sulphate or potassium chloride.

Zinc Deficiency

Zinc acts as a catalyst for chemical reactions in plants, particularly those involving growth hormones. Zinc deficiency in plants causes symptoms known as little-leaf because the leaf development is impaired resulting in a reduced, often almost perfect, stunted leaf. Such symptoms have been reported in palms following pot trials (*Principes* 23, 1979, pp. 171–2). Garden-grown plants have been observed with similar symptoms but it is not positive that these arise from zinc deficiency. Zinc deficiency can be easily corrected by application of balanced trace element mixtures or zinc sulphate applied to the soil or as a foliar spray (3–5 g per litre of water).

OTHER DAMAGING FACTORS

Wind Burn

Developing palm fronds may be damaged by hot, dry winds so that when they open they are disfigured by grey or white papery patches. This is usually a minor problem and is only apparent in sensitive species following long periods of such unfavourable conditions.

Frost Burn

Very many species of palms, especially those of tropical origin, are sensitive to cold spells and especially frost. Young plants are more susceptible than mature specimens with some height, however, there are many recorded cases of frost damage or even death in tall palms following unusual or unseasonable frosts. Roots, especially the growing tips, are very sensitive to freezing and ground temperatures of -2°C may cause considerable damage. A palm has a single growing apex and lacks any secondary measure of survival and once irreparable damage has been done to this apex, death follows. Late spring frosts and black frosts can be particularly damaging. An interesting phenomenon has been noted in palms with a clumping growth habit. Whereas sensitive species with a solitary trunk may be killed, established clumping palms may regrow from the base even though all of the aerial stems have been killed. Such damaged plants may, in a few years' time, completely re-establish themselves.

The symptoms of frost damage are the blackening and collapse of developing leaves and the formation of brown patches on mature fronds. Sensitive species collapse dramatically, usually turning brown or black with the crown becoming a soggy mess. Protection from mild frosts can be obtained by planting close to buildings or large shrubs or under the protective canopies of established trees. In cold areas, sensitive species must be grown in glasshouses.

Salt Burn

In coastal districts, onshore winds deposit salt from sea-water on the leaves of nearby plants. This salt is usually damaging to the leaf tissues and only the most resistant plants tolerate it without any damage. Many species of palm suffer this damage which is known as salt burn. However, this generally causes only a minor setback to growth and is mainly detrimental to the plant's appearance. The margins of the leaflets are burnt and become white and papery. There is no means of control, except perhaps hosing the plants down thoroughly after onshore gusts. Damage from salt burn has been noticed in *Syagrus romanzoffiana* (generally tolerant), *Caryotis mitis* and *C. urens*, *Chrysalidocarpus lutescens* and *Phoenix rupicola*. Species of *Caryota* seem to be especially sensitive to salt burn.

Hail

Palms, because of their large leaves, are very sensitive to damage by hail. The usual symptoms are large chunks torn out of leaflets, leaf segments, inflorescences and any sensitive areas. The damaged areas may be invaded by secondary fungi such as species of *Alternaria*. A wise precaution is to spray badly damaged palms as soon as possible with a fungicide. An insidious side effect of hail is freezing damage. Hail collects in sunken sites such as the top of the crown and around objects such as the base of trunks and, as it thaws, the adjacent plant tissue can be damaged. The thick epidermis will protect the base of the trunk from injury except in the case of young palms.

Sunburn

Young plants of almost any but the very hardiest of palms should be protected from direct exposure to hot sun. In fact, most species of palm need to be protected for their first two to three years of life and then can be gradually hardened to the effects of the sun. Shade-loving palms should be protected at all times.

Premature or unexpected exposure to hot sun results in the leaves becoming sunburnt. This shows up as white or brown papery patches in the leaves and in severe cases whole leaves or even the whole plant may die. Sunburnt plants look tatty and unsightly.

Palms to be planted-out should be hardened to the sun before planting-out and in very hot conditions should be protected by a surround of leafy branches. Water should not be lacking during such times.

Strong Winds

While columnar palm trunks are extremely resistant to strong winds with the long, slender trunks even being renowned for their bending and survival ability during cyclones, their leaves suffer greatly by comparison. The leaflets of feather palms and the leaves of fan palms are usually shredded after strong winds and it is not uncommon for the tips to appear as if they have been flayed. Leaves are frequently broken at the petioles but are usually quickly replaced by new ones. Severe wind damage to young palms may result in a setback to their growth and establishment.

White Oil Damage

The practice of polishing the foliage of indoor palms with white oil to create a glossy appearance is a very common one and can lead to patches of dead brown tissue in the leaves. Although it does create an impression of luxuriance in the plants, the practice is detrimental and should be abandoned. White oil is frequently associated with plant tissue damage and for scale control should not be used any stronger than a dilution of one part in sixty with water and this can be further reduced to one in eighty during hot weather.

Dryness

Like other plants, palms react to dryness by wilting. Palms are generally quite resistant to dry soil but in cases of severe dryness, the fronds and leaflets take on a drooping appearance. Frequently also the crown becomes sparser because the leaf sheaths open and the less turgid petioles hang away from the trunk. The leaves or leaflets characteristically lose any glossy lustre and may fold together or curl inwards. Some palms may stay in a wilted state for quite long periods without any obvious detrimental effect, although constrictions in the trunk or very short internodes may be the result of such dry periods. Following long periods of dryness the tips of the leaflets usually wither and die back.

Twisting

Twisting is an unusual condition of palms whereby the crown takes on a distorted or lopsided appearance, with most fronds seeming to end up on one side. Frequently the trunk may have a bend or kink below the crown. The causes of this condition are not known but it is suspected that it results from damage caused by wind, hail, tree branches or perhaps pest or disease organisms. The condition is most noticeable in old palms which may be on the decline, although sometimes it is seen in younger specimens. It occurs in both species of *Howea* and has also been observed in Bangalow Palm (*Archontophoenix cunninghamiana*).

CHAPTER 7

THE PROPAGATION
OF PALMS

Palms can be propagated sexually—from seed; or asexually—by division, aerial layering or more recently by tissue culture. Most palms are grown from seed and this propagation technique is well within the scope of the average enthusiast. Professional nurserygrowers each year raise thousands of seedlings of the more popular indoor and outdoor palms as well as lesser quantities of collectors items. Propagation by division and aerial layering is mainly useful to enthusiasts and municipal gardeners and has limited application to nurserygrowers. Tissue culture is mainly the province of the research worker and for large-scale enterprises, particularly with commercial species such as the Oil Palm (*Elaeis guineensis*) the Coconut (*Cocos nucifera*) and the Date Palm (*Phoenix dactylifera*).

SEED PROPAGATION

Palms are variously described in the literature as being notoriously difficult and slow to germinate or being easy. This variation is probably a reflection of the facilities and expertise of the person involved but it does highlight the variability that a new grower can expect to meet. The following paragraphs explain some of the procedures used and some of the difficulties that may be encountered. It should be realised, however, that even today there are still considerable gaps in our knowledge of palm propagation. As a rule, however, fresh seed of most species germinates freely in suitable cultural conditions.

HAND POLLINATION

For most palms, pollination is not a limiting factor in seed production, however, it can be a problem in dioecious species. The Date Palm, for example, has been hand pollinated for centuries to enhance production, despite the fact that the trees are planted in groves with male plants in proximity to the females. Cultivated *Chamaedorea* palms may also be tardy in their seed production.

Hand pollination of ornamental palms is not new; in fact, as early as 1749 pollen of *Chamaerops humilis* was transferred from one part of Germany to another to effect pollination. Nowadays, even with modern transport and mailing facilities, the procedure is not practised widely by enthusiasts.

Pollen Collection and Storage

Inflorescences or portions of inflorescences with nearly mature male flowers (preferably with some already shedding pollen) are collected and placed in a sheet of paper in a warm, draught-free room. The pollen is shed onto the paper and can be used for immediate pollination or is collected and stored. The pollen should be freed of all floral fragments prior to storage. It can be stored in gelatin capsules, paper envelopes or in vacuum-sealed envelopes and held in a dry atmosphere. Pollen stored at room temperature will retain some viability for four to six weeks. Pollen stored in a refrigerator will retain its viability for much longer periods.

Moisture and humidity are the enemies of pollen. Flowers wet from rain or dew should not be cut for pollen collection and high humidity quickly destroys pollen viability. Hence, stored pollen must be held over an absorbent material such as silica gel or in a similar dry atmosphere. Containers such as gelatin capsules should only contain a small amount of pollen as large quantities deteriorate quickly as a result of drying difficulties.

Pollinating Female Flowers

Pollen is dusted onto the stigmas of female flowers using a small camel-hair brush. The female flowers should be at their peak of development with the stigmatic lobes appearing moist and receptive (pollen sticking easily to the stigma is a good guide). Female flowers which open widely are easy to pollinate but those which remain tubular or partially closed are more difficult. For these species, careful removal of some of the petals may be necessary to expose the stigma. To eliminate the chance of unwanted hybridisation, the equipment should be sterilised by dipping in alcohol after each use.

Seed Collection

The fruit should be collected as soon as it becomes colourful and ripens. The presence of fallen fruit on the ground is a good indicator and for many species the whole bunch of fruit may be harvested at this stage. However, the fruit of some palms, such as *Borassus* spp. and *Cocos nucifera*, must be harvested individually as the fruit of these palms ripen sporadically throughout the year.

When a whole bunch of fruit is harvested, some fruits will be immature and may not germinate. This is especially problematic in species such as *Howea forsteriana*, *H. belmoreana*, *Hedyscepe canterburyana* and *Lepidorrhachis mooreana*. In these palms it is also very difficult to gauge when the fruit are ripe as they take several years to mature and colour very slowly.

When collecting in the bush or travelling, it is often necessary to gather immature fruit or miss out completely. If the fruit have started to colour or even if they are large and plump, a sample is worth collecting as some can often be induced to germinate. Immature fruit shrivel very readily and immediately after collection they should be stored in moist peat moss in a plastic bag and sown at the first opportunity.

Caustic Fruit

The skin and fleshy of some palm fruits contain needle-like crystals of calcium oxalate called raphides which can enter the skin and cause severe irritation or a burning feeling. Handling such fruits is best avoided but if this is not possible then good protective gloves must be worn. The most familiar of the palms with caustic fruit are the various species of *Caryota* but others include *Arenga australasica*, *A. pinnata*, *Drymophloeus beguinii* and *Gaussia maya*. Under no circumstances should such fruit be eaten. Deaths have been reported following ingestion of the fruit of *Caryota urens* (see also *Stinging Crystals*, p. 72).

Germination Requirements

For successful germination, palm seeds require prolonged exposure to high temperatures ($35°-38°C$) and high humidity. The high temperatures can be supplied by methods such as a heated glasshouse, bottom heat cables or simply an enclosed metal shed which the sun will heat rapidly during the day. However, care should be taken as lethal damage may occur to palm seeds of some species exposed to long periods of temperatures above $38°C$. Humidity is supplied by watering the mixture containing the seeds and also watering the surroundings.

Palm species vary in the amount of exposure to heat and the temperature they require to induce germination. Thus the Oil Palm (*Elaeis guineensis*) has an effective temperature range of $38°-42°C$ and needs exposure for eighty days. On the other hand, the Palmyra Palm (*Borassus flabellifer*) germinates well after thirty-five days at $38°C$ and *Caryota mitis* after twenty-eight days at $35°C$. Temperatures higher than $38°C$ may be lethal to some palms and useful to others. Response to temperature seems to be cumulative and fluctuations, such as occur with overnight cooling, are not detrimental and do not negate the high temperatures of the day.

Moisture is essential for palm seeds to germinate and it has been shown that pre-soaking the seeds of some species may reduce the period of exposure to high temperature needed before germination can occur. The seeds are simply soaked in water for up to seven days before sowing. It is preferable that the water should be changed daily to remove any inhibitors that are leached out. Some growers use a heated water bath with the temperature held constant and the water circulated around the seeds, but this equipment is not essential.

Propagating Mixes

Palm seeds can be sown in a well-structured garden soil and will often germinate satisfactorily and grow quite strongly. Soils, however, commonly contain disease organisms, weeds and pests, so other propagating materials may be safer, especially if large quantities are to be propagated. Nurserygrowers use materials such as coarse sand, peat moss, vermiculite, perlite, pine bark and sawdust.

Coarse Sand Usually obtained from alluvial deposits. It must be washed thoroughly to remove dirt and weed seeds. Drainage is excellent but it dries out rapidly after watering and is best mixed with a water-retentive organic material.

Peat Moss Rotted organic material which has reached a stable point of decay. It is very acid (pH 4.5), sterile and absorbs many times its volume in water. It has excellent aerating properties and mixes well with other materials. It can be used on its own to germinate palm seeds.

Vermiculite A naturally occurring mica which is expanded by subjecting to temperatures in excess of $1000°$C. It is a very light material with a high water-holding capacity and excellent aeration. It mixes well with other materials but must not be over-watered or compacted as it readily becomes cloggy. It is sterile because of the high temperatures used in its production.

Perlite A naturally occurring silicate material which is treated with temperatures in excess of $700°$C. It forms grey, spongy particles which are very light and is best mixed with a water-retentive organic material.

Pine Bark Finely ground pine bark is often called pine peat because of its similarity in appearance and in some properties to peat moss. When fresh it contains toxins and must be stored moist for six to eight weeks before use. It is best mixed with coarse sand or perlite.

Sawdust Like pine bark, fresh sawdust contains plant toxins and hence must be stored moist in a heap for six to eight weeks before use. It mixes well with sand or perlite and has good moisture retention and aeration properties.

Peat moss and vermiculite can be used by themselves as a germination medium for palm seeds but it is more common to use a mixture. Very successful mixtures can be made by combining two parts coarse sand or perlite with one part peat moss, vermiculite, treated pine bark or treated sawdust. The mixture must be moistened before sowing as materials such as peat moss and sawdust are difficult to wet when dry. If weeds, pests or diseases cause problems during germination it may be necessary to sterilise the sowing mixture.

Sowing Techniques

Palm seeds should be covered with 1.5–3 cm of the propagating medium or, as a general rule, covered by at least their own thickness of material. They can be sown quite close together and if recovered and potted soon after germination, do not suffer greatly through competition. Suitable containers for sowing palm seeds must be fairly deep as the roots appear first and grow rapidly downwards before the first leaves show above ground. Pots are very successful but deep trays can also be useful, especially for large quantities.

For optimum rapid germination the seeds should be sown in trays, covered with a plastic bag, and held at high temperatures ($35°$–$38°$C) in a germination cabinet. Few enthusiasts, however, have the facilities available to produce such conditions and fortunately palms prove amenable to other conditions, although the time taken for germination to occur may be considerably prolonged and the percentage that germinate is reduced.

Some nurserygrowers in the tropics sow direct into prepared beds in bush houses or under trees. In these cases the seeds may be sown in a mixture of the above propagating materials or direct into well-prepared soil. Once the seedlings are large enough they can be dug and potted on. This technique is excellent for easy-to-handle species such as *Howea* spp., *Sabal* spp. and

Washingtonia spp. but is not so good with species that resent root interference, such as *Coccothrinax* spp., *Thrinax* spp. and *Archontophoenix cunninghamiana.*

A few palms, the best known of which are *Borassus* spp. *Hyphaene* spp., *Orania* spp. and *Lodoicea maldivica*, produce a premordial growth known as a sinker. This grows down into the soil for lengths up to a metre before the usual leaf is produced above ground. Palms of this type will not tolerate disturbance even when quite young and for success are best sown direct into a large container and planted in the ground when established. They can also be sown direct into their final position in the ground but are then subject to the ravages of various creatures which find the sinker very tasty.

Palm fruits which have only a thin layer of flesh can be sown direct without any prior cleaning; but those fruits that are pulpy—for example, *Butia capitata*—must have the flesh cleaned from the seeds before sowing. If the flesh is not removed fungus disease can rapidly build up, causing the seeds to rot. There is also the strong possibility of germination inhibitors being present in the flesh. The pulp of some fruit can be removed readily—for example, *Jubaea chilensis*—while others are tenacious and may require fermentation (see *The Bag Technique*, below).

Palms take from one month to two years to germinate, depending on the species and the state of the seed when sown. Some species with hard seeds take a long time and the process may be speeded up by first cracking the shell (see *Causes of Poor Germination*, p. 101).

Palms with very large seeds, such as *Borassus* spp. and *Cocos nucifera,* are best sown direct into a large container. Each container holds a solitary seed which is left half exposed on the top of the good, friable mixture. The containers are then placed in a warm, protected, shady place and kept moist until germination occurs. Once they have germinated, the plants can be moved to a suitable growing area until ready for planting out.

Fermentation Technique

The pulp of some palm fruits is fibrous and difficult to remove from the seed. Observations have shown that if moistened fruits are hung in a plastic bag for a couple of weeks the pulp ferments and falls away readily from the seed after hosing. This is a very useful technique and results have indicated that subsequent germination of the seed may be improved by the fermentation. Not only

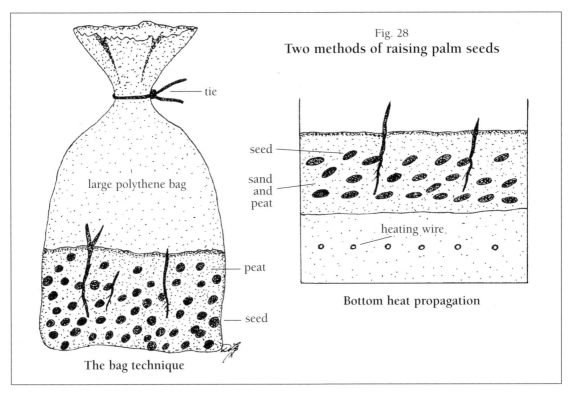

Fig. 28
Two methods of raising palm seeds

tie

large polythene bag

seed

sand
and
peat

heating wire

peat

seed

Bottom heat propagation

The bag technique

is the time for germination reduced but the seedlings appear together, rather than over several months. Seeds should not be left to ferment for more than two to four weeks or they may rot, especially if the prevailing temperatures are high.

The Bag Technique

An extension of the fermentation technique is to actually germinate the seeds in a plastic bag and not in containers such as pots or trays. The seeds are mixed thoroughly with moist peat moss and the whole lot sealed in a sturdy plastic bag. The bag is then placed in a warm, protected, shady position, such as under the bench of a glasshouse or hung from a shady tree. Observation will show when germination takes place and the seedlings can be potted up as needed.

This is a very useful technique for handling large quantities of seed. Little space is taken up and the seedlings can be handled readily without significant damage to the roots, although they may become entangled if the plants are left too long after germination or are too crowded in the bag. Germination also seems to be more uniform if this technique is used.

The bags used may be clear or milky white but must be relatively sturdy. They break down if exposed to excessive ultraviolet light and should be examined fairly regularly as the peat and seeds can dry out rapidly if there is a tear in the bag. For this reason the opening of the bag must be well sealed also. It is a wise precaution to include a label with the seeds before sealing the bag or to write the details on the outside with a permanent marker pen.

Bottom Heat

Some species of palm germinate very slowly and erratically and seeds in any one batch may germinate over two years or more. Germination of such seeds can be greatly speeded up if they are placed in a bottom heat propagation unit and kept continually warm and moist. The bottom heat unit is set at 25°–28°C and the palm seeds are mixed with moist peat and placed just above the heating elements. Exposed seeds are covered with a layer of moist peat and the whole unit is watered daily. Germinated seeds are removed as they reach a certain stage of development and are potted. This technique is especially useful with the Kentia Palm (*Howea forsteriana*) and Curly Palm (*H. belmoreana*) but can be applied to many other species in which erratic germination is a problem.

Causes of Poor Germination

Non-viable Seeds

The most common cause of poor germination in palms is sowing seed that has been harvested while immature, is too old, or has been stored incorrectly. Such seed usually rots fairly quickly after sowing and this in itself is a good indication of the cause. Seeds with a hard shell—for example, *Butia* spp.—may appear normal while in fact they have rotted on the inside. Internal rotting will only show up by cracking a sample lot and examining the contents. The embryos and endosperm of viable seeds fill the cavity and are creamy-white, while non-viable seeds are shrunken and discoloured.

Physical Barrier

The seeds of a number of palms have a hard woody endocarp which is impervious to water and acts as a physical barrier to germination. If the hard endocarp is cracked, sliced open or filed close to the hilum, germination of the seed can begin. These, however, are arduous techniques for large numbers of seeds and if damage occurs to the endosperm or embryo then the seed can rot. Seeds with hard coats will eventually germinate but the seedlings appear very sporadically and may take a long time to do so. Cracking or filing the woody coat has been shown to be beneficial for *Acrocomia* spp., *Arenga engleri* and *Butia capitata*. A technique of using hydrogen peroxide (H_2O_2) has proved useful for seeds of *Licuala grandis* and *Coccothrinax barbadensis*. The seeds are soaked for seventy-two hours in a six per cent solution before sowing.

Chemical Inhibitors

The fruit flesh that surrounds the seeds of many plant species contains germination inhibitors and there is no reason to suggest that palms are exceptions. As a consequence, it is wise to remove any fleshy pulp that surrounds the seed before it is sown. (For seeds with tenacious flesh see *Fermentation Technique*, p. 100).

Fruits that are fed on by animals may require partial digestion in an animal's gut before germination can be effected. Many colourful palm fruits are readily eaten by animals and birds and although information is scant, some observations suggest that digested palm seeds germinate rapidly and uniformly in the dung. Digestion can result in the breakdown of a woody coat as well as assist in the removal of inhibitors.

Leaching experiments have shown that inhibitors in parts of the seed other than the flesh (exocarp) may also affect germination. Thus, defleshed seeds of such diverse species as *Archontophoenix alexandrae*, *Aiphanes erosa*, *Arenga engleri*, *Chrysalidocarpus lutescens*, *Elaeis guineensis*, *Euterpe edulis*, *Caryota mitis*, *Pinanga* spp. and *Ptychosperma* spp., have responded to pre-soaking in changes of water for up to seven days. The water leaches out soluble inhibitors and swells the embryo. The inhibitors may be present in the woody endocarp, the filmy membrane around the kernel (pellicle) or in the endosperm itself. In *Jubaeopsis caffra* (a difficult species to raise from seed) maximum germination was achieved in pure oxygen and excised embryos were inhibited if some endosperm tissue was added to the culture medium.

Potting Seedlings

Palms are ready for potting as soon as the first leaf has expanded. By this stage, the root system will be quite long and well branched, but not yet tangling with its neighbours. As palms are generally fairly slow growing in their early stages, it is best to grow them in pots until they are of a sufficient size to plant out and survive on their own.

A suitable potting mixture must encourage healthy root growth for anchorage as well as strong growth above ground without any disease organisms being present. A safe and useful mix can be made up from five parts coarse sand or perlite; four parts peat moss, vermiculite or milled pine bark; and three parts friable loam.

Some fertilisers should be added to this mixture to promote growth (see also p. 78) and the pH should be adjusted to around 6.0 using lime or dolomite. Slow-release fertilisers that are available in commercial packs are very useful for supplying nitrogen, phosphorus and potassium. They should be used at the rates recommended on the pack. Trace elements can be added as a prepared mixture if needed. Organic fertilisers such as blood and bone, or hoof and horn can also be used at 2–4 kg per cubic metre of potting mix.

Young palms are best potted into fairly small, but deep containers, so that the root system is not too restricted. The young palms should be potted to the same depth as they were in the seedling container. A major cause of loss in palm seedlings occurs when the seedling is potted too deeply, resulting in rotting. Plastic growing tubes about 7 cm across are ideal for the first potting and when the palms are well established in these they can be moved into larger containers. Immediately after potting, the seedlings should be thoroughly watered to consolidate the soil around the root system. Palms can also be potted into large containers and tubs using a similar soil mixture and technique.

After Potting Care

After potting, the palms must be watered regularly to keep the soil mixture moist and the surroundings humid so as to encourage root and top growth. They should be kept in a shaded position, such as in a bush house or under trees, where they will receive filtered sun or direct sun for part of the day only. Pests such as slugs and snails, scales and mealy bugs should be controlled when noticed. Fertilising with liquid preparations will encourage growth and these are best applied during the warm months while the plants are in active growth.

Seed Viability

As a general rule, palm seeds do not retain their viability for very long and are best sown as soon as possible after collection. There are exceptions of course, such as the Coconut which can germinate after many months at sea. Seeds of *Pseudophoenix* spp. still germinate readily after storage for two years and there is some suggestion that these palms will not germinate if sown fresh. However, for most species it is better to be safe and sow the seeds immediately. For example, species of *Oncosperma* and some species of *Pinanga* may lose their viability within a few days of ripening and this short period precludes their distribution by post—at least in the dry state. If it is suspected that the fruit of a species may have a limited period of viability, then it is wise to mix the ripe fruit with moist peat moss and store in a sealed bottle or plastic bag as soon as possible after collection. The seed will not dehydrate in these conditions and may even begin to germinate in the peat moss.

A useful guide to the viability of various palm species can be obtained by considering the climatic conditions in its place of origin in combination with the thickness of the endocarp. Thus palms from areas with a distinct wet/dry or hot/cold climate and with seeds having a thick endocarp can be expected to retain their viability for longest. Seeds of these palms will last for two to four months without any special conditions. On the other hand, palms from the hot, wet, humid tropics generally have a thin endocarp and have a very short viability period of two to six weeks. These must be handled carefully and quickly if they are to be successfully propagated.

The above generalisation holds well for genera with a restricted distribution but considerable variation can be expected in widespread genera, where the species occupy a variety of ecological niches. Even for genera restricted to the tropics those species which grow on high peaks can be expected to behave differently from those on the lowlands. Thus these considerations can only be used as a useful guide.

Seed Storage

Palm seeds lose their viability by loss of moisture through the endocarp. When a significant amount of moisture is lost, the kernel containing the embryo shrinks away from the walls of the

TABLE 6
Viability guide to seeds of some palm genera. Storage time (weeks)

GENUS	2–4	4–6	8–16	GENUS	2–4	4–6	8–16
Aiphanes		X		Latania		X	
Archontophoenix			X	Licuala		X	
Areca	X			Linospadix	X		
Arenga		X		Livistona		X	
Bactris		X		Metroxylon	X		
Bentinckia	X			Normanbya	X		
Borassus			X	Nypa	X		
Brahea			X	Orania	X	X	
Calamus		X		Orbignya			X
Carpentaria		X		Phoenix			X
Caryota		X		Pinanga	X		
Chamaedorea		X		Pritchardia		X	
Chamaerops			X	Ptychosperma		X	
Chrysalidocarpus		X		Raphia	X		
Corypha		X		Reinhardtia		X	
Copernicia			X	Rhapis			X
Cyrtostachys	X			Rhopalostylis		X	
Dictyosperma			X	Roystonea		X	
Elaeis			X	Sabal			X
Euterpe	X			Salacca	X		
Gronophyllum	X			Serenoa			X
Gulubia	X			Syagrus			X
Howea			X	Thrinax			X
Hydriastele	X			Trachycarpus			X
Hyophorbe			X	Trithrinax			X
Hyphaene			X	Veitchia	X		
Jubaea			X	Verschaffeltia	X		
Laccospadix			X				

endocarp and the surface may take on a shrivelled appearance. If a badly shrivelled seed is cut open, the tissue will be discoloured and have a dry texture. Such seeds will not germinate and if sown will rot quickly. The cut-off figure seems to be a moisture loss of about twenty per cent.

Moisture loss is affected by time, the humidity of the atmosphere surrounding the seed and the temperature. Thus in high temperatures and a dry atmosphere the seeds will lose their viability quite quickly. The embryo of a palm seed can be damaged by exposure to low temperatures and thus a successful method of palm seed storage must reduce water loss but avoid injury from cold. The most successful method seems to be to mix them with moderately dry peat moss, seal in a container such as a plastic bag and store at about 20°C. Some palms will withstand temperatures much lower than 20°C but the very tropical ones will not. Seeds properly stored can be held for periods of six to fifteen months.

The technique employed by some commercial seed firms may or may not be suitable. The seeds are freed of all pulp and air-dried for a short period before placing in an aluminium envelope. The air is then evacuated by a suction pump and the envelope is sealed hermetically. Palm seeds of some species will store satisfactorily under these conditions but they are often held at too low temperatures resulting in cold damage to the embryo.

Transport of Seeds

Seed if collected fresh and properly packed will survive air transport to most parts of the world. The seeds should be prepared and packed as outlined above and despatched by the fastest practical method. As air cargo holds are subject to freezing conditions, it is a wise policy to provide some outer insulating cover such as a polystyrene container to prevent cold damage to the embryo. Dry peat moss packing is also a better insulating agent than wet peat moss.

VEGETATIVE PROPAGATION

Propagation from Basal Offsets

Division is a technique of vegetative propagation and the new plants so produced are identical to the parents. In palms, division can only be carried out on those species that produce basal offsets (usually known as suckers) freely, such as *Chamaedorea brachypoda*, *C. cataractarum*, *C. costaricana*, *C. pochutlensis*, *C. siefrizii*, *C. woodsoniana*, *Chamaerops humilis*, *Chrysalidocarpus lutescens*, *Laccospadix australasicus*, *Phoenix dactylifera*, *P. reclinata*, *Ptychosperma macarthurii*, *Rhapis excelsa* and *R. humilis*.

Division is a simple technique of propagation, but if it is to be successful it still requires care both during and after the removal of the sucker. The tools required, such as spades, knives and saws, must be sharp. The soil should be dug away carefully to expose the base of the sucker. If it has good roots it can be severed with the spade or saw or a strong, sharp knife. Damage to the tissues should be kept to a minimum and the cut surface can be sealed by rubbing with garden lime. As much as possible of the offset's root system should be removed with it to aid in its establishment.

Often when the sucker is exposed it is found to have few or no roots and this means that it is deriving all its sustenance from the main stems. This is especially a common feature of *Chrysalidocarpus lutescens* and *Phoenix dactylifera*. Such suckers can be transplanted, however, the loss rate is usually high with the *Chrysalidocarpus*, although the technique can be successful with *Phoenix dactylifera*. In fact, in this latter species, above-ground suckers from the trunk can be removed and successfully established. Existing leaves, however, must be drastically cut back and the suckers given plenty of tender loving care.

A rootless sucker can be induced to form roots by a couple of simple techniques. The easiest is to twist or wrench the sucker downwards so·that its junction with the parent clump is severely disrupted but not completely broken. The sucker can be reburied in soil and roots should form in about two months. Another technique is to slice or notch the sucker on the lower side near its base. Lime should be rubbed into the cut and the sucker reburied until roots form. Suckers respond to the application of root stimulants such as Formula 20.

Another very successful technique with suckers is to employ a plastic bag as in the germination process outlined earlier. This technique is especially successful for establishing divisions of small suckering palms such as *Reinhardtia*. After the division is potted and watered, the entire pot is enclosed in a large, clear plastic bag, the top tied and the lot is then placed in a warm, semi-shaded area. After a few months, if the division is growing the bag can be gradually opened for hardening off. At this stage, make sure excess water drains away by puncturing or upturning the bag.

Watering is not necessary while the bag is tied but plants may need to be watered during the hardening process. With this method there is no need to remove or cut back the leaves from the division. The technique is also useful for establishing seedlings or small transplanted palms with reduced or damaged root systems.

After separation, any sucker which has good roots should be planted directly into position or else it can be potted and held in a semi-shaded, protected situation until sufficiently established to plant. The sucker should be planted slightly deeper than in its original soil depth and thoroughly soaked to consolidate the new soil and encourage new root growth. The fronds should be trimmed back by at least one-third to one-half to reduce transpiration. A mulch applied to the surrounding soil will help the new plant to get established. At no time should it be allowed to dry out.

The optimum time for removing basal offsets from palms is during late spring and early summer when the plants have just started growth or are in active growth. New leaf growth is usually associated with new root growth and if division can coincide with the emergence of new roots then the sucker will establish more easily.

Aerial Layering

Aerial layering, also known as marcotting, is a fairly common technique used to propagate fruit trees and ornamental shrubs and trees that may be difficult to propagate from cuttings. It is interesting to note that it can also be used to propagate some specialised palms such as some *Chamaedorea* spp., some *Pinanga* spp. and *Rhapis* spp. and cultivars. This technique is especially useful for *Rhapis* cultivars such as those with fine leaves and variegated leaves. In *Chamaedorea* it is mainly successful with those species that produce aerial roots on their stems (*Chamaedorea serpens*, *C. elatior*, *C. elegans*, *C. ernesti-augustii*, *C. stolonifera*). The best time for aerial layering is in the warm growing months.

The technique is fairly straightforward and easily mastered. The stem is cleared of leaf sheafs, fibres, etc. in the vicinity of a node and the surface of the stem is lightly wounded. A pad of moist to wet sphagnum moss is wrapped around the wounded node and in turn this is wrapped tightly with a piece of aluminium foil or polythene film. This whole section must be wrapped thoroughly and tied tightly at each end to seal the bundle and prevent the sphagnum moss from drying out. The plant is then placed in a warm, sheltered environment and kept well watered. After six to twelve weeks, roots should show up through the plastic and when a good root system is formed in the sphagnum moss the section can be severed and potted. When potting there is no need to remove the sphagnum moss. After potting, the new plant should be provided with plenty of tender loving care until it is well established.

Stem Layering

The slender stems of *Chamaedorea stolonifera* form aerial roots from just below the crown. The stems are quite pliable and can be bent down to soil level or placed in contact with a container of potting mix. These stems can be anchored in place and when the root system is sufficiently well developed the stem section can be severed as a separate plant.

Bulbils in Palms

In palms, bulbil is the term used for vegetative shoots produced on aerial parts of the trunk. Such shoots are rare but are interesting because they are genetically identical to the parent plant and are capable of growth after separation (by a procedure not dissimilar to that for suckering palms). A technique similar to aerial layering can also be used to induce root formation before the bulbil is severed and thus make separation more reliable.

Bulbils are usually modified inflorescences but may also occur as small growths on the rachillae. Some plants consistently produce bulbils instead of flowers. In *Arenga pinnata*, bulbil formation may occur following damage to the growth apex of the plant. Bulbil formation has been noticed in the Coconut, Oil Palm and other species. A couple of species of *Salacca* (*S. flabellata, S. wallichiana*) form bulbils on the stems of the male inflorescence and when these come in contact with the ground, they form roots and become separate plants.

Tissue Culture

Tissue culture is the propagation of plants from small pieces of tissue which are excised and grown under sterile conditions. The tissue is grown on special culture media which contain a balanced supply of plant nutrients and hormones. This propagation technique must essentially be carried out in a laboratory and requires the use of expensive and specialised equipment. By changing the hormone content of the medium, the tissue can be induced to multiply or to become organised into plantlets, complete with leaves and roots.

The best way to propagate a plant by tissue culture is to remove and grow a bud, either from a shoot tip or a lateral bud in the axil of a leaf. The problem with palms is that each stem contains a single apical bud and therefore the plant must be destroyed if this method is employed. In fact, many plants must be destroyed because of the difficulties of sterilising and establishing a shoot tip in culture. The most common approach for palms therefore is to establish a culture of callus tissue from a developing organ such as a leaf, stem, petiole, inflorescence or root tip. After the callus tissue is growing and multiplying, different hormones in the culture medium can be used to generate shoots and roots.

Tissue culture with palms seems at present to be limited to species with commercial significance, such as superior specimens of Date Palms, Oil Palms and Coconuts. In Malaysia, tissue cultured Oil Palms have been successfully planted in the field for evaluation. In fact, only tissue cultured plants are currently used in propagating this palm. Date Palms have also been successfully propagated by tissue culture and planted out on a fairly large scale. Tissue culture of the Coconut has proved to be extremely difficult.

There is also interest by nurserygrowers in tissue culture as a technique for the rapid propagation of uniform, disease-free plants. To date, this work is in its infancy and has concentrated on high-value species such as *Howea belmoreana* and *H. forsteriana*. Tissue culture may have a significant role to play in the survival of rare species, such as *Hyophorbe amaricaulis*, which are on the verge of extinction.

Embryo Culture

The culture of seedling embryos is known as embryo culture and this technique has had a limited application with palms. Embryo culture begins with the excising of an embryo from a seed under sterile conditions and its transfer to a flask of sterile nutrient medium for germination. The flasks are held under controlled conditions of light, temperature and photoperiod and the embryo germinates and grows in the sterile conditions of the flask.

Embryo culture is a useful practice for overcoming problems in species which are slow or difficult to germinate. By removing the embryo, any influence of chemical inhibitors or physical barriers due to a hard seed coat is avoided. It is also a very useful technique for rare species where only a limited amount of seed may be available. To date, embryo culture of palms is in its infancy and has been carried out on such species as *Caryota urens, Chamaedorea costaricana, Howea forsteriana, Hyophorbe lagenicaulis, H. veschaffeltii, Jubaea chilensis, Jubaeopsis caffra, Pritchardia kaalae* and *Veitchia joannis*.

CHAPTER 8

PALMS FOR CONTAINERS, INDOOR AND OUT

PALMS AS INDOOR PLANTS

Palms make excellent indoor plants and have been used in this way for over 100 years. The large, airy rooms and hallways of the stately homes of England have long been graced by the delicate, drooping fronds of a Kentia or its relative. Today, people are well aware of the decorative value of plants in the home and there is a wide range of palms available for selection. Businesses also realise the importance of indoor plants and palms are a significant component of the range used by plant hire firms.

Although palms as a group are generally considered suitable for indoor decoration and are often promoted as such, a large proportion of species do not respond favourably to indoor conditions. Practical experience shows that only a limited number of species succeed consistently indoors. More species succeed indoors in tropical regions than in the temperate zones because palms as a group are tropical in origin and therefore the range for selection is greater. Also, indoor plants tend to grow more readily in the tropics because of the open, airy design of houses and the warmer climate. Some modern houses are designed for a more open living style and provide excellent conditions for the growth of plants.

Suitable Species

In general, palms that consistently grow well indoors are those that will tolerate fairly dark positions, a usually dry atmosphere and neglect. The best of these are undoubtedly the two species of *Howea* native to Lord Howe Island, which are generally sold as Kentia Palms. These not only look graceful, but will grow indoors and tolerate the conditions mentioned previously. When well cared for and in a position which suits them, they are a magnificent addition to the indoor decor. Unfortunately they are fairly slow growing and an advanced specimen is expensive. Some other species can also be very suitable for indoors and these are listed in Table 7.

Choice of a Plant

Once the choice of a suitable indoor species is made, the next step is to select the plant. As with other plants, a good palm should be a sturdy, healthy specimen in active growth. Look for a plant with healthy, dark green leaves. Avoid those with dull leaves that have a dry appearance, as they have probably been neglected or held too long in the store or nursery. Plants with very lush, soft growth should also be avoided as they may have just been taken out of a glasshouse and will deteriorate when placed indoors.

Buyers should be especially wary of plants which have pests already established. A few holes in the leaves from caterpillars or grasshoppers is acceptable but colonies of scales, mealy bugs or spider-mites can be expected to proliferate in the indoor environment. These pests are difficult to eradicate once established and it is better to start with clean stock. When checking for mealy bugs or scale it is advisable to look under the leaf sheaths, in the petiole grooves and similar places.

TABLE 7
Palms suitable for indoors

SPECIES	LIGHT TOLERANCE	COMMENTS
Aiphanes aculeata	Bright	Good, but prickly
Archontophoenix alexandrae	Bright	Not very good
Arenga caudata	Bright	Interesting clumping palm
Calyptrocalyx micholitzii	Dull–Bright	Attractive foliage
Calyptrocalyx petrickiana	Dull–Bright	New foliage colourful
Carpentaria acuminata	Dull–Bright	Good, cold-sensitive
Caryota mitis	Bright	Good, needs regular spelling
Caryota urens	Dull–Bright	Good, needs regular spelling
Chamaedorea arenbergiana	Dull–Bright	Excellent, broad leaflets
Chamaedorea cataractarum	Dull–Bright	Excellent, dense clumps
Chamaedorea costaricana	Dull–Bright	Excellent
Chamaedorea elegans	Dull	Excellent
Chamaedorea geonomiformis	Dull–Bright	Excellent
Chamaedorea hooperiana	Dull–Bright	Hardy and adaptable
Chamaedorea metallica	Dull–Bright	Excellent
Chamaedorea radicalis	Dull–Bright	Excellent
Chamaedorea sartorii	Dull–Bright	Excellent
Chamaedorea seifrizii	Dull–Bright	Good
Chamaedorea stolonifera	Dull–Bright	Excellent
Chrysalidocarpus lutescens	Dull–Bright	Good
Hedyscepe canterburyana	Dull–Bright	Good, slow growing
Howea belmoreana	Dull	Excellent
Howea forsteriana	Dull	Excellent
Kentiopsis oliviformis	Dull–Bright	Attractive foliage
Laccospadix australasica	Dull	Excellent
Licuala borneensis	Dull–Bright	Neat small palm
Licuala cordata	Dull–Bright	Spectacular small palm
Licuala flabellum	Dull–Bright	Dense clumps
Licuala grandis	Bright	Very tropical
Licuala orbicularis	Dull–Bright	Attractive shiny foliage
Licuala ramsayi	Bright	Good, cold-sensitive
Linospadix minor	Dull	Excellent
Linospadix monostachya	Dull	Excellent
Livistona robinsoniana	Bright	Bright green leaves
Livistona rotundifolia	Dull–Bright	Excellent
Lytocaryum insigne	Dull	Excellent
Lytocaryum weddellianum	Dull	Excellent
Neodypsis decaryi	Dull–Bright	Attractive colouration
Phoenix roebelenii	Dull	Excellent
Pinanga coronata	Dull	Dark foliage, dense habit
Ptychosperma elegans	Bright	Difficult
Ptychosperma lineare	Bright	Clumping species
Reinhardtia gracilis	Dull	Excellent
Rhapis excelsa	Dull	Excellent
Rhapis humilis	Dull	Excellent
Rhapis multifida	Dull–Bright	Excellent
Rhapis subtilis	Dull	Excellent
Rhopaloblaste augusta	Bright	Graceful leaves
Rhopaloblaste elegans	Bright	Attractive
Syagrus romanzoffiana	Dull–Bright	Good, needs regular spelling

Indoor Conditions

Light

Palms growing indoors prefer a position where they receive some light coming in through an archway, window, skylight or doorway. Direct sunlight for part of the day can be tolerated happily by palms, provided it is not long exposure to hot, summer sun. Morning sun or sun filtered through shrubs and trees is ideal. Bright light through coloured or frosted glass provides an attractive background for a group of palms and can be quite suitable for their growth. Such glass, however, may transmit heat and the plants will require more frequent watering and attention to humidity than would be the case in other situations. Solar films applied to windows to reduce heat and glare are very detrimental to indoor palms.

Humidity

Indoor atmospheres are generally of low humidity and tend to fluctuate considerably with changes in the outdoor environment as well as from heating and air conditioning. Palms generally dislike low humidities, although there is a considerable range of response within the group. Species that are intolerant of low and fluctuating humidities lose their lustre and appear dull and are often severely attacked by pests such as spider mite and mealy bug.

The answer to low humidity is to change the atmosphere around the plants. This does not necessarily mean increased watering, although the plants must not be allowed to dry out. Many an indoor palm has been killed by over-watering because its leaves seemingly advertised that it was dry. The dryness, in fact, is caused by the low humidity and cannot be compensated for by increased watering of the potting mixture. Increasing the humidity around the leaves is the most successful solution. This can be achieved by grouping indoor plants so that each contributes to the atmosphere around the other. For the same reason, a number of plants in each pot is more successful than just one. Standing the pot in a large saucer of wet, evaporative material such as scoria is also a useful technique to increase the surrounding humidity, as is misting the plants at regular intervals with a fine spray.

Temperature

The majority of palms are tropical in origin and therefore dislike low temperatures, although some of the successful indoor species grow very well in temperate regions. Tropical palms which are in active growth may be damaged by temperatures around 14°C, but if they are dormant or are growing slowly, they can withstand somewhat lower temperatures. The length of time the plant is exposed to the low temperature also exerts a considerable influence. Low temperatures are not a problem in tropical regions, but in highland districts and temperate zones the winter temperature indoors can drop to a level which can cause damage to sensitive species.

Houses fitted with internal heating avoid the problems of cold damage to indoor palms, but the resulting very dry atmospheres may cause other problems such as excessive water loss. Indoor palms in winter grow slowly, not only because of lower temperatures, but also through reduced light intensity and short daylight hours.

Care of Indoor Palms

Watering

Although watering indoor plants is basic common sense, for various reasons it creates more problems and frustrations than any other aspect of the plant's care. Healthy, actively growing plants need regular watering and the frequency depends upon the prevailing temperatures and humidity. In summer, plants can be safely watered daily, whereas in winter their needs are much less. The same parameters as for other indoor plants apply to palms: vigorous-growing plants will need more water more frequently than those growing slowly or not at all; plants will need watering more regularly in the summer than in the winter; and plants growing in bright light will dry out more quickly than those in dim positions.

Other factors must also be considered, such as the type of potting mix, the pot size and how full of roots it is and the prevailing temperature and humidity. The potting mixture must drain well but should also retain sufficient water for the plant's growth. Heavy soils that become soggy when watered are useless for palms as they lead to rotting of the roots and retardation or death of the plant.

Palms that are kept too dry lose the sheen on their leaves, generally look unhealthy and may even wilt. Palms that are too wet suffer damage to the tips of the leaflets which become brown and die. If they have suffered root damage, such as rotting of the root tips, waterlogged palms may also wilt because the damaged roots are unable to extract water from the soil. Wilting in palms is not as prominent as in other groups of plants.

The ideal watering regime keeps the potting mixture sufficiently moist to maintain an adequate supply of oxygen and water to the roots for growth. Regular topping up can be quite satisfactory but at intervals the potting mixture should be thoroughly soaked so that water flows out the drainage holes. This ensures a thorough wetting of the root system and also leaches out salts which may have accumulated from the breakdown of the fertilisers. This leaching process should be performed out of doors or in a bath or sink.

In any group of indoor plants some specimens are going to require more regular watering than others. It is a temptation to water all of the plants at the same time but this should be avoided and individual needs catered for. If all of the plants are watered each time the most vigorous plant dries out, then the least vigorous ones will receive too much water.

Recuperation

Indoor palms, like any indoor plants, appreciate a 'freshen up' at intervals. This can consist of a hosing down in the garden to wash the dust off the leaves and freshen the plants generally. This simple act can also be important in reducing pest build up, in particular discouraging species such as mites, which like dry conditions. It is also a good policy to put palms outside in rain or drizzle, but they must not be left out if the sun appears. Such sudden exposure can drastically burn plants that have been shielded from it.

Resting indoor palms by moving them to a shady position in the garden or bush house is very beneficial. Here they can be well watered, repotted or fertilised if necessary and generally encouraged to recuperate and put on new growth. Once spelled, the plants can then be moved back indoors. With planning, a series of palms can be circulated in this way and those indoors can always be at their peak. As a guide, palms should be rested outside for two to three weeks after every two-month period indoors.

Fertilisers

Indoor palms benefit from the application of fertilisers but these should only be applied during the warm growing months. Fertilisers applied during winter when growth is slow, or sudden applications of quick-release fertilisers to starved or debilitated palms may be of no benefit and indeed may even cause severe burning. Fertilisers are best applied in small doses at regular intervals and the soil mixture should be watered thoroughly and regularly after their application. Quick-release fertilisers should never be applied to newly potted palms or to those where the root system has been damaged—for example, by waterlogging—because the weakened or new roots can be readily burned.

A wide range of commercial products is available to fertilise indoor plants and most of these will be successful with palms. Fertilisers are usually incorporated in the potting mix to encourage initial growth and these may be supplemented with side dressings when it is felt that the growth is in need of a boost. Complete fertiliser mixtures are usually used in the potting mixes and these may be rapid or slow releasing. Organic manures and fertilisers can be very beneficial, but some, such as blood and bone, have the drawback of being smelly and attractive to dogs. A suitable potting mixture including fertilisers is included on p. 112.

Supplementary fertilising of palms can be carried out using slow-release fertilisers, plant pills or liquid preparations. Liquid fertilisers are very beneficial and are usually safe except where the plant is suffering from over-watering. A useful, cheap, nitrogenous preparation can be made by dissolving one-and-a-half teaspoons of urea or ammonium nitrate in a watering-can of water. Commercial preparations have the application amounts shown on the packet and these recommendations should be adhered to. Some nutrients can be applied through the leaves in a process known as foliar feeding. This is generally a much less satisfactory and more expensive way of boosting growth than applications to the root system.

Pests

Pests are dealt with in detail on pp. 82–90, but it should be mentioned here that container-grown palms may be more susceptible to certain pests than those palms grown in the garden. The three most serious pests of container-grown palms are mealy bugs, spider mites and scale insects. Spider mites revel in dry conditions and are a severe pest of indoor palms. Their effects can be reduced by frequent misting or hosing. Mealy bugs and scale may be present on any palm, but become prevalent and severe on those plants that are weakened or debilitated through neglect. Healthy plants resist pests far better than weakened ones.

PALMS AS OUTDOOR PLANTS

The preceding section deals mainly with palms for indoor decoration. What is not generally realised is that palms also make excellent container plants for outdoor decoration on areas such as terraces, patios, verandahs and around barbecues and pools, to name just a few. Because the plants are outdoors, they do not have to tolerate the restricted environment of indoor plants and a much wider range of species can be grown. Almost any palm will make a suitable container plant for a while, but vigorous species soon outgrow their container and will need regular attention. A selection of suitable container palms is provided in Table 8. Virtually any of the dwarf to small palms will also make excellent container plants. Because they are grown outdoors, the species must be able to tolerate the climatic regime of the area. Once a container-grown palm has become too big, it can always be planted in the garden or sold.

Outdoor Conditions

Sun

Container-grown palms can be used for display in sunny or shady aspects. As a general rule, young palms need protection from direct, hot sun for the first two to three years of their lives. Some palms, such as the various species of *Phoenix* and *Sabal,* will tolerate sun from a very early age, while others such as *Chamaedorea elegans* and *Linospadix monostachya,* need shade even when mature. The majority of palms will tolerate sun when their fronds are about 1 m high and this is an excellent size to start off a palm in a container. Tolerance to sun will also vary depending on how moist the potting mixture is kept. Palms kept well watered will tolerate much more sun that those which are allowed to dry out excessively between watering.

When a new palm is purchased from a nursery, take note of how much sun it is receiving. If it has been held in a shade house or glasshouse it will need to be hardened-off before being placed in full sun. Hardening consists of increasing the exposure to sun the plant receives each day while keeping it well watered. A sudden prolonged exposure, especially in summer, will lead to severe burning by the ultraviolet rays of the sun.

Wind

Palms generally do not like wind and those in containers should not be placed in a windy or draughty position. Cold winds may cause chilling with resultant stunted or deformed growth, while hot winds cause severe desiccation. Damage may also occur to the leaves and leaflets. Tall palms in tubs are also top-heavy and tend to blow over easily.

Frost

Palms in containers can be moved if severe environmental conditions are imminent. Frost is a major enemy of palms and only hardy species should be chosen for cold areas.

Care of Outdoor Container Palms

Watering

Watering outside containers is easier than those indoors because hoses are available and there is less need to worry about the mess. Because they are outside, however, the plants tend to be forgotten and their watering is often neglected. If container-grown palms are to maintain a good appearance they must be regularly watered and at no stage allowed to dry out completely. Watering will vary with the climatic conditions prevailing. In summer, a daily watering may be necessary, whereas in winter once or twice a week may be sufficient. Windy weather dries plants out and extra watering may be needed. Heavy rain will water the plants but do not fall into the trap of thinking that light rain will do the same. Rain needs to be fairly heavy and persistent to penetrate the potting mix and many a plant has died because the owner thought that rain had watered the plant sufficiently.

In addition to normal watering it is a good policy to thoroughly soak the plants every two weeks. This prevents dry spots developing in the potting mixture and also leaches out excess fertiliser salts. Hosing down the foliage is also useful because it reduces the build up of dust and discourages pests.

Potting Mixes

A suitable potting mix is of tremendous importance to the successful growth of a container palm whether it be held indoors or outside. A potting mix must supply anchorage for the roots and encourage their growth by ensuring adequate aeration, moisture and nutrient supply. Two suitable potting mixes are:

Mix 1	Mix 2
5 shovels coarse sand	5 shovels coarse sand
4 shovels milled pine bark or peat moss	3 shovels peat moss
3 shovels friable loam	10 shovels milled pine bark
45 g Osmocote (3–4 months formulation)	70 g Osmocote (3–4 months formulation)
130 g Osmocote (8–9 months formulation)	240 g Osmocote (8–9 months formulation)
40 g Dolomite	60 g Dolomite
3 g iron sulphate	5 g magnesium sulphate
	45 g iron sulphate
	40 g trace element mix

Note that palms like well-rotted animal manures and these can be added to either mix.

Fertilisers

Some of the fertilisers incorporated in the potting mixtures are slow release and will maintain growth for up to nine months. After this time the palm can be repotted if necessary or else treated with additional slow-release or liquid fertilisers as outlined on p. 110.

Repotting

Repotting of container-grown plants is necessary to maintain their appearance and growth. The containers eventually become filled with roots and the potting mix exhausted. Such plants are referred to as being pot-bound and watering them can be difficult.

Repotting will be necessary every one to two years. The plant can be put into a bigger container, or if the present container is still suitable the plant can be put back into it after removal of the old potting mixture and some of the root system. Repotting is best carried out in spring or early summer. The plants should be thoroughly watered immediately after repotting.

Note: Pests and diseases are dealt with in detail in Chapter 6.

General Hints

Certain palms, such as species of *Livistona* and *Phoenix,* make excellent container plants but are very prickly and are best avoided if the containers are to be placed near thoroughfares.

Containers should not be placed directly on the ground but on concrete or bricks or supported above the soil surface. This is to prevent diseases and pests (grubs and worms) from gaining entry via the drainage holes and damaging the palm's root system. It also prevents the palm roots from growing through the drainage holes and becoming entrenched in the soil.

Some palms have a prodigious root system that quickly fills a container. If these palms are not repotted regularly their roots are quite capable of bursting the container. This is a particular problem with plastic containers but may also apply to terracotta pots or even the large concrete containers used in shopping centres.

TABLE 8
Palms suitable for outside containers

SPECIES	POSITION	CLIMATIC CONDITIONS*	NOTES
Aiphanes aculeata	sun or shade	Tr–STr	Prickly
Archontophoenix alexandrae	sun or shade	Tr–STr	Strong grower
Archontophoenix cunninghamiana	sun or shade	Tr–STr–Te	Strong grower
Arenga caudata	sun or shade	Tr–Te	Interesting clumping
Arenga engleri	sun	STr–Te	Cluster
Brahea armata	sun	STr–Te	Bluish leaves
Butia capitata	sun	STr–Te	Excellent
Calyptrocalyx micholitzii	shade	Tr–STr	Lovely foliage
Calyptrocalyx petrickiana	shade	Tr–STr	New leaves purple-brown
Caryota mitis	sun	Tr–STr	Nice foliage
Caryota urens	sun	Tr–STr	Nice foliage
Chamaedorea brachypoda	shade	Tr–STr	Dense clumps
Chamaedorea cataractarum	shade	Tr–Te	Dark green foliage
Chamaedorea costaricana	shade	Tr–STr–Te	Small cluster palm
Chamaedorea elegans	shade	Tr–STr–Te	Small palm
Chamaedorea geonomiformis	shade	Tr–STr	Small palm
Chamaedorea metallica	shade	STr–Te	Lovely foliage
Chamaedorea pochutlensis	shade	Tr–STr	Shiny foliage
Chamaedorea seifrizii	shade	Tr–STr	Small palm, fast growing
Chamaerops humilis	sun	STr–Te	Very hardy, prickly
Chambeyronia macrocarpa	shade	Str–Te	Colourful new leaves
Chrysalidocarpus lutescens	sun	Tr–STr–Te	Excellent, good colour
Coccothrinax alta	sun	Tr–Te	Hardy
Coccothrinax argentata	sun	Tr–Te	Hardy
Coccothrinax argentea	sun	Tr–Te	Hardy
Cyrtostachys renda	sun	Tr	Colourful
Dictyosperma album	sun	Tr–STr	Nice foliage
Hedyscepe canterburyana	shade	STr–Te	Slow growing
Howea belmoreana	sun or shade	STr–Te	Excellent
Howea forsteriana	sun or shade	STr–Te	Excellent
Hydriastele wendlandiana	shade	Tr–STr	Slow growing
Hyophorbe lagenicaulis	sun	Tr–STr	Excellent accent plant
Hyophorbe verschaffeltii	sun	Tr–STr	Slow growing
Laccospadix australasica	shade-some sun	Tr–STr	Excellent
Licuala cordata	shade	Tr–STr	Spectacular small palm
Licuala glabra	shade	Tr–STr	Large rounded leaves
Licuala grandis	sun	Tr	Spectacular
Licuala orbicularis	shade	Tr–STr	Shiny leaves
Licuala rumphii	shade or sun	Tr–STr	Large clumps
Licuala ramsayi	sun	Tr–STr	Excellent
Linospadix minor	shade	Tr–STr	Small palm
Linospadix monostachya	shade	STr–Te	Small palm
Livistona australis	sun	Te–STr	Hardy
Livistona chinensis	sun	Tr–STr–Te	Spreading, prickly
Livistona decipiens	sun	Te–STr	Hardy, drooping segments
Livistona rotundifolia	sun or shade	Tr–STr	Attractive foliage
Lytocaryum insigne	shade	Tr–STr	Excellent small palm
Lytocaryum weddellianum	shade	Tr–STr	Excellent small palm

*Tr = tropical; STr = subtropical; Te = temperate.

SPECIES	POSITION	CLIMATIC CONDITIONS*	NOTES
Nannorrhops ritchiana	sun	Te–STr	Very hardy
Neodypsis decaryi	sun	Te–Tr	Interesting trunk
Normanbya normanbyi	sun or shade	Tr–STr	Attractive foliage
Phoenix pusilla	sun	Tr–STr–Te	Attractive foliage
Phoenix roebelenii	sun	Tr–STr–Te	Excellent
Phoenix rupicola	sun	Tr–STr–Te	Attractive
Pinanga aristata	shade	STr	Mottled foliage
Pinanga copelandii	shade	Tr–STr	Dark foliage
Pinanga coronata	shade	Tr–STr	Dense habit
Pinanga densiflora	shade	Tr–STr	Colourful mottled foliage
Pinanga disticha	shade	Tr–STr	Marbled foliage
Ptychosperma lineare	shade or sun	Tr–STr	Clumping species
Ptychosperma macarthurii	sun	Tr–STr	Excellent
Ptychosperma microcarpum	shade or sun	Tr–STr	Interesting leaves
Ravenea rivularis	sun	Tr–Te	Lovely foliage, fast grower
Reinhardtia gracilis	shade	Tr–STr	Excellent small palm
Reinhardtia latisecta	shade	Tr–STr	Spectacular leaves
Rhapis spp.	sun or shade	STr–Te	Excellent
Rhopaloblaste singaporensis	shade or sun	Tr–STr	Attractive clumps
Roystonea regia	sun	Tr–STr	Large container
Sabal etonia	sun	STr–Te	Hardy
Sabal minor	sun	Tr–STr–Te	Very hardy
Siphokentia beguinii	shade or sun	Tr–STr	Attractive leaves
Syagrus romanzoffiana	sun	Tr–STr–Te	Large container
Thrinax excelsa	sun	Tr–Te	Attractive leaves
Thrinax morrisii	sun	Tr–Te	Hardy
Thrinax radiata	sun	Tr–Te	Long lived
Trachycarpus fortunei	sun	STr–Te	Very hardy
Wallichia caryotoides	shade or sun	STr–Te	Attractive foliage
Wodyetia bifurcata	sun	Tr–STr	Spectacular foliage

*Tr = tropical; STr = subtropical; Te = temperate.

Part Two

ALPHABETICAL ARRANGEMENT OF PALMS

Livistona australis (*Botanical Magazine*, Vol. XXXIII of the
3rd series, plate 6274)

The following species descriptions detail a range of palms which are commonly grown in various countries throughout the world. Also included are many collector's items—which are mainly grown by enthusiasts—and many other species which await introduction to cultivation. Because of the tremendous enthusiasm shown by palm collectors, the species described here can in no way be considered an exhaustive list of the palms being grown.

ACANTHOPHOENIX

(from the Greek *acantha*, thorn, and *phoenix*, a palm)

A monotypic genus which is endemic to the Mascarene Islands of Mauritius and Réunion, where it grows in forests from near sea level to high altitudes. It is a solitary, spiny, feather-leaved palm with a well-developed crownshaft. The inflorescence arises below the crownshaft and bears unisexual flowers of both sexes. This palm has become extremely rare in the wild with the once extensive populations being reduced to fragmentary remnants, mainly as a result of collection of its cabbage.

Cultivation: Popular in cultivation and grows readily in a range of locations from tropical to warm–temperate regions. Plants need good drainage and will tolerate sun from an early age. Single plants are capable of producing fertile seeds.

Propagation: Fresh seed germinates readily within two to four months of sowing. Each fruit contains a single seed.

Acanthophoenix rubra
(reddish)

Palmiste Rouge, Barbel Palm

MASCARENE ISLANDS

Leaves of young plants of this elegant palm are dark green with prominent red veins, but in older plants the veins lose this attractive colouration. When young most parts of this palm are covered with stout spines and prickles, but these are often shed as the plants mature, although young plants may stay spiny. The leaves spread in an attractive crown and the nearly pendulous leaflets may be green or whitish beneath. White to red flowers are followed by clusters of ellipsoid black fruit each about 1 cm x 0.7 cm. Plants grow to about 15 m tall and are suitable for tropical, subtropical and warm–temperate regions. The species is extremely variable but no varieties or subspecies have been named.

ACOELORRHAPHE

(from the Greek, *a* without, *coelos*, hollow, *raphe*, seam; the seed lacks an impressed seam)

A monotypic genus which occurs in southern Florida, the West Indies and adjacent coastal parts of Central America (Honduras and Guatemala). It forms thickets in brackish swamps. It is a clumping, prickly fan palm with the slender stems enclosed in persistent fibrous sheaths. The inflorescences arise among the leaves and bear bisexual flowers.

Cultivation: A popular palm which has proved to be adaptable and highly ornamental when well grown. It is a sun-loving species which responds well to fertilisers and watering. Plants grown in dry situations are very slow growing. Single plants are capable of producing fertile seed.

Propagation: Removal of established suckers is relatively easy. Seed is small but germinates mostly within two or three months if sown fresh. Each fruit contains a single seed.

Acoelorrhaphe wrightii.

Acoelorrhaphe wrightii
(after Charles Wright, 19th century American botanist)
Silver Saw Palmetto, Paurotis Palm, Everglades Palm

FLORIDA—CENTRAL AMERICA—WEST INDIES

This tough fan palm which extends from Florida to the Caribbean coast of Central America and the West Indies, usually grows in damp, sandy soil, often in brackish swamps. It is a clumping species, the most distinctive features of which are its prominent, silvery fan leaves and an intricate fibrous sheath around the slender brown trunks. A few trunks may dominate the clump and are surrounded by a dense growth of suckers. This species makes a highly ornamental palm and in Florida is often planted around municipal buildings. It is also suitable for planting in parks and on large acreage. Plants are tropical or subtropical in their requirements and will tolerate full sun when quite small.

Coastal conditions suit them admirably. They need plenty of water and are best grown where their roots can reach ground water. Once established, plants are quite hardy. The species has been variously known as *Paurotis wrightii*, *P. androsana* and *Acanthosabal caespitosa*.

ACTINOKENTIA
(from the Greek *actino*, radiating from the centre and *Kentia*, another genus of palm)

A small genus of two species of palms endemic in New Caledonia. They grow in wet rainforests on serpentinic soils. They are solitary, unarmed, feather-leaved palms with a crownshaft and few leaves in the crown. The inflorescence arises below the crownshaft and bears unisexual flowers of both sexes.
Cultivation: Generally these palms are rare in cultivation, although *A. divaricata* was grown in European glasshouses last century. Generally a sheltered position is essential, especially for young plants. Single plants are capable of producing fertile seed.
Propagation: Fresh seed germinates readily two to five months after sowing. Seedlings are slow to establish. Each fruit contains a single seed.

Actinokentia divaricata
(spreading widely apart)

NEW CALEDONIA

A choice, slender palm from the lowland rainforests of New Caledonia where it grows at low to moderate elevations on soils derived from serpentinite. Its very slender trunk to about 6 m tall is crowned by only four or five spreading, pinnate leaves, each of which has curiously arched leaflets. The new leaves are bright red and the mature ones a dark, glossy green. The petiole maintains a reddish colouration and contrasts with the shiny greenish-brown crownshaft. Pinkish to reddish flowers are followed by purplish ellipsoid fruit about 2.5 cm long. In cultivation the species likes a shady aspect and deep, well-drained soil. It appears to be moderately adaptable and has been grown successfully in California, Australia and Honiara. Previously known as *Kentiopsis divaricata* and *A. schlechteri*. This palm would probablymake a very attractive container specimen.

ACTINORHYTIS

(from the Greek *actino*, radiating from the centre and *rhytos*, wrinkled, in reference to the seeds)

A small genus of two species of palms found in New Guinea and the Solomon Islands. Both grow in rainforest. They are tall, solitary, unarmed feather-leaved palms with a slender crownshaft and large fruit. The inflorescence arises below the crownshaft and bears unisexual flowers of both sexes.
Cultivation: One species is widely grown, while the other is virtually unknown. They are attractive palms which are very tropical in their requirements. Single plants are capable of producing fertile seed.
Propagation: Fresh seed germinates readily two to four months from sowing. Each fruit contains a single seed.

Actinorhytis calapparia

(from the Malay name, *kelapa*, coconut)

Calappa Palm, Pinang Penawar

NEW GUINEA—SOLOMON ISLANDS

Widespread in New Guinea and the Solomon Islands, this palm, which is an inhabitant of dense rainforest has become commonly cultivated in South-East Asia for its supposed magical properties. The plants grow tall (12 m or more) with a slender, grey-brown trunk topped with a bright green crownshaft and an arching crown of finely divided fronds. The pinnae are glossy green and arch out from the rachis in an attractive manner. The large ovoid fruit, which are about 8 cm long, are reddish when ripe and are carried on a large, much-branched, complex infructescence. They may be used as a betel nut substitute. This species has excellent ornamental features which give it great potential for cultivation in the tropics. Previously known as *Areca calapparia*.

Actinorhytis poamau

(a local native name)

SOLOMON ISLANDS

A tall palm from the Solomon Islands (Loyalty and Treasury Islands south-east of Bougainville). Plants have a slender trunk to about 24 m tall and spreading fronds with curved, linear segments drawn out

Actinorhytis calapparia and *Areca catechu* in habitat. J. Dransfield

Actinokentia huerlimannii

(after Dr. H. Hurlimann, original collector)

NEW CALEDONIA

A small palm with a slender trunk to about 4 m tall and a sparse crown of four or five fronds each about 1 m long. The fronds have shiny green leaflets and a reddish petiole. This palm is of restricted occurrence in New Caledonia, where it grows in rainforest on soils derived from ultrabasic rocks at moderate elevations. It appears to have been rarely tried in cultivation and may have specialised requirements.

into an acuminate tip. The leaflets have transparent scales on the underside. Much-branched inflorescences bear white flowers followed by large, ovoid to ellipsoid fruit (to 6 cm x 4 cm) which are dark green when ripe. These fruit are used locally as a substitute for betel nut. A handsome palm for tropical regions.

AIPHANES

(from the Greek *aiphanes*, jagged, ragged, torn, in reference to the leaflet tips)

A relatively large genus of thirty-eight species of palm found in the West Indies, Central America and South America with a strong development in Colombia. They grow as understorey palms in rainforest. They are solitary, pinnate palms which have long, dark spines on the slender trunks, leaf-bases, petioles, rachises and even the leaflets of some species. The leaves are coarsely divided into broad, jagged leaflets and produce a distinctive silhouette. The inflorescence arises among the leaves and produces unisexual flowers of both sexes. Many species were previously well known in the genus *Martinezia*. For a recent study of the genus in Ecuador see F. Borschenius and H. Balslev, 'Three new species of *Aiphanes* (Palmae) with notes on the genus in Ecuador', *Nordic Journal of Botany* 9(4) (1989): 383–93.

Cultivation: Novelty palms which despite their spiny nature have tremendous horticultural appeal. They are especially popular with palm enthusiasts. Most species need tropical conditions but some succeed in the subtropics. Spines should be removed from eye level and below from plants sited near paths. Single plants are capable of producing fertile seeds.

Propagation: Fresh seeds germinate readily two to four months from sowing. Seedlings establish quite quickly. Each fruit contains a single seed.

Aiphanes acanthophylla

(with thorny leaves)

Coyure Palm

PUERTO RICO

This Puerto Rican palm is a slender species with a graceful crown of pinnate leaves that have moderately crowded segments. Each segment is irregular at the end and a much darker green above than

Aiphanes aculeata inflorescence.

below and with black spines on both surfaces. Its trunk is adorned with rings of long, black spines and these are the only drawback to the cultivation of an otherwise very decorative palm. It succeeds best in tropical and to a lesser degree subtropical regions and needs protection from direct sun when small. Plants thrive in organically rich soils and need plenty of water during dry periods.

Aiphanes acaulis

(without a stem)

COLOMBIA

An interesting dwarf palm from northern Colombia, where it is only known from two small populations at low to moderate altitudes (170–700 m). It is a stemless palm with eight to ten leaves arising in a crown from ground level. Each leaf, 0.6–1.2 m long, has forty to sixty narrow leaflets arranged in the one plane. These are to 27 cm x 2.5 cm, green and glabrous on both surfaces and the upper margin is prolonged into a narrow tail about 4 cm long. Purple flowers are borne on a bristly spike which is subtended by a bract of similar length to the spike, the whole inflorescences being about as long as the leaves. An ornamental collector's palm for the tropics.

Aiphanes aculeata. G. Cuffe

Aiphanes aculeata

(bearing thorns)

Ruffle Palm, Chonta Ruro

SOUTH AMERICA

Although formidably armed with long, black, brittle spines on the trunk, petioles and leaflets, this palm has been a firm favourite in cultivation for well over 100 years. It was a popular glasshouse palm in England during the nineteenth century. In the garden it can be grown outside in subtropical and tropical regions. Although small plants need protection from direct sun, larger specimens will happily tolerate positions from partial shade to full sun. Its solitary trunk is ringed with black spines which are quite sharp and should be removed from ground to eye level. The crown consists of many graceful pinnate leaves, the leaflets of which are broad and jagged as in those of the genus *Caryota*, but with irregularly spaced leaflets, each of which has spines on the lower surface. Ruffle Palm likes a

rich, damp, well-drained soil and plenty of water during dry periods. It is native to Columbia, Venezuela and Ecuador and grows naturally in rainforests. Clusters of orange-red to scarlet fruit on the tree are decorative and are reportedly edible. *A. caryotifolia* is a synonym.

Aiphanes chiribogensis

(from Chiriboga, Ecuador, the type locality)

ECUADOR

The very slender, pendulous inflorescences with very few rachillae are an immediate guide to the identity of this species. It is a solitary palm with a slender, spiny trunk to 4 m tall and arching fronds with narrowly wedge-shaped leaflets clustered in groups of up to three along the rachis. The species originates in Ecuador where it grows in dense rainforest at about 2000 m altitude.

Aiphanes chocoensis

(after the Choco region of western Colombia)

COLUMBIA

This species can be recognised by a combination of two unique features, an undivided, spike-like inflorescence and entire leaves which are notched at the apex. Plants are almost stemless and the leaves, to 65 cm x 25 cm, have irregularly toothed margins. The petioles and rachis are densely covered with short, brown spines and the main lateral veins on the upper surface are similarly adorned. This species has excellent ornamental potential but is unknown in cultivation. It is native to western Colombia, where it is reported to be locally common along the Rio Mutata River.

Aiphanes concinna

(neat, trim)

COLOMBIA

A clumping species with very attractive leaves, with the crowded, overlapping leaflets presenting a ruffled appearance. Plants have slender stems to 5 m long and about 7.5 cm wide which are ringed with spines. The leaf sheaths, petioles and rachises are also densely armed with spines to 9.5 cm long. The leaves, which are about 75 cm long, have about sixty-six leaflets borne in irregular groups along the rachis and arranged more or less in two planes. The wedge-shaped leaflets have irregularly erose outer margins and a prominent yellow midvein. The globose fruit, about 1.5 cm across, are red when ripe. This highly ornamental palm originates in Colombia where it grows in humid forests at about 3000 m altitude.

Aiphanes eggersii

(after M. Eggers, original collector)

ECUADOR

This species is known from dry deciduous forests in Ecuador where it is reported to be common in the province of Manabi. It is a slender clumping palm growing to about 5 m tall with the frond bases and rachises bearing a mixture of black spines to 5 cm long and short, fine spines. The fronds have wedge-shaped segments (to 45 cm x 9 cm), green above, whitish beneath, irregularly arranged in groups along the whitish rachis and also in several planes. The male flowers are yellow and the fruit, 1.5–1.7 cm across, are red and globose. These are eaten by the local people where it occurs naturally. A highly ornamental palm which awaits introduction to cultivation.

Aiphanes erinacea

(prickly, like a hedgehog)

SOUTH AMERICA

A clumping palm with up to fifteen stems to 5 m tall. The leaves have yellow spines on the rachis and about forty leaflets which are irregularly grouped and arranged in several planes. The inflorescences have prominent brown to purple spines on the rachillae and white to purple flowers. The fruit mature through red to black. The species is native to south-western Colombia and Ecuador where it grows in moist rainforests at moderate to high altitudes.

Aiphanes erosa

(as though bitten or chewed)

WEST INDIES

The long, narrow, triangular leaf segments with their irregularly lacerated margins and the long, brown spines on the petioles and veins give imme-

diate clues to the identity of this palm which hails from the islands of Barbados, Gualeloupe and Martinique in the West Indies. It is a slender species with a crown of bright green pinnate leaves of delightful symmetry. As well as being spiny, the leaf petioles are covered with a white, mealy powder. Inflorescences are crowded with cream, fragrant flowers and are followed by red fruit. The species is not common in cultivation and is mainly to be found in enthusiasts' collections. It is strictly tropical in its requirements, being very cold-sensitive.

Aiphanes fosteriorum

(after M.B. & R. Foster, original collectors)

SOUTH AMERICA

This clumping palm originates in south-western Colombia and north-western Ecuador where it grows at about 1000 m altitude in moist forests. The stems (usually two develop, the rest are suckers) to 10 m high, are densely spiny and the large leaflets (to 45 cm x 14 cm) are glossy dark green on the upper surface and with brown hairs beneath. These are irregularly clustered along the rachis and also arranged in several planes. Robust infructescences about 3 m long carry rose-red, globose fruit about 2 cm across. An interesting palm for tropical gardens.

Aiphanes gelatinosa

(gelatinous, jelly-like)

SOUTH AMERICA

The flowers and fruit of this palm are often covered in a jelly-like substance, hence the specific epithet. The species is native to south-western Colombia and north-western Ecuador where it grows in very moist rainforest at about 900 m altitude. It is not known to be in cultivation but appears to be an interesting palm with ornamental qualities. Plants have a clumping habit (one or two stems dominant) and slender stems ringed with spirals of downward-pointing spines. The narrow wedge-shaped leaflets (to 50 cm x 13 cm) are dark green above and whitish beneath. There are about thirty on each leaf, with the terminal pair united and much broader than the rest. They are regularly arranged along the rachis. The fruit, about 12 mm across are red to black when ripe.

Aiphanes gracilis

(graceful, slender)

PERU

This species is native to Peru where it grows in rainforest in mountainous regions at about 1400 m altitude. Plants have a slender, short trunk to about 0.5 m tall and fronds 1–1.2 m long. The petioles, which are very slender, are densely armed with black spines about 4.5 cm long. Frond segments (about forty per frond) are arranged in irregular clusters along the rachis. They are glaucous green on the upper surface and paler beneath, with black spines on the margins. The longest segments, to 15 cm x 3.5 cm, occur in the centre of the frond. By all accounts this is a very attractive palm which merits introduction to cultivation.

Aiphanes hirsuta

(hairy)

COLOMBIA

This species forms robust clumps with thick stems to about 8 m tall which are well armed with black spines. Fronds are about 1.5 m long and carry about forty leaflets which are irregularly arranged along the rachis in clusters of three to five. The leaflets, which are broadly wedge-shaped (to 35 cm x 7 cm) have irregularly jagged outer margins and bear yellow bristles, at least when young. The branched inflorescence is covered with dark brown bristles. Native to Colombia where it grows in highland rainforests at about 1600 m altitude.

Aiphanes kalbreyeri

(after W. Kalbreyer, original collector)

COLOMBIA

An interesting palm which originates in Colombia where it grows in rainforests at about 1700 m altitude. Plants form multi-stemmed clumps to about 5 m tall, with the trunks densely armed with black spines to 20 cm long. The fronds are divided into about eighty, narrow, rigid segments, to 35 cm x 5 cm, which are dark green and shiny on the upper surface and with dense, long, white hairs over the lower surface. The stems of the branched inflorescence are covered with dark brown spines.

Aiphanes leiostachys

(with smooth spikes)

COLOMBIA

A clustering species from Colombia which forms dense clumps of stems to 5 m tall, each with an arching crown of fronds to 3 m long. The leaf sheaths, petioles and rachises are densely armed with long black vicious spines. The fronds are divided into about forty large, narrowly wedge-shaped leaflets (to 37 cm x 6 cm) which are some-what irregularly arranged along the rachis. The species grows in forests at about 850 m altitude.

Aiphanes linearis

(linear, of uniform width)

COLOMBIA

A clumping species which has narrow, rigid leaflets to 45 cm x 3.5 cm, arranged in clusters of two to four along the rachis. Plants develop thick trunks to 6 m or more tall which bear long, black, reflexed spines. The fronds which are about 2.5 m long, have up to eighty crowded leaflets. The fruit have a covering of yellow hairs and it is recorded that the seeds are edible. A very attractive palm from high-land rainforests in Colombia, growing at about 2300 m altitude.

Aiphanes luciana

(from St Lucia in the Windward Islands)

WEST INDIES

A solitary species with a trunk to 8 m tall and 13 cm thick, ringed with upward-pointing spines. The leaves, to 2 m long, have about forty alternate leaflets to 50 cm x 5 cm. These are bright green and glossy on the upper surface and slightly paler beneath, with the outer margins appearing as if irregularly chewed. The inflorescence is a simple hanging spike to about 1 m long. The globular, red fruit, about 1.5 cm across, have a thin layer of edible flesh. This species grows in dense rainforest on the island of St Lucia in the West Indies.

Aiphanes macroloba

(with large lobes)

SOUTH AMERICA

A dwarf species from highland rainforests in Colombia and Ecuador, growing at about 1000 m altitude. A prickly trunk (often decumbent) to about 2 m tall is topped with a crown of six to eight slender, graceful fronds, to 1 m x 40 cm, which are more or less elliptical in outline. Each frond may be entire (in Colombian specimens) or is divided into about eight large segments (in Ecuadorian specimens) which have wavy, irregularly toothed outer margins with the upper margin produced into a narrow tail. The apical segment is the largest of all and it has a short central notch. The inflorescence is a simple spike. The red to orange, ellipsoid fruit, each about 1 cm long, have a small apical beak. An attractive small palm which awaits introduction to cultivation.

Aiphanes monostachys

(single spike)

COLOMBIA

A clumping species with slender trunks armed with brown spines and with a crown of fronds each about 1.5 m long. The rachis is covered with brown, scurfy scales and the broadly wedge-shaped leaflets are dark green and shiny on the upper surface and whitish beneath. Each frond has about twenty of these segments irregularly arranged in groups of two or three along the rachis. The apical margin of the segments is irregularly lobed with the margins drawn out into long tails. The inflorescence is a long, arching, simple spike to 1.2 m long. The fruit are red when ripe. An interesting palm originating in Colombia where it grows in mountainous forests at about 1000 m altitude.

Aiphanes pachyclada

(with thick branches)

COLOMBIA

Another species of *Aiphanes* which originates in highland rainforests of Colombia, growing at about 1800 m altitude. Plants should grow well in subtropical regions in a sheltered location. It is a clumping species with spiny stems to about 5 m tall and fronds about 1.5 m long. The fronds have about fifty, dark green, shiny, very rigid leaflets to 33 cm x 7.5 cm, which are irregularly clustered along the rachis. The inflorescence is about 60 cm long, with pendulous branches.

Aiphanes parvifolia

(with small leaves or leaflets)

COLOMBIA

A novelty palm which has a slender, prickly stem (about 10 cm across) to 1.5 m tall and fronds about 45 cm long. The petioles and rachises bear brown, scurfy scales and short, black spines. The fronds are divided into about twenty wedge-shaped segments to 10 cm x 4 cm, which are more or less evenly arranged along the rachis. The species occurs naturally in Colombia where it grows as an understorey palm in highland rainforests at about 2000 m altitude.

Aiphanes schultzeana

(after H. Schultze-Rhonhof, original collector)

SOUTH AMERICA

A solitary species to 6 m tall with a slender trunk only about 5 cm across. The arching leaves, to about 2 m long, have the leaflets arranged in several planes. These are large (to 35 cm x 20 cm) and broadly wedge-shaped with the apex wavy and irregularly incised. The inflorescences are held erect above the leaves and the small, red fruit (about 7 mm across) are prominently beaked. An interesting palm from the Amazonian lowlands of eastern Ecuador and south-eastern Colombia extending up the slopes of the Andes to about 1850 m altitude.

Aiphanes simplex

(simple, of one piece or series, in reference to the inflorescence)

COLOMBIA

A clustering palm which develops slender, prickly stems 2–3 m tall, each with a crown of small (less than 1 m long) graceful fronds. The rachis and petiole are covered with white bristles interspersed with black spines on the underside. The segments (20–24 per leaf) are broadly wedge-shaped, to 12 cm x 4 cm, green on both surfaces and irregularly arranged in clusters of two or three along the rachis. The simple, spike-like inflorescence is about 50 cm long. A highly attractive palm from the Colombian Andes where it grows at 1500–2000 m altitude.

Aiphanes tessmannii

(after G. Tessmann, original collector)

PERU

This species has a cluster of fronds to about 2 m long, which arise directly from the ground as the trunk is subterranean. The stiff petiole has a mixture of long, black spines and fine, yellowish spines. The fronds have numerous (about forty) narrowly wedge-shaped segments to 32 cm x 3 cm. These are evenly arranged along the rachis, green on both surfaces and the upper margin is drawn out into a narrow tail. The inflorescence, which is about 45 cm long, is densely covered with yellowish hairs. An attractive novelty from the lowland forests of eastern Peru.

Aiphanes verrucosa

(warty, in reference to the apex of the fruit)

ECUADOR

An interesting feature of this species is its large fruit (about 3 cm across). These are green to whitish at maturity and the mesocarp cracks while still attached to the infructescence, allowing the seed to fall free. It is a clumping palm with each very spiny stem having a sparse crown of graceful, arching fronds. The narrow lanceolate leaflets (to 40 cm x 2.5 cm) present a crowded impression, being arranged in several planes and in irregular groups along the rachis. This unusual palm is known only from highland forests in southern Ecuador and is not known to be cultivated.

Aiphanes vincentiana

(from St Vincent in the Windward Islands)

WEST INDIES

A large species in the genus with a trunk to about 50 cm across armed with rosettes of flattened spines to 10 cm long. The leaves are about 3 m long and each has about sixty pairs of crowded, bright green leaves. These are long and narrow (to 1 m x 9 cm) and the upper surface has numerous prominent veins. Narrow, branched infructescences about 1 m long carry bright red, globular, shortly beaked fruit, about 1.5 cm across. This robust palm grows in woodlands and forests on the island of St Vincent in the West Indies.

Aiphanes weberbaueri

(after A. Weberbauer, original collector)

PERU

A small species with a slender trunk to about 30 cm tall and fronds to about 1.5 m long. Leaf sheaths, petioles and rachises are armed with dark brown to black spines to 6 cm long. The leaflets, which are clustered in irregular groups along the rachis, are wedge-shaped (to 24 cm x 4.5 cm), green on the upper surface, whitish beneath, with the apex irregularly toothed and the upper margins produced into a tail. An interesting palm from dense highland (1700 m altitude) rainforests in Peru.

ALLAGOPTERA

(from the Greek, allagos, alternate, on the opposite side, and pteron, a wing)

A genus of four species of palms restricted to South America (Brazil, Bolivia, Argentina and Paraguay) where they grow in sparse, open habitats. They are small, unarmed clumping palms with subterranean trunks and pinnate leaves. The inflorescences, which arise among the leaves, are erect and unbranched and carry large, chunky, unisexual flowers of both sexes. The genus Diplothemium is synonymous.

Cultivation: Interesting palms which have proved to be versatile in cultivation, tolerating moderately harsh conditions but also responding readily to watering and fertilisers. Single plants are capable of producing fertile seeds.

Propagation: Well-established suckers can be successfully removed. Fresh seed germinates three to five months after sowing. Seed apparently loses its viability quickly and should be sown soon after collection. Each fruit contains a single seed.

Allagoptera arenaria

(growing in sandy places)

BRAZIL

This palm, is tolerant of extreme coastal exposure and has proved to be a very useful species for coastal tropical and subtropical planting. In nature it grows on sand just above the high-tide mark of Brazilian beaches. Plants in cultivation have performed very well but require freely draining soil in a position exposed to some sun. The plants form a

Allagoptera arenaria inflorescence.

Allagoptera arenaria.

clump of erect to arching pinnate fronds to about 2 m tall and have a branching, subterranean trunk. The pinnae are bright green above and silvery beneath and are in an attractive whorled arrangement. The small, greenish-yellow fruits are edible.

Allagoptera brevicalyx

(with a short calyx)

Cachando Palm

SOUTH AMERICA

A recently described species (1993) from Brazil, Bolivia, Argentina and Paraguay which grows in short, sparse vegetation on coastal sand dunes and low, nearby scrub (Cerrado vegetation). Plants have a subterranean stem and a sparse crown of up to eight leaves each to about 1 m long. The leaf sheath and petiole is densely covered with woolly hairs as also are young leaflets. Each leaf has fifty to eighty lanceolate leaflets which are grouped irregularly along the rachis. These are lobed on the apex, glossy green on the upper surface and

glaucous beneath. Simple unbranched spikes to 90 cm long, after flowering carry ovoid to top-shaped fruit 1.5–2 cm long. This species is rarely cultivated, if at all.

Allagoptera campestris

(growing in the fields)

BRAZIL

A tough palm occurring naturally in the Cerrado vegetation of Brazil. Plants have a subterranean trunk and a sparse crown of leaves with leathery, pointed leaflets arranged along the rachis in clusters of two to four. Young fronds have a covering of soft, white hairs as does the ovoid to ellipsoid fruit. This palm, which is rarely grown, demands excellent drainage in a sunny location.

Allagoptera leucocalyx

(with a white calyx)

BRAZIL

A small palm from Brazil which occurs naturally in low scrub (Cerrado vegetation) which may be burnt on a regular basis. Plants have a subterranean trunk and a crown of greyish-green to dull green fronds to 1.5 m long with leathery, pointed leaflets clustered along the rachis in groups of two or three. Young fronds lack prominent hairs. The fruit, however, which are ovoid to ellipsoid in shape, bear loose, cottony hairs. This species appears to be rarely cultivated.

125

Archontophoenix alexandrae in flower.

ARCHONTOPHOENIX

(from the Greek *archi*, chief, first and *phoenix*, a palm)

A genus of six species of palm endemic to eastern Australia where they grow in moist forests, swamp forests and rainforest, often in dense colonies. They are tall, slender, solitary, unarmed feather-leaved palms with a prominent crownshaft. The inflorescence arises below the crownshaft and bears unisexual flowers of both sexes.

Cultivation: Very popular palms which are widely grown in tropical and subtropical countries. They are valued for their graceful ornamental appearance and ease of culture. Single plants are capable of producing fertile seeds. Palms of this genus hybridise freely and this should be borne in mind by growers when collecting seed from mixed plantings.

Propagation: Fresh seed germinates readily one to three months from sowing. Each fruit contains a single seed.

Archontophoenix alexandrae

(after Princess Alexandra)

Alexandra Palm, King Palm

AUSTRALIA

A popular, fast-growing palm which is widely grown in tropical and subtropical countries. Plants usually have a swollen base to the trunk, a light green crownshaft, leaflets which are dark green above, silvery grey beneath and oriented in a vertical plane towards the leaf apex. Flowers are creamy white and are followed by clusters of bright red fruit, each about 1.4 cm x 1 cm. The species is endemic to Queensland where it is distributed between Gladstone and the Melville Range near Bathurst Bay on Cape York Peninsula. It commonly grows in lowland swamp forests but is also found up to 600 m altitude in the ranges. An excellent, handsome garden palm which looks especially decorative when in fruit. Plants are often sold for indoor decoration but are generally unsuitable for this purpose, having high light requirements and disliking the often dry atmosphere inside offices and dwellings. They do make a very attractive tub plant and are suitable for verandas or patios. The variety *beatriceae*, which has a prominently stepped trunk, is no longer regarded as being botanically distinct but is still commonly propagated and sold in the nursery trade.

Archontophoenix cunninghamiana

(after Alan Cunningham, early Australian botanist)

Bangalow Palm, Piccabeen Palm

AUSTRALIA

This, the most cold-tolerant member of the genus, can be grown successfully in subtropical and temperate regions. Plants are relatively fast growing and are well suited to planting in groups or clumps. They are valued for their slender habit, graceful crown of arching, dark green fronds (leaflets green beneath) and displays of lilac-purple flowers and bright red fruit. The species is native to eastern Australia where it is widely distributed from the Eungella Range near Mackay to just north of Batemans Bay in southern New South Wales. It grows in rainforest and moist open forest from the lowlands to about 1000 m altitude. This palm is often sold for indoor decoration but is generally unsuitable for this purpose.

Archontophoenix maxima

(large, great)

AUSTRALIA

A recently described species from moist forests on the Atherton Tableland in north-eastern Queensland, Australia, where it forms colonies close to streams. Plants have affinities with *A. alexandrae* but can be distinguished by their stiffly erect to obliquely erect, straight leaves with a very short petiole. This is an impressive palm which grows very well in subtropical regions and would be worth trying in temperate zones.

Archontophoenix myolensis

(from Myola, the type locality)

AUSTRALIA

A recently described species from the Atherton Tableland in north-eastern Queensland, Australia, where it grows as a rheophyte in rainforest along streams, at about 400 m altitude. It is of restricted distribution and numbers less than 100 plants in the wild. Plants are similar in general appearance to *A. alexandrae* but the crownshaft has a blue-green tinge and the leaflets become pendulous. The mesocarp fibres of the fruit unravel—probably as an aid to dispersal by water. Seedlings grow where they are regularly submerged by floodwater. This species is cultivated on a limited scale.

Archontophoenix purpurea

(purple, in reference to the colour of the crownshaft)

Mount Lewis King Palm

AUSTRALIA

This species has some similarities with *A. alexandrae* but is recognised by its distinctive reddish-purple crownshaft and larger, dark red fruit (2–2.6 cm x 1.8–2.2 cm). It grows as individuals and in

scattered, loose colonies in highland rainforests (400–1200 m altitude) of Mount Lewis, Mount Spurgeon and Mount Finnigan in north-eastern Queensland, Australia. It is an excellent palm for cultivation with its colourful crownshaft and spreading fronds. Plants tend to be slower growing than other *Archontophoenix* species but are more cold tolerant than Alexandra Palm and succeed better in temperate districts. They also need more protection from direct sun and excessive wind when small.

Archontophoenix tuckeri

(after Robert Tucker, original collector)

AUSTRALIA

This species is common on Cape York Peninsula, Queensland, Australia, where it grows in rainforest and swamp forest from sea-level to about 500 m altitude. It is similar in general appearance to *A. alexandrae*, but has smaller leaves, prominently zigzagged rachillae in the inflorescence and larger fruit (to 2.5 cm x 1.5 cm). It has been introduced into cultivation but is mainly grown by enthusiasts.

ARECA

(from a name used by local people on the Malabar coast of India)

A moderately large genus of about sixty species of palm widely distributed from southern China and India through Malaysia, the Philippines, Indonesia and the Solomon Islands to New Guinea. Some species grow in open habitats but most occur as understorey palms in rainforest. They are small to medium-sized, solitary or clumping, unarmed feather-leaved palms with a crownshaft. Some species have a slender, underground trunk and are regarded as being acaulescent. The inflorescence arises below the crownshaft (interfoliar in acaulescent species) and bears unisexual flowers of both sexes.

In the early days of taxonomy the genus *Areca* was the dumping ground for many palms which are now placed in other genera. Some of these errors are perpetuated by horticulturists, such as the use of *A. lutescens* for *Chrysalidocarpus lutescens*.

Cultivation: The genus includes some species which are commonly grown and many others which have great potential but await introduction

Areca catechu group planted.

to cultivation. The Betel Nut Palm (*A. catechu*) is widely cultivated for its fruit which have narcotic properties. This and similar species will grow in sunny conditions but many other species of *Areca* are shade-lovers. Single plants are capable of producing fertile seed.

Propagation: Fresh seed germinates readily one to three months from sowing. The seed of some species loses its viability rapidly and should be sown quickly after collection. Each fruit contains a single seed.

Areca catechu

(from a Malayan name *caccu*, used for the palm)

Betel Nut Palm

MALAYSIA—PHILIPPINES

A tropical palm the seeds of which form the basis of a huge industry (see *Edible Seeds*, p. 55). As a consequence, this palm, which probably originated

in either Malaysia or the Philippines, is a familiar sight in tropical regions around the world. It has become naturalised in many countries. It is characteristically a very tall, slender palm with a crownshaft and a small, crowded crown of semi-erect, silvery pinnate fronds. The fruit, if they are allowed to ripen, are quite large and colourful, varying from orange to scarlet. The species is reputedly cold-sensitive and will only thrive in the warm tropics although healthy plants are known from gardens in subtropical regions. It likes deep, well-drained soils and plenty of water during dry spells. Young plants will tolerate considerable exposure to sunshine and will also grow in the shade. Although very wind-resistant, the crown becomes very tattered following strong blows. Seedlings grow rapidly in good conditions and may flower at six or seven years of age. Betel Nut Palms are variable and many different variants are recognised.

Areca concinna

(neat, elegant)

SRI LANKA

This species is very rare in cultivation and most plants sold by nurseries are in fact *A. triandra*. The true *A. concinna* is restricted to lowland rainforest in Sri Lanka, whereas *A. triandra* is much more widespread. *A. concinna* succeeds best in a shady situation in the tropics and likes an abundance of moisture. The fruit are borne in clusters and are

Areca concinna inflorescence and supporting bract.

scarlet when ripe. In Sri Lanka they are used as a betel nut substitute. It is a very attractive palm well worth growing if true seed of the species can be obtained.

Areca guppyana

(after H.B. Guppy, plant collector in the Pacific)

NEW GUINEA—SOLOMON ISLANDS

An attractive, small, slender palm from New Guinea and the Solomon Islands which grows in rainforests. Its thin, solitary trunk may grow to about 3 m tall and is supported at the base by stilt roots and terminated by a light crown of short fronds, each of which has about five pairs of widely spaced, broad pinnae. The palm also has a slender, smooth crownshaft and bears bright red fruits which may be used by the natives as a substitute for betel nut. A highly ornamental species, this palm is rarely grown. In the Solomon Islands it is regarded as a sacred plant and may be planted on graves.

Areca hutchinsoniana

(after W.I. Hutchinson, botanical collector in the Philippines)

PHILIPPINES

An attractive palm from the Philippines which is useful because of its short, compact growth habit. The slender trunk grows no more than 4 m tall and about 15 cm thick and has prominent, pale, annular rings. The crownshaft bulges somewhat and the fronds are dark green and spread in a graceful crown. The fruits are about 3 cm long, elliptical and an attractive yellow when ripe. Responses in cultivation indicate that this palm grows best in the tropics and the plants are quite cold-sensitive. Small plants should be protected from direct, hot sun.

Areca ipot

(from a native name *bunga ipot*, used in the Philippines)

PHILIPPINES

In some respects this palm resembles a reduced version of the Betel Nut Palm (*A. catechu*). It is a solitary species which does not grow to more than

Areca guppyana.

4 m tall and with a trunk 8–12 cm in diameter. This trunk is ringed and topped by a short, but somewhat inflated crownshaft. The infructescence is densely crowded with large, ovoid fruit which measure about 5 cm x 3 cm and are red when ripe. Highly ornamental in all respects, this palm is becoming much sought after by enthusiasts. Native to the Philippines, it succeeds best in a shady situation in tropical gardens.

Areca macrocalyx

(with a large calyx)

Highland Betel Nut Palm

NEW GUINEA

As the common name suggests this palm is cultivated around villages in the highlands of New Guinea and is used as a betel nut substitute. It is a solitary, fairly tall palm with a thick, dark green crown and bears dense, club-like clusters of fruit. Although it is rarely grown, it appears to be an attractive palm worthy of trial in highland tropical and subtropical areas.

Areca macrocarpa

(with large fruit)

PHILIPPINES

A cluster palm from the Philippines which forms clumps of slender, elegant stems and overlapping crowns of dark green leaves. The large, elongated fruit (up to 7 cm x 3.5 cm) are sometimes used as a betel nut substitute in the area where it grows. Although it is highly ornamental, this palm is mainly to be found in the collections of enthusiasts. It needs hot, humid conditions in deep, organically rich soils and plenty of water for good appearance.

Areca ridleyana

(after N.H. Ridley, English botanist who collected in Malaysia)

MALAYSIA

A dwarf, highly ornamental palm that is native to Malaysia where it grows as an understorey plant on hillsides in rainforest. Each palm has a solitary, slender trunk that grows no more than 1 m tall and

Areca vestiaria colourful crownshaft and fruit.
G. Cuffe

has a crown of dark green leaves about 30 cm long. These are deeply notched at the tip and may be entire or divided into one or two broad leaflets. Brilliant-red, elliptical fruit about 0.5 cm long are carried on a short, sparsely branched infructescence. This is a very ornamental little palm that makes a delightful addition to any tropical garden and requires a sheltered, shady position.

Areca triandra

(with three anthers)

INDIA—SOUTH-EAST ASIA

A delightful palm that forms a dense clump of slender, pale green stems and deep green pinnate fronds. It is very tropical in its appearance and is the perfect neat, cluster palm for the home garden, growing well in the tropics and subtropics. Clusters of fruit are an additional decorative feature, being bright orange-red when ripe and following pale coloured flowers which have a strong lemon perfume. Plants require a situation sheltered from wind and direct sun when small. Deep, rich, organic soils and plenty of water ensure fast growth and an attractive appearance. This species is distributed from India to Borneo and is extremely variable with notable variants deserving recognition.

Areca vestiaria

(clothed)

Pinang Yaki

SULAWESI—MOLUCCAS

The most attractive feature of this slender palm is its crownshaft which is invariably of a brilliant orange to reddish colouration. Native to the Celebes, Sulawesi and the Moluccas, it is a clumping palm with prominent stilt roots. Plants are becoming widely distributed and well known in cultivation. Although succeeding best in the tropics, plants have also been successfully grown in subtropical regions. They require a shady position and like an organically rich soil. In a good position they can be quite fast growing. Trunks may be 10–12 cm thick and the leaves have very broad, dark green leaflets. Fresh seed germinates quite rapidly. Fruit are deep orange, about 2 cm long and carried in tightly packed bunches. *A. langloisiana* is a synonym of this species.

Areca whitfordii

(after H.N. Whitford, plant collector in the
Philippines)

PHILIPPINES

A Philippine species which grows in semi-swampy
areas of the lowlands. It develops trunks to 20 cm
thick and more than 5 m tall and these may be
conspicuously ringed. The crownshaft is prominent, somewhat swollen and the fronds arch
upwards and outwards in the crown. Fruit are 4–5
cm long and up to 2 cm thick, elliptical in shape
and brownish when ripe. This species is mainly to
be found in enthusiast's collections in the tropics.

ARENGA

(from the Javanese name *aren*, used for a palm of
this genus)

Sugar Palms

A genus of seventeen species of palm found mainly
in India, southern China, Malaysia and South-East
Asia, with one occurring in Christmas Island,
Australia, a few in New Guinea and one in northern Australia. The larger species grow in open situations or may be emergent above the forest canopy,
whereas the smaller ones grow as understorey
plants within forests. Palms of this genus are solitary or clumping, unarmed or lightly spiny, feather-
leaved species with the trunk of the solitary species
being monocarpic and that of the clumping ones
being either hapaxanthic or pleonanthic. Some
species are of local commercial significance for the
production of starch, sugar, wine and fibre. The
genus *Didymosperma* is synonymous.

Cultivation: All species are worthy of cultivation,
and those with a clumping growth habit have
tremendous horticultural appeal. Some have proved
to be quite hardy in warm–temperate regions,
whereas others are very tropical in their requirements. All need good drainage and the smaller
species need shelter from excess sun and wind.
Solitary plants are capable of producing fertile seed.
Mature trunks should be removed as they decline
from clumping species and new plants should be
grown to replace those which are monocarpic.

Propagation: Removal of suckers can be successful
but the divisions may be slow to establish. Seeds of
Arenga species are difficult, being commonly slow
and erratic in their germination and with a low percentage. They probably contain chemical inhibitors.
Presoaking for three to five days promotes germina-

Areca triandra.

tion. *Arenga* fruit contain calcium oxalate crystals
and should be handled with care. Each fruit contains one to three seeds.

Arenga australasica

(from Australia)

AUSTRALIA

This species grows in coastal districts of north-eastern Queensland, Australia, and adjacent islands,
and also Elcho Island in the Northern Territory. It
is a large clumping palm (trunks to 20 m tall) with
usually one to three stems dominant and a fringe
of immature suckers. Plants have large (2.5–3.5 m
long) fronds which are mid-green to dark green
and the leaflets have irregularly lobed margins.
Purplish-black, globose fruit are about 2 cm across.
Although too large for the average home garden,

this species makes an interesting lawn specimen for large gardens and parks. It grows well in tropical and subtropical regions.

Arenga caudata

(tailed, bearing a tail)

THAILAND

A delightful clumping *Arenga* from Thailand which has distinctive wedge-shaped leaflets with variously lobed and lacerated margins and a long, drawn-out tail at the apex. The leaflets are glossy green above and silvery-white beneath and provide a pleasant contrast when stirred by the wind. The palm itself forms dense, bushy clumps to about 2 m tall with slender trunks, and is a decided acquisition to any garden. It will succeed in tropical and subtropical areas in a semi-protected position where it will receive some sun. A dwarf-growing, compact form

is sometimes available and is sold as 'Nana Compacta'. The palm was previously known as *Didymosperma caudata*.

Arenga engleri

(after H.G.A. Engler, German botanist who collected widely in the tropics)

Formosa Palm

TAIWAN—RYUKYU ISLANDS

This palm has proved to be exceedingly adaptable in cultivation, growing well in both tropical and temperate regions. It originates in Taiwan and the Ryukyu Islands to the north and will withstand exposure to light frosts. It is an attractive clumping palm that rarely grows more than 3 m tall but may spread as much as 5 m. It has long, graceful, pinnate leaves which are frequently partially twisted. These have numerous, crowded leaflets which are dark green on top and silvery beneath. It is an excellent palm for a large garden providing it has room to spread. It is sufficiently tough to be grown under established trees and will tolerate a semi-shady or sunny position. Good drainage is

Above: *Arenga caudata* infructescence.
Left: Clump of *Arenga caudata*.

Above: *Arenga engleri* ripe fruit.
Left: *Arenga engleri* underside of leaf.

necessary and the plants respond to fertiliser applications. Propagation is from seed or by division which can be difficult. Seeds germinate slowly and erratically, taking up to two years to appear.

Arenga hastata

(spear-shaped)

BORNEO—MALAYSIA

An interesting palmlet which grows in the understorey of humid forests in Malaysia and Borneo. A suckering species with very slender, cane-like stems to about 1.5 m tall and spreading pinnate leaves. These have relatively few, broad, stalked leaflets (with shallow-lobed or coarse-toothed margins) and the distal pair diverge widely, like a fishes' tail. They are green on the upper surface and whitish beneath. Simple, pendulous spikes arise in clusters from the nodes. Globular, purplish fruit are about 5 mm across. An excellent clumping palm for a shady position in the tropics. Also a useful container palm. *A. borneensis* and *Didymosperma borneense* are synonyms.

Arenga hookeriana

(after J.D. Hooker, 19th century English botanist)

MALAYSIA—THAILAND

This species can have either simple leaves or short pinnate leaves with a few leaflets. The leaflets are

Arenga engleri flowers.

Arenga hookeriana.

Young plants have an attractive crown of arching fronds with the long, slender leaflets spreading widely from the rachis. Plants die after flowering and the dead stalks with limp spent inflorescences tower above the surrounding vegetation. The fruit are small (about 12 mm long), ellipsoid and red when ripe. Seeds of the species have been successfully germinated in north-eastern Queensland and plants are growing in Darwin. This palm is tropical in its requirements of warmth, moisture, excellent drainage and a sheltered position.

Arenga longicarpa

(with long fruit)

SOUTHERN CHINA

A clumping species from southern China which grows to about 3 m tall, with many slender (about 7 cm thick) crowded stems in a clump. The fronds have long petioles (2–2.5 m) and the unusually shaped leaflets are stiff and leathery in texture. They are more or less triangular in shape but with an extended, apical, tail-like portion which has irregularly toothed margins. The upper surface is dark green and shiny and the lower surface silvery. Leaflet dimensions are 20–40 cm x 1.5–5 cm. The inflorescences are much-branched and mature fruit are purple, about 1.8 cm long and ovate to oblong in shape. This palm appears to be unknown in cultivation.

Arenga microcarpa.

characteristically shaped like the blade of a paddle but with lobed and sharply toothed margins. They are bright green and shiny on the upper surface and silvery beneath. Plants develop into relatively slender clumps and have narrow, cane-like stems to about 0.5 m tall. Simple, erect spikes arise in clusters from the nodes. Globular fruit are about 1 cm across and are subtended by a persistent orange calyx. This is a highly ornamental small palm which is excellent for a shady position in the tropics. Also useful as a tub plant. It is native to Thailand and Malaysia. Formerly known as *Didymosperma hookerianum.*

Arenga listeri

(after J.J. Lister, who collected on Christmas Island in 1887)

CHRISTMAS ISLAND

An interesting palm that is endemic to Christmas Island where it grows on weathered basalt. Although still reasonably prevalent on the island, it is considered threatened because of a significant reduction in habitat from phosphate mining and the activities of seed-eating crabs that gather beneath fruiting trees and consume most of the seeds. The palm is a solitary, slender species of a medium size (12–20 m tall) and should be ideal for garden cultivation.

Arenga micrantha
(with small flowers)

SOUTHERN CHINA

This poorly known species is native to southern China where it grows on forested slopes at about 1600 m altitude. Plants have similar vegetative features to those of *A. engleri* but the inflorescences have more numerous branches and form a narrow panicle. The male flowers are yellow. This palm is not known to be in cultivation.

Arenga microcarpa
(with small fruit)

Aren Sagu

NEW GUINEA

A rather coarse clumping palm with slender trunks which may reach more than 7 m tall. They are prominently ringed, dark green and bear a crown of spreading leaves which are dark glossy green above and silvery-white beneath. The clumps are usually quite dense with a few stems dominating and other stages present down to new suckers on the periphery. Interestingly, the flowers are dark purple. The red, spherical fruit are about 15 mm across. The species is native to Irian Jaya and Papua New Guinea where it grows in lowland areas near streams. Sago is prepared by the local people from the trunks of this palm and the species has potential for wider planting in tropical regions for this purpose.

Arenga obtusifolia
(with blunt leaves or leaflets)

Lang Kap

INDONESIA—MALAYSIA

A handsome clumping palm which in tropical regions will thrive in an open sunny situation. Plants develop stout, fibrous trunks to 8 m tall which, while young, are clothed to the base with large, spreading pinnate fronds (each to 6 m x 3 m); with age the trunks become bare and are noticeably stepped. Leaflets, which are crowded, leathery, long and narrow (to 1.6 m x 8 cm), are bright green on the upper surface and silvery beneath. Flowering begins from the lower nodes and continues upwards. The large, blackish fruit, reported to reach the size of a small apple, have a prominent sunken depression at the apex. A native of coastal districts of Malaysia and Indonesia, this palm grows on steep slopes in the hills. Plants are cultivated on a limited scale but deserve to become more widely planted.

Arenga pinnata
(divided like a feather)

Sugar Palm, Gomuti Palm, Aren

INDIA—SOUTH-EAST ASIA

This palm, originally from India and South-East Asia, is now widely cultivated throughout tropical Asia for its many useful products, some of which are saleable items. These include sago, sugar, vinegar and alcoholic beverages. It is a very distinctive

Arenga pinnata.

monocarpic palm with a thick, black fibrous trunk, long spines on the leaf-bases and a dense, erect crown of finely pinnate leaves with drooping, dark green leaflets which are satiny white beneath. The flowers and fruit are carried in large, drooping panicles and the purple flowers have an unpleasant odour. The palm succeeds best in the tropics where it can be quite fast growing, reaching maturity within ten years. It can also be induced to grow in cooler areas of the subtropics and warm–temperate regions but is much slower to mature. It needs rich, well-drained soil and plenty of water throughout the year. Plants will tolerate direct sun when quite small. The species was previously known as *A. saccharifera*.

Arenga porphyrocarpum
(with purple fruit)

JAVA

A small palm which forms dense clumps to about 1 m tall. Individual stems are very thin and cane-like and the leaves have few leaflets. These are variable in shape but may be entire or with several lobes

Arenga tremula.

and the margins are finely toothed. The upper surface is bright green and the underside is greyish white to glaucous. The sparsely branched inflorescences usually bear unisexual flowers, with the female flowers (purplish in colour) being followed by fruit. These are about 1.5 cm long, reddish-purple to orange and shiny. An excellent small palm which is ideal for a shady location under established trees. It is native to Java. *Didymosperma porphyrocarpum* is a synonym.

Arenga retroflorescens
(flowering backwards)

SABAH

Flowering in this species commences from the lowest nodes of a mature stem and proceeds upwards. The palm is native to Sabah where it is restricted to a few coastal sites growing on the landward side of mangrove communities. It is a clumping species with slender, fibre-covered stems to 80 cm tall and erect to arching pinnate leaves to 3 m long. Each leaf has about forty-four narrow leaflets which are green above and brownish, scaly beneath. These are regularly spaced along the rachis except for the lowest six to ten which are clustered. The inflorescencs are short, erect, unbranched unisexual spikes and the fruit are distinctly three-sided. An interesting palm which would be a very decorative addition to tropical gardens.

Arenga tremula
(trembling, jelly-like)

PHILIPPINES

A handsome cluster palm which may form clumps 3–4 m tall. It has proved to be quite adaptable, surviving well in warm–temperate regions and also succeeding in tropical areas. The trunks are slender and green with prominent pale rings. Each supports a handsome spreading crown of relatively broad fronds which are dark green above and dull, glaucous green beneath and with a prominent pale rachis. The leaflets are quite narrow, of nearly uniform width throughout and with a few small teeth along the margin. The inflorescences are quite large and held well above the foliage. *A. tremula* is native to the Philippines and is an excellent garden palm. It succeeds in partially protected situations where it receives some sun during the day.

Large clump of *Arenga westerhoutii*. J. Dransfield

Arenga undulatifolia

(with wavy margins on the leaflets)

Aren Gelora

BORNEO—PHILIPPINES

One of the most ornamental of all palms, this species, which originates in Borneo and the Philippines, forms a neat, dense clump of fronds which are a lustrous, blue-green colour. These arch out from a short, stocky trunk or cluster of trunks and may reach more than 4 m tall. The leaflets are fairly long (to 70 cm), crowded on the fronds and have toothed, characteristically wavy margins which give rise to the appropriate specific epithet. They are dark green above and silvery white beneath. Large plants are very decorative indeed. For cultivation, a tropical climate is most suitable, although plants can be grown in a sheltered position in the subtropics. It makes a handsome lawn specimen and can also be mixed with other palms in a collection. *A. ambong* is a synonym.

Arenga westerhoutii

(after J.B. Westerhout, 19th century Dutch collector)

SOUTH-EAST ASIA

This palm is a solitary species, although it often occurs in large groups. Each plant has a massive crown. The stems are covered with black fibres which arise on the leaf sheaths and the individual fronds may be nearly 8 m long. These are flat, with stiff, oblong leaflets and are green above and greyish-white beneath. Fruit are oblong, 5–7 cm across and blackish when ripe. Like most *Arenga* species, this is a handsome palm well worthy of cultivation. It is widely distributed, being found in Malaysia, Thailand, Burma and southern China and grows as a common palm in rainforests. Plants thrive in warm, tropical conditions and are sensitive to cold.

Arenga wightii

(after Robert Wight)

Wight's Sago Palm

INDIA

An Indian palm which grows in sheltered forests at moderate elevations in northern ranges. It forms small clumps consisting of a few crowded, black, fibrous trunks, each with a crown of arching leaves 2–3 m long. The leaves have long, narrow, linear leaflets (to 30 cm x 5 cm), dark green above, white beneath, with a pair of prominent auricles at the base and small teeth and lobes along the margins. These leaflets spread in almost a flat plane on either side of the rachis. The blackish, nearly round fruit, to 2 cm across, often contain 3 seeds. This species is a collector's palm which will grow in a sheltered position in the tropics.

ASTROCARYUM

(from the Greek *astron*, star, *caryon*, a nut or kernel of a nut; from unusual star-like markings on the seed)

A genus of forty-seven species of palm distributed from Mexico to Brazil and Bolivia and also occurring on the island of Trinidad in the West Indies. Some species grow in open habitat such as savanna, some on forest margins and others occur within the canopy of lowland forests. They are solitary or clumping palms (sometimes acaulescent) with simple or pinnate fronds, the leaflets being arranged regularly or unevenly. Many species are heavily armed, with long sharp spines being distributed on most organs. The inflorescence arises below the leaves and bears unisexual flowers of both sexes. In their natural state these palms are used for a variety of purposes, particularly in fibre production. Many species also have an edible cabbage.

Cultivation: This genus is not greatly popular in cultivation and some species are only moderately ornamental. They are generally easy to cultivate with most species being best suited to tropical regions. All require free drainage. Some species tolerate full sun whereas others need a sheltered position. Solitary plants are capable of producing fertile seeds.

Propagation: From fresh seed which germinates erratically and may respond to cracking or soaking in warm water prior to sowing. Fruit may contain one or two seeds. The fruit of some species, such as *A. standleyanum,* have unusual, star-like markings on the endocarp.

Astrocaryum aculeatum

(bearing prickly thorns)

SOUTH AMERICA

A graceful palm from South America which has a slender, grey trunk copiously armed with black spines and an almost round crown of bright green leaves. These are quite long (about 5 m) and are finely divided into numerous leaflets which are plumose and arranged in groups. The leaf stalks, rachises and midribs of the leaflets are all armed with sharp, black spines. Clusters of large, orange fruit are a decorative feature and as a bonus are edible. This species is best suited to the tropics in large gardens or parks.

Astrocaryum mexicanum

(from Mexico)

CENTRAL AMERICA

This tough-looking palm is formidably armed with prickles and yet has an interesting appearance. Young plants in particular are quite appealing with their large, simple or sparingly divided leaves and spiny petioles and rachises. In older plants the fronds are distinctly pinnate with a marked variation in the size of the leaflets. The terminal pair tend to be united and fishtail-like. All leaflets are dark green above and silvery beneath. The spathe

subtending the inflorescence is interesting, being deeply cup-shaped and densely spiny. The fruit are also spiny, about 3 cm across and the seeds have an edible kernel. Native to Mexico and adjacent Central American countries, this palm is mainly grown in botanical collections. Plants are of a very manageable size, rarely exceeding 2.5 m in height. They grow well in subtropical regions.

BACTRIS

(from the Greek, *bactron*, a walking stick)

A large genus of 239 species distributed from Mexico to Paraguay and the West Indies, with a major development of the genus in Brazil. Species of this genus grow in a range of habitats including savanna, swamps, mangroves and rainforest. They are solitary or clumping (sometimes acaulescent) palms, usually with pinnate fronds but in some species these are entire with a notched tip. The inflorescence arises among the leaves and bears unisexual flowers of both sexes. A number of species have edible fruit and are also a source of cabbage and fibre.

Cultivation: The genus includes some ornamental palms but few have become popular in cultivation, possibly because of their spiny nature. Most species require well-drained soils but some will tolerate

Nearly mature fruit of *Bactris gasipaes.*

wet conditions. Single plants are capable of producing fertile seed.

Propagation: Clumping species can be propagated by division but are slow to establish. Seeds of some species have a very short period of viability and must be sown fresh. Germination is usually rapid, within two months of sowing. Fruit are usually one-seeded.

Bactris gasipaes

(a native name)

Peach Palm or Pejibaye

This palm, which is widely grown in some tropical countries for its fruit which are eaten after cooking and also for its edible cabbage, is unknown in the wild. The ripe fruit are yellow, with the attractive appearance of a peach and hang in pendulous clusters. Two crops are produced each year and seedless forms are known. The palm is clumping with slender, prominently ringed, spiny trunks and a crown of deep green pinnate fronds which are also liberally covered with spines. It is best suited to hot, humid, tropical regions, although plants have been successfully grown in temperate regions. In the tropics, plants can be very fast growing and are appreciative of regular watering and feeding. Young plants need protection from direct sun. Seeds germinate readily within two months but should be sown very soon after collection, as they quickly lose their viability. The species was previously known as *Guilielma gasipaes*.

Bactris jamaicensis

(from Jamaica)

JAMAICA

This clumping palm, as the name suggests, comes from Jamaica where it may grow in thickets. Plants have slender, spiny trunks, spiny leaf-bases and pinnate leaves. They are very tropical in their requirements and succeed best in a partially protected position.

Bactris major

(larger than normal)

SOUTH AMERICA—WEST INDIES

The clustered trunks of this palm may grow to 8 m tall and are no more than 5 cm thick. They are

Bactris major.

spiny when young but the spines are shed with age and the older trunks are smooth with prominent white rings. Each trunk is crowned with a cluster of dull green, pinnate fronds. Clusters of purple fruit are quite showy and have a juicy flesh which is edible or can be made into wine. The species succeeds best in the tropics. Once established, plants will tolerate full sun, but need protection while small. In nature this palm grows on the margins of estuaries at the extreme of tidal influence.

BALAKA

(from *Mbalaka*, a Fijian native name used for the palm)

A small genus of about seven species of palm, five of which occur in Fiji and two in Samoa. All species grow as understorey plants in humid forests. They are solitary, unarmed, slender to wispy palms with arching, feathery fronds and a

Balaka seemannii. C. Goudey

prominent crownshaft. The inflorescence arises below the leaves and bears unisexual flowers of both sexes. The kernel of the seed is eaten locally and the hard straight stems have been used for spears and walking sticks.

Cultivation: These small, slender palms deserve to become more widely grown. They are excellent for a shady position or one exposed to partial sun. Soil drainage must be excellent and plants respond to mulches and regular watering. Solitary plants are capable of producing fertile seed.

Propagation: Fresh seed germinates readily within two to four months of sowing. Each fruit contains a single seed.

Balaka longirostris

(with a long beak)

FIJI

A palm from the rainforests of Viti Levu, Fiji, growing in moist areas at moderate elevations in the mountains. Plants have a very slender trunk (4–5 cm in diameter) and scattered, wedge-shaped leaflets similar to those of *B. seemannii*. A true collector's palm, this species is very rare in cultivation and stringent in its requirements of a shady, warm position in well-drained acid soil.

Balaka macrocarpa

(with large fruit)

FIJI

A little-known palm which is distinctive for having the largest fruit in the genus (about 4 cm x 1.5 cm). It is native to the islands of Viti Levu and Vanua Levu, Fiji, where it grows in the understorey of rainforests at moderate elevations. It is rarely cultivated and would appear to have similar cultural requirements to those of *B. longirostris* and *B. seemannii*.

Balaka microcarpa

(with small fruit)

Spear Palm

FIJI

Young leaves of this distinctive, slender palm are covered with grey wool which is shed as the leaf expands and matures. A native of the lowland rainforests of Viti Levu, Fiji, the plants typically have a slender trunk to 10 m tall and a sparse crown (about five leaves) of arching fronds with narrow, more or less erect leaflets which are broadest near the middle and taper to each end. The slender, hard, straight trunks of this palm were prized as spear shafts by the Fijians and were later harvested by Europeans and made into walking sticks. Plants of this palm succeed best in the tropics and require well-drained acid soil in a shady to semi-shady position. Once established, plants are tolerant of considerable exposure to sun. Growth can be fast, especially if plants are supplied with plenty of water.

Balaka seemannii

(after Dr Berthold Seemann, botanist, explorer, pioneer and author of *Flora Vitiensis*)

FIJI

Often confused in cultivation with *B. microcarpa*, this species can be readily distinguished by its dark green, wedge-shaped leaflets which are irregularly jagged on the end. Both species are similar in having a slender trunk and a sparse crown of fronds. Those of *B. seemannii* number five to seven and surmount a small, narrow crownshaft. An attractive palm which is unfortunately rare in cultivation. It is native to the islands of Vanua Levu and Taveuni, Fiji, and is stringent in its requirements of warmth, shade and well-drained acid soil. Young plants make attractive container specimens. Also known as *Ptychosperma seemannii*, *B. gracilis* and *B. cuneata*.

BASSELINIA

(after Olivier Basselin, 15th century French poet)

A genus of eleven species of palm all endemic to New Caledonia. One species, *B. gracilis*, is widespread in a range of conditions but most of the others are narrow endemics occupying specific habitats and soil types. All species have a crown of stiffly spreading, pinnate fronds, a prominent crownshaft and unarmed trunk, but about five species have a clumping habit and the remainder are solitary palms. The inflorescences arise below the crownshaft and bear unisexual flowers of both sexes.

Cultivation: Palms of this genus have tremendous horticultural appeal but they are rarely encountered in cultivation and remain the province of collectors. Many species appear to have specific requirements and as a result are not adaptable and have proved to be difficult to grow. Experience is also limited by difficulty in obtaining supplies of seed. All species grow in well-drained soils which may be either acidic or ultrabasic. Single plants are capable of producing fertile seed.

Propagation: Fresh seeds germinate readily but seedlings can be difficult to handle. Fruit contain a solitary seed. Division of the clumping species is apparently difficult.

Basselinia deplanchei

(after Dr. M. Deplanche, French naval surgeon and botanist)

NEW CALEDONIA

A small clumping palm which grows in ultrabasic soils at moderate altitudes, usually in open situations and along forest margins. Plants have very slender (about 5 cm across) trunks to 3 m tall, a dark red to black crownshaft and leaves which may be entire and deeply notched at the apex or irregularly divided. This species is very specific in its cultural requirements and appears to be difficult to grow.

Basselinia gracilis

(slender, graceful)

NEW CALEDONIA

A choice collector's palm from New Caledonia where it grows in dense rainforest from sea-level to high altitudes in the mountains. It is exceptionally colourful, rivalling even the brilliant Sealing Wax Palm for gaudiness. The leaves are dark green and glossy, the sheaths and crownshaft are pink to dark red with blue or grey tinges, the petioles purplish and the rachis yellowish-green. The species is usually a clumping palm with several slender stems arising in close proximity and in nature growing to about 7 m tall. It is extremely rare in cultivation and reports indicate that it is somewhat difficult to grow, although it is certainly the most adaptable in the genus. It appears to require cool to warm, shaded, moist conditions in well-drained soil. Selection of provenances may play an important role in its cultivation in different climates. Seed germinates easily within two to three months of sowing. Also known as *B. eriostachys*, *B. billardieri*, *B. heterophyl-la*, *Microkentia gracilis* and *Kentia gracilis*.

Basselinia pancheri

(after M. Pancher, French collector in New Caledonia)

NEW CALEDONIA

An elegant palm which is readily distinguished by its black, two-lobed fruit which are almost kidney-shaped. It grows in drier habitats than do other members of the genus and plants are often exposed to full sun. The solitary, slender trunk grows to about 8 m tall and has a colourful crownshaft which may be in shades of orange, red or purple. The pinnate leaves are irregularly divided with some segments being very broad. Plants require excellent drainage, shelter and may need alkaline soils. Previously known as *Kentia pancheri* and *Microkentia pancheri*.

BENTINCKIA

(after William Henry Cavendish Bentinck, 18th & 19th century Governor-General of the East Indies)

A small genus of two species of palm found in India and the adjacent Nicobar Islands. They grow in forests at low to moderate elevations. They are solitary, unarmed feather-leaved palms with a conspicuous crownshaft. The inflorescence arises below the leaves and bears unisexual flowers of both sexes.
Cultivation: Interesting palms which are grown to a limited degree in tropical regions. Plants require well-drained soil in an open sunny position and respond to watering, mulches and fertilisers. Single plants are capable of producing fertile seeds.
Propagation: Solely from fresh seed which germinates readily within two to four months of sowing. Each fruit contains a single seed.

Bentinckia condapanna

(an Indian name)

Lord Bentinck's Palm

INDIA

This species develops a slender, ringed trunk to 10 m tall and 15 m across, and has a sparse, arching or rounded crown of fronds 1–1.6 m long. The leaflets, which may reach 60 cm x 3 cm, are held

rigidly erect and split near the apex. The inflorescence has prominent colourful bracts which may be bright red or violet blue. The dark brown, egg-shaped fruit contain a seed which has a deep groove in the testa. This palm, which was once locally common growing on steep cliffs at moderate to high elevations in the Travancore Hills, India, has been reduced to rarity by clearing and the ravages of feeding elephants (which also relish the cabbage). Plants are relatively easy to grow in tropical and warm subtropical regions.

Bentinckia nicobarica

(from the Nicobar Islands)

NICOBAR ISLANDS

Superficially, this palm closely resembles the widely cultivated Bangalow Palm but can be immediately distinguished by the longer crownshaft, the intricately branched inflorescence and the leathery leaflets which have two blunt lobes at the apex. It is a very tall, slender species reaching upwards of

Bentinckia nicobarica.

15 m and has a graceful crown of dark green pinnate fronds which often twist along the rachis. The small, round fruit are scarlet when ripe. The species is endemic to the Nicobar Islands in the Indian Ocean and is widely grown in tropical countries. Being very cold-sensitive it is best suited to the hot, humid lowlands and once established is fast growing.

BISMARCKIA

(after Prince Otto Von Bismarck, first German Chancellor)

A monotypic genus of palm which is endemic to Madagascar. It is a very conspicuous palm which grows in open habitats, particularly savanna, on the western side of the island. The species is a fan palm with a solitary, unarmed trunk and striking costapalmate leaves. The inflorescence arises among the leaves and bears unisexual flowers. The trunks and leaves of this palm are used for the construction of dwellings.

Cultivation: A popular species which requires unimpeded drainage and a sunny location in tropical and warm subtropical regions. Plants are dioecious and separate male and female trees are needed for the production of fertile seed.

Propagation: Fresh seed germinates readily. Each fruit contains a single seed.

Bismarckia nobilis

(noble, stately)

Bismarck Palm

MADAGASCAR

A magnificent fan palm prized for its large, heavy crown of blue-green leaves which creak in the slightest breeze. The lamina may be in excess of 3 m across and is strongly costapalmate with rigid, waxy segments. It is supported by a very thick petiole which is covered with a waxy, woolly material. The petiole splits at the base, where it is attached to the trunk. The trunk itself is stout, clear of petiole bases and may reach 10 m tall in nature. The fruit is about 3 cm across and brown when ripe. In cultivation, the palm is reputed to be fast growing. It is ideal for group planting and seen at its best in an open situation where it can develop to its potential. It requires a sunny aspect in well-drained soil and grows best in the tropics.

Bismarckia nobilis.

BORASSUS

(from the Greek *borassos*, an immature spadix of the date palm)

Palmyra Palms

A genus of seven species of tall palm found in drier tropical regions. They are remarkably widespread for a small genus being distributed in Africa, Madagascar, Arabia, India, Indonesia and New Guinea. Most occur in open situations such as savanna and littoral forests but they are also found in riverine forests and in mountainous regions. They are solitary fan-leaved palms with an unarmed trunk covered in the upper part with a lattice of leaf-bases. The inflorescence arises among the leaves and bears unisexual flowers. At least one species (*B. flabellifer*) is widely used for a variety of purposes and others are of local significance.

Cultivation: *Borassus flabellifer* is widely planted for its economic products and less commonly for its ornamental features. Other species appear to be rarely grown. Being large palms they require plenty of room to develop fully and also must have excellent drainage. Male and female trees are essential for fertile seed production. Borassus palms are sensitive to Lethal Yellowing disease and also root rotting fungi.

Propagation: *Borassus* seeds are large and because they germinate with a long sinker, they are commonly sown into a large container or directly into the ground in the position where they are to grow. Seeds take six to nine months to germinate and true leaves emerge two or three months later. Each fruit contains one to three seeds embedded in tough, reddish fibre. Germination is sporadic and may be hastened by pre-soaking the seeds in water for up to two weeks or filing a nick in the outer coat.

Borassus aethiopum

(from Africa)

African Palmyra Palm, Black Rhun Palm, Ronier Palm

A native of Africa where it is distributed disjunctly from tropical regions to the northern Transvaal Province. Plants develop a characteristic bulge in the trunk at maturity and have smaller fruit (about 9 cm across) than those of *B. flabellifer* which it closely resembles. Outside Africa this palm appears to be rarely cultivated. In tropical Africa, a drink is made from the sap of this palm and the immature fruit are eaten. Natural populations of this species are suffering from interference and are shrinking.

Borassus flabellifer

(bearing fans)

Palmyra Palm, Toddy Palm, Talauriksha Palm

INDIA—MALAYSIA

This tall palm is frequently noticeable in drier tropical regions where it grows to perfection. It is native to India and Malaysia and is widespread and common in open situations, frequently growing in dry, sandy soils near the coast and often forming communities. The hard, black trunks are often curved and bear a dead skirt beneath the crown of large (3 m across), rigid, blue-green, fan-shaped leaves. In its native state the very old plants are cut for their hard, black timber and also the sap is tapped to yield palm sugar and the leaves may be made into paper. When fresh, the large, black fruit (almost as large as coconuts) contain one to three large seeds which are surrounded by a layer of orange, fibrous flesh which is sweet and juicy. Palmyra Palms greatly resent disturbance and the seeds are best sown in their permanent position in the ground. They like a sunny aspect in well-drained soil and are very sensitive to cold. Seed takes two to six months to germinate. This palm has been erroneously recorded from Australia, this record being based on a planted specimen in the garden of Somerset, Cape York Peninsula, Queensland.

Borassus aethiopum.

Borassus flabellifer young plant.

Borassus flabellifer mature plant.

BRAHEA

(after Tycho Brahe, 16th & 17th century Danish astronomer)

Hesper Palms

A small genus of about sixteen species of hardy palm distributed in Baja California, Mexico and Central America, where they grow in colonies in open, rocky situations. They are solitary or, less commonly, clumping palms with unarmed or spiny trunks and large, fan-shaped leaves. The inflorescence arises among the leaves and the flowers are bisexual. The leaves are commonly used for thatching. All species of the genus *Erythea* are now included in *Brahea*.

Cultivation: These palms have proved to be highly adaptable and will succeed from tropical to temperate areas. Their main requirements are unimpeded drainage and an open position exposed to sun and air movement. Single plants are capable of producing fertile seeds.

Propagation: Fresh seed germinates readily within two to twelve months of sowing, although sometimes germination may be sporadic. Each fruit contains a single seed.

Brahea armata in flower.

Brahea aculeata

(prickly, with thorns)

Sinaloa Hesper Palm

MEXICO

The large, yellowish-green, fan leaves of this Mexican palm (the fans are up to 1 m across) are supported by a slender, glaucous petiole which appears too thin to carry the weight. This species is a robust, squat-growing palm with a trunk to about 5 m tall and 20 cm thick and a relatively sparse, open crown. Plants bear masses of small, brown to black fruit (20–25 mm across) in heavy, hanging clusters. They will grow in temperate, subtropical and drier tropical regions and need an open, sunny position in well-drained soil.

Brahea armata

(armed with thorns)

Blue Hesper Palm

MEXICO

The most distinctive feature of this palm is its crown of stiff, blue-green, fan-shaped leaves which radiate in a crown from the apex of the trunk. These leaves persist after death, hanging as a brown skirt. The relatively stout trunk (40–50 cm across) grows up to 12 m tall and the species produces masses of cream flowers in arching inflorescences, 4–6 m long, which extend well clear of the foliage. The shiny, brown fruit are flecked or striped with white, 18–24 mm long and ovoid in shape. This palm, very decorative with its bluish leaves and impressive inflorescences, has proved to be very hardy and is best suited to temperate and subtropical regions. Plants withstand severe frosts without damage and for best leaf colour should be grown in an open, sunny situation. They will grow in poor soils but will not tolerate bad drainage. The plants are relatively slow growing but may flower while quite small. The species is native to northern Baja California, north-western Mexico, where it grows in rocky canyons.

Brahea aculeata.

146

Brahea bella

(pretty)

MEXICO

A species from northern Mexico where it grows in crevices on calcareous cliffs and among boulders at altitudes up to about 1000 m. Its leaves are bright green on both surfaces, somewhat shiny and the leaflets appear limp and drooping. The blades are about 1.5 m or more wide and divided into about fifty leaflets that split towards the end. The inflorescence is densely covered with white, woolly hairs and the broadly oblong fruit, to 13 mm x 11 mm, are green to reddish when ripe and flattened on one side.

Brahea berlandieri

(after M. Berlandier)

MEXICO

A solitary palm from north-eastern Mexico where it grows on rocky, limestone hills. Plants have a stout trunk to 3 m tall and a rounded crown of stiff, green to glaucous leaves. The branches of the inflorescence are smooth and lack any hairs. An interesting sun-loving palm which demands excellent drainage.

Brahea brandegeei

(after T.S. Brandegee, botanist and original collector)

San José Hesper Palm, Palma de Taco

MEXICO

Brahea clara.

This species is native to San José del Cabo on the southern end of the peninsula, Baja California, where it grows in canyons near streams and on sheltered slopes. Populations have suffered as the trees are cut for wood and edible cabbage and the leaves used for thatch. This palm was reputed to be the tallest in the genus with trunks growing in excess of 25 m tall, but it is now believed that these dimensions were the result of confusion with *Washingtonia robusta* which also grows in the area. *B. brandegeei* seems to grow to about 12 m tall and has a moderately large, dense crown. Frequently a petticoat of dead leaves persists below the crown and the leaf-bases clothe most of the trunk. The leaves are dull green above and whitish beneath from a waxy layer. Fruit are shiny brown with lighter flecks or bands, 15–22 mm long and oblong to globose in shape. When yellow they are edible, tasting somewhat like dates. This palm is now fairly widely grown and is well suited to temperate and subtropical regions. Plants need well-drained soil and will tolerate direct sun from an early age.

Brahea clara

(clear)

MEXICO

This species is similar to *B. armata* but with a more slender trunk, shorter inflorescences, smaller flowers and smaller fruit. The species is native to Mexico (Sonora and Baja California) where it

147

Brahea decumbens, an attractive dwarf palm.

grows in canyons and cliffs, often close to the sea. Plants have a dense crown and the leaves are usually blue-green but are often green. The inflorescences hang out beyond the leaves and after flowering bear clusters of brownish, striped fruit (to 20 mm x 16 mm) which are flattened on one side. A very decorative palm which is ideal for avenue planting. It grows well in subtropical zones, particularly in areas with a drier climate. Plants must have unimpeded drainage. *Erythea roezlii* is a synonym.

Brahea decumbens

(prostrate with the apex erect)

MEXICO

A delightful small palm which has a decumbent trunk terminated by an erect and rounded crown of striking, bluish-green leaves. The trunks grow in excess of 2 m long and produce offsets. The leaves of these offsets and those of seedlings of this palm are characteristically green, contrasting markedly with the leaves of the mature plants. The ascending inflorescences have a few short, crowded branches and the fruit, about 1.5 cm long, are densely hairy. As well as being highly ornamental, this palm is tolerant of adversity as its natural habitat is on exposed, rocky, limestone hills in north-eastern

Mexico. Plants are rare in cultivation and have proved to be slow growing. They need a sunny aspect and excellent drainage.

Brahea dulcis

(sweet tasting)

Rock Palm

MEXICO

The small, yellow fruits of this palm hang in large clusters up to 2 m long. They have a layer of succulent flesh which is quite sweet and tasty and were eaten by the Indians of Mexico where the palm is native. As its common name suggests, the species grows in poor, skeletal soils in rocky areas. It is a stout palm with a large crown of deeply divided, stiff, fan-shaped leaves which are covered with a mealy white powder, at least while young. Although relatively slow growing, the species is adaptable, succeeding in temperate and tropical regions. It needs well-drained soil and once established is hardy and long-lived.

Brahea dulcis.

Brahea edulis

(edible)

Guadalupe Palm

MEXICO

When in fruit, this palm bears prodigious quantities of small (25–35 mm across), round, black fruit which have a fleshy, sweet pulp around the seed. This pulp is edible and in some specimens is very acceptable and tasty. Native to Guadalupe Island off the west coast of Mexico where it grows in deep, warm ravines. This palm is now promoted for its large, handsome, fan leaves, its hardiness and its tasty fruit. Mature specimens are very ornamental, with a large crown of heavy, green fronds atop a stout trunk to 10 m x 40 cm. Plants have proved to be very adaptable in cultivation, succeeding from subtropical to temperate areas and even withstanding frosts. Young plants, which are generally slow growing, tolerate exposure to full sun, but well-drained soils are essential for success.

Brahea elegans

(elegant, graceful)

Franceschi Palm

MEXICO

An elegant fan palm which has a graceful crown of leaves. Plants have a short, stout trunk (about 1 m tall) and a large, open crown. Each leaf is borne on a long, slender, spiny petiole and the blade (about 75 cm across) is deeply divided into about fifty, stiff, thin-textured, spreading segments. Inflorescences about as long as the leaves bear showy, white flowers followed by oblong to pear-shaped, yellowish fruit, 18–22 mm long. This species was described in 1907 from a plant cultivated in Los Angeles and believed to have originated in Sonora, Mexico. It grows well in subtropical regions.

Brahea moorei

(after Professor L.H. Moore, Jr, noted 20th century palm botanist)

MEXICO

A native of north-eastern Mexico where it grows in shady forests on rocky, limestone slopes. It is a small palm with a subterranean trunk and a sparse crown. The leaves are dark green on the upper surface and chalky white beneath from a dense powdery coating. The slender petioles are unarmed. The erect inflorescence extends well above the leaves. The fruit are about 1 cm across and it is interesting to note that the seeds are capable of germination after being exposed to hot sun for many months. Some populations of this palm apparently produce suckers whereas others do not. This species has only recently been introduced into cultivation.

Brahea edulis.

Brahea nitida

(shiny)

MEXICO

A robust palm with an attractive rounded to elongated crown of bright green, somewhat shiny fronds. These have broad, stiff blades with the tips of the segments lax or conspicuously drooping. Long, much-branched inflorescences arch out beyond the lower leaves and after flowering carry small, ovoid, blackish fruit, each with a prominent apical beak. The trunk of this palm is relatively

Brahea nitida ornamental leaf-bases and trunk fibres.

stout and is covered by persistent, pale brown, fibrous leaf-bases. This is an ornamental, but slow-growing palm which has been successfully grown in subtropical and warm temperate regions.

Brahea pima
(a native Mexican name)

MEXICO

A little-known species which grows on basaltic loams at about 1500 m altitude in Mexico. Plants develop a moderately stout trunk to about 4 m tall and have a rounded crown of stiff, dull green leaves. The ovoid, beaked fruit, about 1.2 cm long, are brown to blackish when ripe. This species, which appears to be rarely grown, is probably suitable for subtropical and warm temperate regions.

Brahea prominens
(prominent)

MEXICO

This species was described from a striking specimen growing in a part in Oaxaca, Mexico. Its origin is uncertain but plants have a stout trunk to 10 m tall and over 30 cm thick and large leaves (about 1 m across) which are light green on the upper surface and silvery beneath with a glaucous coating.

Each leaf is divided into about fifty narrow segments. The inflorescences are longer than the leaves and after flowering carry clusters of yellowish fruit (about 17 mm x 10 mm) which are flattened on one side.

Brahea salvadorensis
(from El Salvador)

CENTRAL AMERICA

This species is widespread being native to El Salvador, Guatemala, Belize, Honduras and Mexico. It grows in open areas on rocky slopes of igneous formation. It is a robust palm with a relatively stout trunk and a crown of large, bright green, somewhat shiny fronds. These are deeply divided into narrow segments which hardly droop. Small, whitish flowers are followed by ovoid, brown to blackish fruit. This species is rare in cultivation. It has proved to be slow growing and is suitable for subtropical and warm–temperate regions.

BRASSIOPHOENIX
(after L.J. Brass, original collector and *phoenix*, a palm)

A small genus of two species both endemic to New Guinea where they grow in lowland rainforest. They are solitary palms with a slender unarmed trunk topped by a crownshaft and spreading pinnate leaves, with unusual wedge-shaped leaflets which have three prominent apical points. The inflorescence is borne among the leaves and carries unisexual flowers of both sexes.
Cultivation: Although rarely grown, these palms are highly ornamental and deserving of wider planting. They are shade-loving species which are tropical in their requirements. Single plants are capable of producing fertile seed.
Propagation: From fresh seed. Each fruit contains a single seed.

Brassiophoenix drymophloeoides
(resembling a *Drymophloeus*)

NEW GUINEA

Rainforests of eastern New Guinea are the home of this very attractive, slender palm which may grow as much as 10 m tall. It bears a neat crown of fronds (which have unusually shaped leaflets) above a prominent crownshaft. The inflorescence is densely covered with white, woolly tomentum and this feature, together with the dark red fruit, serves to distinguish this species from the following. Attractive features, including interesting foliage and colourful fruit, make this palm a very desirable subject for cultivation and it must surely become more widely grown in the future. Partial shade in a well-watered, tropical garden are its requirements.

Brassiophoenix schumannii

(after K. Schumann, original collector)

NEW GUINEA

This palm occurs in north-eastern and eastern Papua New Guinea where it grows in rainforest. Like the previous species it is a slender palm with excellent ornamental prospects but unfortunately remains virtually unknown in cultivation. From the climate where it grows and reports about its performance in America, it would appear to need a partially shaded position in the tropics. It can be distinguished from the previous species by its pale, yellow–orange fruit and dark, scaly hairs on the inflorescence.

BUTIA

(from a local Brazilian name for one species)

A genus of about eight species distributed in southern Brazil, Paraguay, Uruguay and Argentina. They grow in open situations, often forming colonies, in habitats such as savanna and sparse woodland. They are relatively stout, solitary, unarmed or spiny palms with prominent leaf-bases retained over most of the trunk and arching, feathery fronds with narrow, stiff leaflets. The inflorescence arises among the leaves and bears unisexual flowers of both sexes. Rare hybrids are formed with *Jubaea* and *Syagrus*.
Cultivation: Popular palms for their hardiness, ease of culture and ornamental appeal. They can be grown from the tropics to temperate regions and require unimpeded drainage and a sunny position. Single plants are capable of producing fertile seed.
Propagation: Seed may be slow and sporadic to germinate because of a hard, woody coat. Cracking the seed coat prior to sowing and pre-soaking in warm water may aid germination. Each fruit contains one to three seeds.

Butia archeri

(after W.A. Archer, original collector)

BRAZIL

A small palm from Brazil which grows in grassland or in the shrubby vegetation known as Cerrado. Plants either have a subterranean trunk or a short, emergent trunk to about 1 m tall and arching fronds about 1 m long. The petiole lacks spines and the inflorescence is prominently glaucous with smooth (not woolly) spathes. Yellowish fruit (about 20 mm x 14 mm) are more or less globose with a short beak.

Butia arenicola

(growing on sandhills)

SOUTH AMERICA

The glaucous leaves of this species appear to erupt out of the ground since the short stem is subterranean. They grow to about 1.2 m long and have spiny petioles. The spathes of the inflorescence are glaucous or bear short, brown hairs and the yellow fruit (to 23 mm x 10 mm) have a short beak. This species is native to Paraguay and Brazil, growing in savanna and surviving, after clearing, in pasture. There is some speculation that this species may be a dwarfed variant of *B. capitata*.

Butia bonnetii

(after M. Bonnet, 19th century French horticulturist)

Plants are grown under this name in California but appear to be variants of *B. capitata*. The species was named from a cultivated plant and no type specimens exist. The description of the species matches *B. capitata* but is smaller growing with shorter leaves and smaller, ovoid fruit.

Butia capitata

(head-shaped or in heads)

Wine Palm, Jelly Palm

SOUTH AMERICA

An eye-catching and distinctive palm readily recognised by its stout, woody trunk to 5 m tall and graceful crown of arching, bluish-green pinnate fronds. It is native to South America (Brazil and Uruguay) where it is widespread in the drier regions and is sometimes locally common. The fruits are very decorative, being 2–3 cm across, somewhat flattened and yellow or reddish when ripe. They are edible with a fruity flavour but rather fibrous consistency. This palm is very hardy and will thrive in temperate and subtropical regions as well as inland districts. It requires a sunny position in well-drained soils and will tolerate well-structured clay or limey soils. Young plants make very decorative and hardy tub specimens. The species can be distinguished from its close relatives by the glaucous, glabrous spathes of the inflorescence, the long spines on the petioles (8–11 cm long) and the small flowers (3–8 mm long) and fruit (1.5–2.5 cm long).

B. capitata is variable and some plants in cultivation appear quite distinctive. Some of these have been named as varieties but may just represent oddities. B. capitata var. strictior appears distinctive because of its stiffly erect fronds which shed completely and leave the trunk clear.

Butia capitata.

Butia eriospatha
(with a woolly spathe)
Woolly Butia
BRAZIL

Although basically similar to *B. capitata,* this species can be distinguished by the spathes which are densely covered with short, brown wool. Plants can have green or glaucous fronds. The fronds have a stout, spiny petiole which may reach 1 m long while the rachis can be 2–2.5 m long. Fruit are globular, about 2 cm across and yellow when ripe. The plants grow up to 6 m tall and are very hardy in cultivation, requiring a sunny aspect in well-drained soil. The species is native to Brazil and is uncommonly grown.

Butia microspadix
(with a small spadix or inflorescence)
BRAZIL

An apparently trunkless species (the trunk is subterranean) which grows in the grasslands of Brazil, being locally common in some areas. Plants have a cluster of erect to arching, glaucous leaves to about 1 m long with the petiole lacking spines. The inflorescence is very short and the spathes are densely covered in dark brown wool. The relatively narrow, yellowish fruit (about 20 mm x 11 mm) have a prominent, long beak. This would be an interesting palm for cultivation.

Butia paraguayensis
(from Paraguay)
SOUTH AMERICA

As the specific name suggests, this species is native to Paraguay, but it also extends to Argentina and Brazil. It grows in open vegetation (Cerrado) including savanna and is reported to survive some grazing by cattle in pasture. Plants may have either a subterranean trunk or an emergent trunk to about 1.5 m long and have a relatively long petiole which is armed with short spines. The fronds are gracefully curved and have glaucous to green leaflets. The spathes are covered with short, brown hairs and the ovoid fruit, to 37 mm x 27 mm, have a prominent beak and often contain two seeds.

Above Left: *Butia eriospatha*.
Above Right: *Butia eriospatha* flowers.

Butia purpurascens

(purplish)

BRAZIL

This species grows in the vegetation known as Cerrado, and is native to the state of Goias, Brazil. It is locally common, with the plants withstanding regular fires. Plants develop a woody trunk to about 4 m tall and are similar in general appearance to *B. capitata* but with smooth (not toothed) petiole margins and purplish to purple inflorescences, flowers and fruit (yellow in *B. capitata*). The fruit of *B. purpurascens* are about 28 mm x 13 mm and ovoid in shape. An attractive palm which has recently been introduced into cultivation.

Butia yatay

(a native name)

Yatay Palm

SOUTH AMERICA

This handsome palm from northern Argentina and Uruguay is little-known in cultivation, being mainly grown by enthusiasts. In nature it forms extensive forests, growing in sandy soils. Plants will

Butia paraguayensis.

153

Butia yatay.

grow in tropical, subtropical and temperate regions and also seem to be hardy enough for inland districts. Its trunk tends to be taller (to 12 m) and more prominent than in most Butias and is adorned with dark, persistent leaf-bases. Its bluish fronds are in an attractive crown and the large, yellowish fruit (3–4 cm long) are reportedly edible. Well-drained soils in a sunny position are essential for success. Plants are generally slow growing and seed is slow and erratic to germinate, taking from six months to two years. The species can be distinguished from its close relatives by the glaucous, hairy spathes of the inflorescence, the short spines on the petioles (about 3 cm long) and the larger flowers (10–16 mm long) and fruit (3–4 cm long).

X BUTIAGRUS

(a hybrid name between the genera *Butia* and *Syagrus*)

A hybrid genus of palm originating from crosses between the genera *Syagrus* and *Butia*. Sporadic plants occur in batches of seed obtained where the parents grow in close proximity.

X *Butiagrus nabonnandii*

(after Paul Nabonnand, who hybridised the species prior to 1908)

This palm is an intergeneric hybrid between *Butia capitata* and *Syagrus romanzoffiana*. It was apparently produced originally as a deliberate cross by a French horticulturist using the *Syagrus* as the pollen donor. Crosses also occur in the wild where mature plants of the parents grow in close proximity and these hybrids show up sporadically in batches of seed raised from these plants. Seedlings and larger plants of the cross have characters intermediate between the parents. A couple of large plants occur in municipal gardens of Brisbane, Australia and others are also in the collections of enthusiasts. Plants from the original crosses are growing in gardens in France and Italy. They are quite adaptable to cultivation but are sterile. *Syagrus* X *fairchildiana* is a synonym.

CALAMUS

(from the Greek *calamos*, a reed, in reference to the slender, cane-like stems of these palms)

A large genus of about 375 species of palms widely distributed in tropical regions, including Africa, India, South-East Asia, Malaysia, Indonesia, Fiji, New Guinea and Australia, with the major zone of diversity centred on Indonesia and New Guinea. Many species are climbing palms with thin, tough, spiny stems; others are shrubby with erect stems; and some are even stemless or have solitary trunks. The spreading leaves are pinnate and often have thorns on the rachis. *C. arborescens* from Burma forms erect, unsupported clumps with stems to more than 7 m tall. Climbing is aided by thorns on the rachises and petioles and spines on the young stems, all of which catch in surrounding vegetation. Plants also produce long, specialised, hook-bearing extensions of the rachis (cirri) or flagella which are modified inflorescences and arise in the leaf axils.

As a group, these palms are known as rattans because the slender, pliable stems are harvested for furniture construction. Some species have very thin, whippy stems while others are quite stout. They mainly climb into breaks in the forest canopy and are common along roads, trails, and so on. Many species have fruit with a thin layer of edible

Calamus stems (rattan) drying. G. Cuffe

or thirst-quenching flesh surrounding the seed. Drinking water can be collected from the cut stems of some species and young suckers and their starchy base can be eaten after cooking.

Cultivation: Because of their prickly nature very few species of *Calamus* are grown and they are mainly to be found in botanical collections and those of enthusiasts. Some species make striking pot plants and are suitable for indoor decoration. Young plants of many species are very sensitive to sun damage and may be readily killed by over-exposure. As a group, they are best grown in semi-shady conditions and appreciate organic mulches and litter.

Propagation: Male and female plants are essential for fertile seed production. Seed seems to germinate readily provided that it is fresh. Seed of some species may lose its viability quite rapidly and for best results should be sown very soon after collection. The fruit usually contain a single seed but in some species can contain up to three seeds. Division of clumps can be successful but the divided pieces are often very slow to establish.

Calamus aruensis

(from the Aru Islands)

AUSTRALIA—NEW GUINEA—ARU ISLANDS

Young plants of this vigorous climber have handsome, dark green leaves (30–40 leaflets per frond) and make attractive pot specimens. As they tolerate dark conditions they could be useful for indoor decoration. They are, however, extremely tropical in their requirements, being very sensitive to periods of cold and will not withstand even the slightest frost. The species is native to northern parts of Cape York Peninsula, Queensland, Australia; New Guinea; and the Aru Islands.

Calamus australis

(southern)

Hairy Mary, Lawyer Cane, Wait-a-while

AUSTRALIA

A robust climbing palm with a rather unfriendly nature, its leaf sheaths, petioles and rachises being liberally coated with long, brittle spines. The leaves have 25–56 dark green, uncrowded leaflets and

Bundles of rattan ready for processing.
G. Cuffe

Calamus moti quiescent clump on rainforest floor.

white, scaly fruit are borne on long, hanging clusters. The species, which is locally common in the rainforests of north-eastern Queensland, Australia, is grown on a limited scale. Seedlings make interesting pot plants and could be suitable for indoor decoration.

Calamus caryotoides

(resembling the genus *Caryota*)

Fishtail Lawyer Cane

AUSTRALIA

This, the most slender of the Australian species of *Calamus*, is a desirable palm for cultivation. With its thin, cane-like stems and distinctive fishtail leaflets, it makes an interesting plant for shady conditions. Plants can be grown under established trees and like plenty of water, especially during dry periods. They are slow growing in their early years of establishment but may then increase in height quite rapidly. Some have a single stem for many years while others produce suckers from an early age. A cold-tolerant palm which can be successfully grown in sheltered positions in temperate regions. In its native state, it grows in drier rainforests and along forest margins and stream banks in lowland and highland areas of north-eastern Queensland, Australia.

Calamus ciliaris

(fringed, as with eyelashes)

INDONESIA

A slender climbing palm native to Java and Sumatra where it climbs by hooked extensions to the leaves. Although the species lacks garden appeal, the young plants are graceful and make interesting pot plants. The species has been cultivated for display in English glasshouses for many decades. In this species, the pinnate leaves are verdant green and the narrow segments are quite hairy. In pots it likes a well-drained mixture rich in organic matter and as it is rather sensitive to cold may need to be kept in a glasshouse. The plants must not be allowed to dry out as they may easily die.

Calamus discolor

(of different colours)

PHILIPPINES

This is a rather attractive climbing palm as the pinnate leaves are green above and silvery white beneath. It is not as vigorous as many climbing species and plants can be held in a pot for considerable periods. The slender stems are spiny and the thorny extensions can be removed if a nuisance. Native to the Philippines, it is best suited to tropical and perhaps subtropical regions.

Calamus hollrungii

(after Udo Max Hollrung, German collector in New Guinea)

AUSTRALIA—NEW GUINEA

An unusual rattan which occurs in north-eastern Queensland and New Guinea. It grows in monsoonal rainforests and along stream banks, often in relatively open situations. Plants consist of a single climbing stem and the fronds have large, widely spaced, dark green leaflets. The stem is supported by long, hooked cirri which tangle in the surrounding vegetation. This palm is poorly known in cultivation but seedlings have handsome, arching leaves and could have potential for use indoors or in glasshouses and conservatories. Plants are very sensitive to frost damage.

Calamus merrillii

(after E.D. Merrill, specialist in Philippine flora)

Rattan Palm

PHILIPPINES

A vigorous climbing palm with slender, sinuous stems which in nature may exceed 60 m in length. Its pinnate fronds bear many hooks which aid in climbing and the young stems are clothed with spines. The white fruit are about 1 cm across, scaly and with a thin layer of edible flesh. This is a fast-growing palm native to the Philippines and its stems are harvested for rattan cane. It is hardly suitable for garden culture, nevertheless its seeds are frequently offered for sale. It succeeds best in tropical conditions.

Calamus moti

(an Aboriginal name)

Yellow Lawyer Cane

AUSTRALIA

A robust, climbing palm with an attractive general appearance but a very unfriendly nature. The leaf sheaths are well armed with rows of yellow, bayonet-like thorns and the petioles and underside of the rachis bear shorter, often curved, yellow to reddish thorns. Leaves are dark green, with numerous segments (up to 100) and long, pendulous flagella (to 4 m long) are very thorny. Seedlings are very handsome and young plants can be successfully used indoors and make an interesting novelty item.

The species is common in the rainforests of north-eastern Queensland, Australia.

Calamus muelleri

(after Baron F. von Mueller)

Southern Lawyer Cane

AUSTRALIA

A relatively cold-tolerant species which can be grown in temperate regions. Plants, however, have very limited appeal and are mainly found in botanical collections. The species is native to eastern Australia, being found in the rainforests of south-eastern Queensland and north-eastern New South Wales. It is a slender climbing palm with thorny stems and relatively few leaflets (15–20) per leaf.

Calamus ornatus

(ornate, decorated)

Limuran

SOUTH-EAST ASIA

This climbing palm from the Philippines and South-East Asia bears clusters of red, scaly fruit each up to 3.5 cm across. These have edible flesh and are sometimes sold in the local markets. The palm itself has large, dark green pinnate leaves and spiny stems. It is an attractive species but prickly.

Calamus moti spines on leaf sheaths.

It probably fruits best in tropical areas and should be planted in groups in a shady position in well-drained soil.

Calamus radicalis

(arising from the root)

Vicious Hairy Mary

AUSTRALIA

An impressively armed Australian palm which forms untidy thickets and ascending clumps in the rainforests of north-eastern Queensland. Leaf sheaths, petioles and rachises are covered with long, brown, brittle spines and the long, pendulous flagella and inflorescences are also armed. This species is rarely cultivated, although seedlings would make an interesting conversation item.

Calamus reyesianus

(after C. Reyes, forester in the Philippines)

Apas Palm

PHILIPPINES

A vigorous climbing palm from the Philippines, this species suckers freely and forms spreading clumps. Its long, pinnate leaves are bright green and finely divided and bear a long, prickly cirrus at each tip. When young, the leaves are bright pink and are most decorative. Large white fruit up to 2 cm across are borne in long, hanging clusters. Although vigorous, this palm makes an attractive pot subject as it suckers early in its development. Prickly extensions can be removed if a nuisance. It should be planted in a shady situation where it can climb and is probably best grown in tropical regions.

Calamus viminalis

(with long flexible shoots)

INDIA—MALAYSIA

A vigorous climbing palm from India and Malaysia which has succeeded very well in the Royal Botanic Gardens, Sydney, Australia. It has moderately stout, prickly stems with woolly white leaf sheaths and spreading, pale green leaves about 1 m long which are divided into numerous narrow leaflets. These leaflets are not spread evenly but rather are clustered in small groups along the rachis. The stems climb with the aid of flagella. The fruit are about

1 cm across, yellow when ripe and are globular with a prominent beak.

Calamus warburgii

(after Otto Warburg)

AUSTRALIA—NEW GUINEA

Although appearing to be less prickly than some other members of the genus, this species is still formidable as the long spines which clothe the leaf sheaths and petioles are very sharp and brittle. The species is found in New Guinea and north-eastern Queensland, forming untidy clumps. It is very rarely cultivated, although seedlings may have some appeal to enthusiasts.

CALYPTROCALYX

(from the Greek *calyptros*, veil or covering and *calyx*, the outer whorl of the perianth)

A genus of thirty-eight species with the majority occurring in New Guinea and one species in the Moluccas. They are small understorey palms found mainly in rainforest at low to moderate elevations, occasionally growing in wet or boggy soil. Plants are usually clumping, rarely solitary, with slender unarmed stems, a small crownshaft and entire or pinnate leaves. The inflorescence, which is undivided, arises among the leaves and bears unisexual flowers of both sexes. The genus *Paralinospadix* is synonymous.

Cultivation: All species have excellent ornamental qualities but are rarely encountered in cultivation. Most would appear to be shade-loving palms which need well-drained, loamy soils and an abundance of moisture. Many are probably excellent subjects for containers. Single plants are capable of producing fertile seeds.

Propagation: Suckers can be transplanted fairly readily. Fresh seeds germinate two to five months from sowing. Each fruit contains a single seed.

Calyptrocalyx angustifrons

(with narrow fronds)

NEW GUINEA

The fronds of this species are like elongated fans and they form an attractive spreading crown atop a slender stem. They are mostly undivided, have a deep apical notch and taper to a slender petiole about 50 cm long. This attractive species, from mountainous regions (about 1000 m altitude) in the East Sepik District of Papua New Guinea awaits introduction to cultivation.

Calyptrocalyx bifurcatus

(notched twice)

NEW GUINEA

A delightful palm with a trunk to about 2 m tall and 6 cm across. The rigid fronds (lamina to 30 cm long) are fan-shaped in outline, notched apically, taper like a wedge to the base and are borne on slender petioles about 12 cm long. They overlap to form an attractive crown. The simple, undivided inflorescence is shorter than the leaves. This species is native to montane rainforests of the East Sepik District, Papua New Guinea, growing at about 1000 m altitude.

Calyptrocalyx caudiculatus

(with a small tail)

NEW GUINEA

A graceful small palm with a slender trunk as thick as a finger and irregularly pinnate fronds, to 60 cm x 30 cm, which are ovate in outline. The fronds are divided into eighteen, curved, very narrow segments (to 8 cm x 1 mm) which are unevenly arranged along the rachis. Broadly ellipsoid, red fruit, to 11 mm x 6 mm, are borne on simple, slender spikes about 1.2 m long. An interesting species from forests on the south coast of West Irian.

Calyptrocalyx geonomiformis

(resembling a *Geonoma*)

NEW GUINEA

An elegant, graceful, small palm with a very slender stem about 1 m tall. The fronds are usually simple, although sometimes they are lobed at the base. They are obovate in outline, to 55 cm x 30 cm, with a notched and toothed apex and are narrowed to the base, tapering into short petioles. Simple inflorescences are about as long as the leaves. The species is native to lowland forests of West Irian.

Calyptrocalyx hollrungii

(after Udo Max Hollrung, 19th century German collector)

NEW GUINEA

A slender, clumping palm found on the Huon Peninsula of Papua New Guinea where it grows as an understorey plant in rainforest. It is remarkable for the variability in its leaves with some being scarcely divided into coarse, broad segments while others are finely divided into narrow segments. Whatever the leaf shape, it is a highly desirable palm for a shady position in a tropical garden and would make an extremely decorative plant for a large container.

Calyptrocalyx lauterbachianus

(after C.A.G. Lauterbach, German collector in New Guinea)

NEW GUINEA

A New Guinea gem which rarely grows to 3 m tall and which is found in shaded rainforest at moderate altitude. The plants may have either olitary or multiple stems and have erect, dark green leaves, the pinnae of which radiate in many directions to give a plumose appearance. Newly emer-ging leaves are reported to be reddish. The inflorescences are long, unbranched, pendulous spikes carried in small groups and after flowering bear moderately large, bright crimson fruit. All in all, this is a very decorative palm which would make an ideal subject for a shady nook in a tropical garden.

Calyptrocalyx lepidotus

(covered with small scales)

NEW GUINEA

A little-known species from the Western District of Papua New Guinea where it grows in lowland forests. Plants grow to about 5 m tall and the new fronds emerge brownish and are covered with loose, brown scales. Each frond (about 80 cm long) is divided into about twenty, fairly stiff leaflets which are irregularly clustered along the rachis. They are dark green on the upper surface, brownish beneath and the apex is drawn out into a narrow point.

Calyptrocalyx merrillianus

(after E.D. Merrill, botanist who collected widely in Melanesia)

NEW GUINEA

A clumping species which originates in lowland forests of the Fly River region in the Western District of Papua New Guinea. Plants have slender (1.3 cm across) trunks to about 4 m tall and a loose crown of fronds, each about 1.5 m long. About thirty, narrow, curved leaflets (to 30 cm x 2.5 cm) are regularly disposed along the rachis, these being dark green on the upper surface and paler beneath. Dark purple-black, beaked fruit (about 1.8 cm x 8 mm) are borne on slender spikes.

Calyptrocalyx micholitzii

(after W. Micholitz, original collector)

NEW GUINEA

This extremely beautiful palm which originates in the dense rainforests and jungles of New Guinea, has been grown in England as a glasshouse plant for many years. The plants form a crown of arching, broad, bright green pinnate fronds and when well grown, make delightful tub specimens. The young leaves are deep purplish and this colour gradually fades as they expand. The plants like a rich, well-aerated, organic soil mix and should not be allowed to dry out. They can be grown outside in tropical areas but require shady, moist conditions.

Calyptrocalyx pauciflorus

(with few flowers)

NEW GUINEA

A palmlet from shady forests in the East Sepik District of Papua New Guinea, growing at about 850 m altitude. Plants have a subterranean trunk and a crown of five or six fronds each about 1 m long. The fronds are undivided, oblong to obovate in outline and about 14 cm across near the apex which is deeply notched. Ovate to oblong red fruit (about 13 mm long) are carried on slender, string-like spikes about 1 m long. This species would be a delightful acquisition to any collection.

Calyptrocalyx petrickianus

(after a collector named Petrick)

NEW GUINEA

An elegant New Guinean palm which was apparently described from plants growing in Kew Gardens early this century. Its pinnate leaves are deep purplish-brown when young and expand to over 2.5 m long and nearly 1 m across. The plants form an erect clump of many fronds and are an eye-catching tub specimen for glasshouse decoration. In tropical regions, they can be grown outside but require a very shady position and plenty of water during dry periods. They also like plenty of mulch and organic matter in the soil.

Calyptrocalyx polyphyllus

(with many leaves)

NEW GUINEA

The fronds of this species are finely divided into about forty narrow, dark green segments, each to 45 cm x 18 mm. Plants are solitary with a slender (finger thickness) trunk to 1.5 m tall and a crown of sessile fronds each about 1 m long. Ovate, beaked red fruit, to 1 cm x 6 mm, are borne on very slender, string-like spikes about 90 cm long. An elegant small palm from the East Sepik District of Papua New Guinea, growing at about 600 m altitude.

Calyptrocalyx spicatus

(spike-like, in a spike)

MOLUCCAS

A rarely grown collector's palm from the Maluku Islands and Ambon in the Moluccas. It has a handsome crown of drooping, green, pinnate fronds reminiscent of the Kentia Palm (*Howea forsteriana*) and a long, pendulous, simple spike as its inflorescence. The trunk, which may grow to more than 12 m tall, is prominently ringed. The rounded fruit are bright red when ripe and are used as a betel substitute in its native country. This species is essentially a palm for the tropics and needs shade for best appearance. It deserves to be far more widely grown and should be tested for its suitability as an indoor plant.

CALYPTRONOMA

(from the Greek *calyptron*, a veil or covering and *nomos*, a cap, in reference to the petals forming a dehiscent cap)

A small genus of three species restricted to the West Indies and adjacent islands. They are moisture-loving palms which grow in swamps and near streams, sometimes in fairly deep water. They are solitary, unarmed palms which lack a crownshaft and with an arching crown of pinnate leaves. The inflorescences, which arise among the leaves, are at first erect then later pendulous and carry unisexual flowers of both sexes. The leaves are used for thatch.
Cultivation: These palms are rarely seen in cultivation and are mainly of interest to collectors. Plants need an abundance of moisture and will grow in a sunny position in heavy soils. Single plants are capable of producing fertile seeds.
Propagation: Fresh seed germinates readily two to four months from sowing. Each fruit contains a single seed.

Calyptronoma dulcis

(sweet-tasting)

CUBA

A Cuban palm with a solitary trunk and large, spreading crown of dark green fronds. In general appearance, plants resemble those of the Coconut

Carpentaria acuminata in fruit.

but with stiffer leaflets. The leaflets droop towards the margins and have a prominent midrib. Fruit are about 2.5 cm long. The species grows well in wet situations and is most suitable for the tropics.

Calyptronoma rivalis

(growing by streams)

PUERTO RICO

This Puerto Rican palm is considered to be endangered in its natural state as it is known only from about three localities on the island. It grows close to streams on permanently moist, alkaline soils derived from degraded limestone. Seed was distributed to palm enthusiasts some years ago and the resulting plants have grown very well in cultivation. They require a shady aspect when young, with plenty of water and succeed in tropical and subtropical regions. This fast-growing palm is attractive with finely pinnate leaves in a spreading to rounded crown and clusters, of small, bright red fruit.

CARPENTARIA

(after P. de Carpentier, former Governer-General of the Dutch East Indies)

A monotypic genus of palm endemic to the Northern Territory, Australia, where it grows in rainforest near streams. It is a tall, slender, unarmed feather-leaved palm with a crownshaft. The inflorescence arises below the leaves and bears unisexual flowers of both sexes.
Cultivation: A desirable ornamental which has become widely planted in northern Australia and southern Florida. It is a fast-growing palm which is well suited to tropical and warm subtropical conditions. Single plants are capable of producing fertile seed.
Propagation: Fresh seed germinates readily two to four months from sowing. Each fruit contains a single seed.

Carpentaria acuminata

(tapering to a long point)
Carpentaria Palm

AUSTRALIA

A distinctive palm which can be recognised by its slender, grey trunk and graceful crown of distinctly

curved or arching fronds with the leaflets obliquely erect. A fast-growing species with a highly ornamental appearance, aided by large, colourful clusters of bright scarlet fruit. The species has become deservedly popular in northern Australia and is being planted on an increasing scale elsewhere. While this palm can be grown as a specimen, it is also well suited to planting in groups. Plants need well-drained soil and an abundance of water during dry periods, especially when small. Young plants need some protection for the first two or three years but once established they will tolerate considerable sun.

CARYOTA

(from the Greek *caryon*, a nut)

Fishtail Palms

A small genus of twelve species of very distinctive palms distributed in India, Sri Lanka, South-East Asia, southern China, Malaysia, the Philippines, Solomon Islands, New Guinea and Australia. They grow in rainforests and monsoon forests from the lowlands to high altitudes. They are solitary or clumping palms and their most distinctive feature is the bipinnate fronds with unusually shaped leaflets (fishtail-like). All species have monocarpic stems, the solitary species dying after maturing the fruit on the lowermost infructescence, the clumping species regularly producing new replacement stems. The inflorescence arises among the leaves and carries unisexual flowers of both sexes. Palms of this genus are used for a range of purposes, including sago, palm wine, fibre and sugar, as well as various minor uses. The taxonomy of *Caryota* is rather confused and the genus is presently under study, with changes in the concepts of cultivated species highly likely to occur.

Cultivation: All species have excellent horticultural appeal being prized mainly for their distinctive bipinnate fronds. They grow readily in tropical and subtropical regions (some also succeeding in temperate regions) and can be fast growing in suitable conditions. Plants like an abundance of moisture and respond well to regular applications of fertilisers. Single plants are capable of producing fertile seeds.

Propagation: Clumping species can generally be divided fairly readily. Fresh seeds germinate quite readily if sown fresh but seedlings often appear sporadically, sometimes taking twelve months or more. The fruit contain caustic, oxalate crystals and

Above: Caryota cumingii.
Below: Caryota cumingii fruit. J. Dransfield

Two clumps of *Caryota mitis*.

Above: *Caryota mitis* inflorescence.
Above Right: *Caryota mitis* infructescence.

should be handled with care and on no account eaten. Each fruit contains one or two seeds.

Caryota cumingii

(after H. Cuming, 19th century British botanist who collected in the Philippines)

Fishtail Palm

PHILIPPINES

In general appearance this palm resembles a single-stemmed form of *C. mitis*, but has a deeply ringed trunk and different male flowers. It is native to the Philippines and is not commonly encountered in cultivation. It is very tropical in its requirements and succeeds best in the hot, humid lowlands. Cultural requirements are as for *C. urens*. Mature plants have been recorded at about 12 m tall. Germination of this species is very haphazard with some seedlings taking ten to twelve months to appear.

Caryota maxima

(biggest, largest)

Giant Mountain Fishtail Palm

INDONESIA—MALAYSIA

This giant palm from the mountains of Java, Sumatra and Malaysia is truly an imposing palm. Plants may grow to more than 30 m tall, with a solitary trunk and a huge, elongated crown of fronds. Individual fronds may be up to 3 m long and 2 m wide and are dull green with prominently drooping pinnae and pendulous leaflets. The fruit are bright pinky-red and most contain only one seed. These are carried on a massive infructescence which may be more than 2 m long and is composed of numerous pendulous spikes. Unfortunately this majestic palm dies after maturing its final bunch of fruit, however it still makes an impressive addition to a park or large garden. Taxonomically this species has had a very confused history and has been known under the names *C. aequatorialis*, *C. obtusa* var. *aequatorialis* and *C. rumphiana* var. *javanica*.

Caryota mitis

(mild, gentle, unarmed)

Clustered Fishtail Palm, Tukas

INDIA—SOUTH-EAST ASIA

A very widespread palm which extends from India to the Philippines, and Sulawesi and Java in Indonesia. It is also widely cultivated in many tropical and subtropical regions of the world and is a familiar garden palm. It forms a characteristic cluster of closely placed stems and the crowns together produce a crowded mass of the attractive, bipinnate leaves with their unusual fishtail leaflets. It is an excellent garden palm and while plants in the tropics may reach more than 5 m tall, those in cooler districts rarely achieve these proportions. It is sufficiently hardy to be grown in a warm position in temperate regions. It is quite a long-lived palm and although each stem is monocarpic and dies when the fruit matures on the lowermost inflorescence, each is replaced by new basal suckers. Requirements are well-drained, rich soil and plenty of water. The plants are especially responsive to applications of nitrogenous fertilisers. They will tolerate full sun when quite small and are fast growing. Young specimens make excellent indoor plants but dislike a dry dusty atmosphere. They can be propagated by division of the suckers or from seed which usually takes four to six months to germinate.

Caryota no

(a native name)

Giant Fishtail Palm

BORNEO

This is one of the biggest of all the fishtail palms and is a true majestic giant in all respects. Plants may grow to more than 25 m tall with a trunk 50 cm or more across and have a crown of obliquely arching to horizontal fronds which have stiff pinnae and pendulous leaflets. The crowded leaf-bases usually hide the trunk. Individual fronds may grow to 4 m long and be nearly 3 m wide. The grey trunk is stout and bulging in a manner reminiscent of *Roystonea regia*. Inflorescences which may measure over 2.5 m long, carry cream flowers or large, black fruit, each of which contain two seeds. This species, native of Borneo, would be an excellent palm for the tropical lowlands. It has been reduced to great rarity in the wild by the collection of its cabbage which is described as being delicious. It has also been known as *Caryota rumphiana* var. *borneensis*.

Caryota obtusa

(blunt)

Mountain Fishtail Palm

INDIA—THAILAND

This palm, from mountainous regions of northern India and Thailand, where it grows up to about 1000 m altitude, is a solitary species similar in general growth habit to *C. urens*, but its wedge-shaped, leathery leaflets are blunt and shallowly toothed at the tips. The leaves are borne on a stout trunk 40–60 cm across. The blackish fruit are 2–2.5 cm across. It is not commonly encountered in cultivation and, if anything, is more cold-tolerant than most of the other species in this genus, with the possible exception of *C. ochlandra* and *C. urens*. It is a solitary monocarpic palm and requires similar cultural conditions to *C. urens*. Propagation is from seed which may germinate erratically. This species has also been known as *C. rumphiana* var. *indica* and *C. obtusidentata*.

Caryota ochlandra

(with yellow stamens)

Chinese Fishtail Palm, Canton Fishtail Palm

CHINA

This is probably the most cold-tolerant of all the species of *Caryota*. It is a solitary palm with a large crown of bluish-green, bipinnate leaves and may grow to about 8 m tall. The leaflets tend to be narrower and blunter than in most other species and with jagged margins. They are densely clustered at the base of each frond. Like other solitary *Caryota*

Caryota rumphiana in flower.

Above: *Caryota urens* inflorescences in various stages of development. Right: *Caryota urens.*

species it is monocarpic, usually beginning to flower after ten to fifteen years. As the common name suggests, it is native to China, being found in the region of Canton. Young plants appreciate protection from direct sun but, once established, will tolerate considerable exposure. Light to medium frosts cause little obvious damage.

Caryota rumphiana

(after G.E. Rumpf (Rumphius), 17th & 18th century German botanist who collected in Indonesia)

Fishtail Palm

MOLUCCAS—NEW GUINEA—AUSTRALIA

This is a variable palm which extends from the Moluccas through New Guinea to Australia. The typical variant is from the Moluccas and is a large, solitary fishtail palm with stiffly spreading fronds and round fruit, each containing a solitary seed. In New Guinea the species is represented as the var. *papuana* which is basically similar to var. *albertii*, the variant found in Australia. A botanical study is badly needed to sort out these relationships. Forms

of *C. rumphiana* from countries west of the Moluccas are referable to other species (see *C. maxima* and *C. no*). Whatever the taxonomic position, these Fishtail Palms all have ornamental qualities and are deserving of cultivation. They grow well in tropical and warm subtropical regions and have similar cultural requirements to other species of *Caryota.*

Caryota urens

(stinging)

Solitary Fishtail Palm, Jaggery Palm, Toddy Palm

BURMA—INDIA—SRI LANKA

Although native to the tropics (India, Burma and Sri Lanka) and widely grown there, this large, impressive palm has proved to be sufficiently adaptable to survive in warm, temperate regions. It is a solitary species with a large, shiny, thick, ringed, grey trunk and a head of gracefully arching, bipinnate fronds (each to 6 m x 4 m). The large, wedge-shaped leaflets are bright green and shiny, with one margin sharply and irregularly toothed

165

and the other produced into a tail. The reddish to blackish fruit are 15–20 mm across. Unfortunately these majestic palms are monocarpic and die once the fruit on the lowermost inflorescence matures. They are, however, an excellent and impressive garden plant and add that tropical flavour. They will tolerate sun when quite small and like a deep, rich soil with plenty of water. They are also very responsive to applications of nitrogenous fertilisers. Once established in good conditions they can be quite fast growing and can easily reach 20 m tall. Young plants make very useful tub specimens for indoor and foyer decoration. In their native countries of India and Malaysia the sap is tapped via the inflorescence and used to make alcoholic drinks or palm sugar. The trunk also contains an edible sago.

CHAMAEDOREA

(from the Greek *chamai*, on the ground and *dorea*, a gift; a probable reference to the small stature of these palms)

A genus of about 100 species of small palms distributed in Mexico, Guatemala, Costa Rica, Panama, Honduras, Belize, Ecuador, Columbia, Brazil, Peru and Bolivia. They grow mainly in wet, undisturbed forests, particularly rainforests and cloud forests. These palms are extremely diverse, ranging from true trunkless dwarfs to solitary or clumping species about 10–15 m tall. Some even have prostrate stems and one species is a true climber. The slender, unarmed stems are prominently ringed and the leaves may be entire, have a notched apex or be variously pinnately divided with few to many pinnae. The inflorescence arises among the leaves or below the leaves and carries unisexual flowers. The genera *Collinia*, *Eleutheropetalum* and *Neanthe* are now included with *Chamaedorea*.

Cultivation: As a group these plants have tremendous horticultural appeal and they are probably the most significant group of palms available for general ornamental use, both indoors and out. Major features include small to moderate size, slender stems, lustrous foliage in a range of interesting shapes and colours and intriguingly colourful contrasts between flowers, fruit and rachillae. Response in cultivation is variable. Many species have proved to be moderately adaptable and easy to grow if provided with their basic requirements, whereas others are much more difficult because

of stringent specialised need. Basically these plants are shade-loving palms which need protection from excesses of sun and wind, require moderate to high humidity and an abundance of water throughout the year. Soils must be well-drained, acidic for most species, with organically rich loams producing excellent results. Many species make highly decorative subjects for containers and some are suitable for indoor and parlour use. The parlour palm, *C. elegans*, is possibly the most widely grown palm in the world, being propagated annually in the hundreds of thousands by the nursery trade in America, Europe and Australia.

Chamaedorea palms are unisexual and male and female plants are essential for fertile seed production. In gardens these palms lend themselves well to planting in groups or massed planting and this arrangement also assures seed production. Hybridism can occur between related species, so due consideration must be given when planning an area where these palms are to be grown.

There has been considerable confusion among horticulturists as to the correct names of some of the *Chamaedorea* species that are grown. Fortunately this situation has now been addressed with the publication of an excellent, detailed monograph of the genus (see Donald R. Hodel, 'Chamaedorea Palms, The Species and their Cultivation', (1992) *The International Palm Society*, Allen Press, Lawrence, Kansas).

Propagation: Clumping plants can be divided successfully and some species such as *C. stolonifera* are suitable for layering as they produce adventitious roots from their stems. Seed germinates readily when fresh, within one to four months. Each fruit contains a single seed. The flesh of the fruit contains caustic stinging crystals and should be handled with care.

Chamaedorea allenii

(after Paul H. Allen, American botanist and original collector)

COLOMBIA—PANAMA

The bright yellow male flowers crowded in a simple spike are a useful guide to distinguish this species from its congeners. Plants grow to about 2 m tall and the leaves have seven to nine pairs of narrow, glossy, dark green leaflets (rarely are they entire and notched). The black fruit, about 1 cm across, are crowded together in the short infructescence. This species grows naturally in the humid forests of Panama and Colombia at about 1000 m altitude. It is introduced into cultivation but is still essentially a collector's item.

Chamaedorea alternans

(alternating, in reference to the female inflorescence)

MEXICO

This species has been cultivated in European glasshouses since about 1875 and recently seed has been distributed through palm societies. It is a handsome palm, somewhat similar to *C. tepejilote*, but with fewer narrower leaflets and prominent, cream margins on the leaf sheaths. Up to four inflorescences can arise from a node, presenting a very crowded impression. The species is native to Mexico where it grows in moist to wet forest at moderate altitude.

Chamaedorea amabilis

(lovely, attractive)

CENTRAL AND SOUTH AMERICA

A beautiful palm, the populations of which have been greatly reduced in the wild due to exploitation by collectors. Plants are prized for their large, butterfly-like, thin-textured leaves which are of a rich green with numerous prominent, parallel veins, toothed margins and a deeply notched apex. An excellent species which is suitable for cultivation in containers or in a shady garden. Although native to moderate elevations (500–1000 m altitude) in Costa Rica, Panama and Colombia, this palm has proved to be surprisingly adaptable, succeeding even in temperate regions. *C. coclensis* is a synonym.

Chamaedorea angustisecta

(with narrow segments or divisions)

SOUTH AMERICA

This species has characteristic green and white mottles on the leaf sheath and petioles. It is a solitary palm, to 4 m tall, with attractive, dark green pinnate leaves, each having numerous (about 80) narrow leaflets (to 45 cm x 3 cm). These spread widely and the leaves are broadest near the middle. The male flowers have a strong, pleasant perfume.

The relatively large, black fruit (about 18 mm long) are supported on orange rachillae. The species, which is native to Bolivia, Peru and Brazil, is cultivated on a limited scale. *C. leonis* is a synonym.

Chamaedorea arenbergiana

(after Duke d'Arenberg-Nieppen)

CENTRAL AMERICA

Widespread from Mexico to Honduras, this elegant *Chamaedora* may grow in excess of 3.5 m tall. It is a solitary species with a slender, prominently stepped, ringed trunk about 2.5 cm thick, ending in a sparse crown of about six pinnate leaves which may be nearly 2 m long. Each leaf has up to ten pairs of widely scattered, broad leaflets which can grow to 60 cm long and over 15 cm wide. These are a dull, deep green above and yellowish-green beneath. Fruiting spikes are orange and the ovoid fruit, about 1 cm long, are black when ripe. A shady situation in the tropics or subtropics is most suitable for this species.

Chamaedoria arenbergiana (*Botanical Magazine*, Vol. XLI of the 3rd series, plate 6838)

Chamaedorea brachypoda.

Chamaedorea brachyclada

(with short stems)

CENTRAL AMERICA

An interesting species which has long pinnate leaves (to 2 m long) which arise from a very short stem. Each plant has only two or three leaves and each leaf has widely spreading (to 30 cm x 3 cm) dark green leaflets. Arching inflorescences arise from the stem at ground level and are intricately branched near the apex. The fruit are extremely small, being only about 5 mm across when ripe. Although this species has been cultivated since the late 1800s it remains rare and is essentially a collector's item. Some growers report that plants of this species resent dryness and prefer cool to warm humid climates rather than hot tropical climates. It is native to Panama and Costa Rica.

Chamaedorea brachypoda

(with a short foot)

CENTRAL AMERICA

A slender, clumping palm from Guatemala and Honduras which, at least in the tropics, develops fairly rapidly into spreading, though compact and crowded clumps (it is slow elsewhere). It likes a shady, moist position in organically rich soil and grows to about 2 m tall. Spread is by underground shoots which emerge at intervals from the parent plant. The stems are very slender (about 7 mm thick) and bear entire, satiny, bright green leaves which are deeply notched at the apex. This is a lovely garden palm that at present is mainly found

in the collections of enthusiasts. It deserves to become widely grown but will probably be restricted by lack of propagating material. Plants of this species are sometimes incorrectly sold in nurseries as *C. stolonifera*, a distinct species from Mexico.

Chamaedorea cataractarum

(growing by cataracts)

Cascade Palm, Cataract Palm

MEXICO

A trunkless, Mexican palm which grows as a rheophyte along the margins of streams and is often covered by floodwaters during peak flow. It has a unique growth habit with each growth splitting and separating to form two growths and eventually plants develop into dense, slowly spreading clumps by this means of division. Each crown bears two to six upright to arching, relatively broad, dark green, pinnate leaves, the terminal leaflets of which are undifferentiated from the rest. The inflorescence,

Chamaedorea cataractarum.

which grows to about 50 cm tall, arises from the basal axils and bears yellowish flowers. The green to black fruit are about 10 mm x 8 mm across and have a powdery bloom at maturity. A well-grown plant of this palm resembles a group of small Kentia Palms and looks most attractive in a large container. As the species grows well inside houses, it should become a very popular palm for home and office decoration. Outside it needs shady, moist conditions in well-drained, organically rich soil and is an ideal garden palm. This species is sometimes wrongly sold in nurseries as *C. atrovirens*.

Chamaedorea correae

(after Mirey Correa, botanist in Panama)

PANAMA

An intriguing little palm which has prostrate stems which produce roots from those nodes which are in contact with the soil and a sparse crown of grey-green, prominently veined, fishtail-like leaves which occasionally have an extra pair of small leaflets towards the base. Although the stems produce roots when in contact with the soil, the plants remain solitary and do not branch. The fleshy, red female inflorescences persist on the plants for a time after the fruit mature. The fruit are broadly ellipsoid, about 8 mm long and black when ripe. This collector's palm occurs in cloud forests on mountain tops in Panama at about 1000 m altitude.

Chamaedorea costaricana

(from Costa Rica)

CENTRAL AMERICA

In general appearance this clumping palm from Costa Rica, Honduras, Panama and Nicaragua, closely resembles both *C. microspadix* and *C. seifrizii*. The former species has bright orange fruit while those of the others are black. When compared with *C. seifrizii*, *C. costaricana* has longer, more closely spaced leaflets that are dull rather than shiny dark green. It can be distinguished from *C. pochutlensis* by the prominent ligules on the leaf-base. Plants of *C. costaricana* tend to be more cold-tolerant than the other species and survive quite happily in a protected situation in temperate regions. They will grow in full shade (in fact young plants demand this

Chamaedorea costaricana.

protection) but once established, plants will withstand considerable exposure to sun, especially if kept well watered and mulched. Plants can be propagated readily by seed or division of the clumps. One interesting characteristic of this species is the formation of suckers or lateral growths above ground level. As a tub plant it is quite handsome and is useful for indoor decoration. *C. seibertii* is a synonym.

Chamaedorea crucensis

(from Las Cruces Tropical Botanical Garden, Costa Rica)

COSTA RICA

A slender species (to 2 m tall) with arching pinnate fronds each having 10–22 relatively narrow, spreading leaflets. These are slightly curved and shiny. The simple or sparsely branched inflorescences have densely crowded flowers and on female plants these are followed by crowded fruit which ripen through red to black. This species, which is native to Costa Rica, has been introduced into cultivation, but is mainly a collector's item.

169

Chamaedorea deckeriana

(after G.H. Decker, German botanist who grew the
type plant)

COSTA RICA

The flowers of this impressive small palm release
a strong, spicy scent which is very noticeable on
warm days. Its crown seems out of proportion to
the slender stem, for it consists of about six large,
arching, simple leaves, each deeply notched at the
tip. These are bright green with a paler midrib. The
inflorescences have densely crowded flowers and
on the female plants these are followed by crowded
fruit which ripen through green to red and then
black. This choice collector's palm is native to
shady, wet forests of Costa Rica.

Chamaedorea elatior

(lofty, higher)

Climbing Chamaedorea

CENTRAL AMERICA

Unusual for this genus, this palm is a climber
with the slender trunk threading its way through
surrounding vegetation and being supported by
spreading, dark green pinnate leaves which have
the upper leaflets reflexed and hooking for support.
It is native to Mexico and Guatemala where it
extends from the lowlands to the highlands, growing in rainforest. Two variants are known, one

Chamaedorea elatior in fruit.

Chamaedorea elegans.

having solitary, thicker stems (about 2 cm across)
and the other being clumping with very slender
stems (about 8 mm across). Stems may reach to
about 20 m in length and will creep along the
ground if they cannot find support. The yellow
flowers emit a very strong, rather unpleasant
fragrance and on female plants are followed by
globose, black fruit which have a powdery bloom.
Plants seem to be best suited to a shady position in
tropical and subtropical regions. They can be quite
fast growing and for seed production should be
planted in clumps. As the stems need support,
plants are best grown among shrubbery or over a
suitable frame or tree. This species has a number
of synonyms, including *C. scandens*, *C. affinis*,
C. montana, *C. resinifera*, *C. bambusoides* and
C. desmoncoides.

Chamaedorea elegans

(neat, elegant)

Parlour Palm, Neanthe Bella, Good Luck Palm

CENTRAL AMERICA

This slender, dwarf palm is very popular for indoor
decoration and is propagated in the thousands each
year by the nursery trade. It is native to Mexico,
Guatemala and Belize, growing in moist, dense
rainforest at moderate elevations, often on limestone. Plants have a solitary, woody, slender trunk

which may grow to 3 m tall with numerous closely spaced growth rings. The pinnate leaves are dark green and ascend in an upright to spreading crown. The leaf sheaths frequently have a prominent, white margin. The fruit, which are about 5 mm long, are black when ripe, contrasting with the orange rachillae. As well as being ideal for indoor decoration, this palm can also be grown outdoors in a cool, moist position protected from hot sun and drying winds. It is suitable for areas from temperate to tropical regions, although severe frosts may damage the leaves. A fine-leaf variant with narrow pinnae is much in demand. *C. elegans* is frequently, but wrongly, sold in the nursery trade as *Neanthe bella* and is sometimes also placed in the genus *Collinia*. Synonyms include *C.humilis*, *C. deppeana* and *C. helleriana*.

Chamaedorea ernesti-augustii

(after Ernest August, 19th century German monarch)

CENTRAL AMERICA

Distributed from Mexico to Honduras, this small palm grows as an understorey plant in shady rainforest. The slender, solitary trunk grows to about 2 m tall and has numerous, prominent, pale, annular rings. Its leaves are simple, pleasantly pleated and, like those of *C. geonomiformis,* are deeply notched apically like a fishes' tail. When in flower and fruit it is quite attractive, with cream flowers and black, ovoid fruit embedded in an orange to red rachis. This dwarf, slow-growing palm is rare in cultivation and is a collector's gem. It can be grown in warm–temperate and subtropical regions and likes a shady, moist situation. Plants of this species are similar to those of *C. geonomiformis* but can be distinguished by the thin, deciduous petals of the flowers. *C. glazioviana* is a synonym.

Chamaedorea erumpens

(breaking out, erupting)

CENTRAL AMERICA

This species has been reduced to synonymy with *C. seifrizii*. Although plants grown as this species have broader leaflets than the commonly grown form of *C. seifrizii*, research has shown that the latter species is variable and encompasses a wide range of variants. *C. erumpens* is merely a broad-leafleted form at one end of the scale of variation.

Chamaedorea ernesti-augustii male plant in flower.

Chamaedorea fragrans

(with fragrant flowers)

PERU

A highly ornamental clumping palm which is readily recognised by its slender, cane-like stems and simple, deeply divided leaves, with the two narrow lobes spreading in a deep vee. Another very distinctive attribute is the male flower which exudes a strong, spicy fragrance in warm weather. Although mainly a choice collector's palm, this Peruvian species deserves to become much more widely grown. It grows best in shady, humid conditions in tropical and subtropical regions. As seed is scarce, plants are mainly propagated by division. Synonyms include *C. pavoniana* and *C. ruizii*.

Chamaedorea geonomiformis

(similar to the genus *Geonoma*)

Necklace Palm

CENTRAL AMERICA

This delightful small palm consists of a single, slender trunk that may grow to 1.5 m tall and has a crown of dark green, simple leaves notched at the apex, each resembling a large fishes' tail. The leaves are shiny, thinly textured and have prominent veins. Small branching racemes or spikes are borne at regular intervals from the lower leaves and carry

Chamaedorea geonomiformis.

cream flowers followed by bluish-black fruit. This palm makes an attractive pot plant and also looks good planted in a shady spot, especially among ferns. Once established, the plants are quite hardy. It originates in Mexico, Guatemala, Belize and Honduras and has proved to be adaptable, growing in warm–temperate regions as well as the tropics. Seed germinates easily but early plant growth is very slow. Plants of this species are similar to those of *C. ernesti-augustii* but can be distinguished by the thickened petals which persist when in fruit.

Chamaedorea glaucifolia

(with blue-green leaves)

MEXICO

This slender, single-trunked, Mexican palm may get quite tall, reaching to over 5 m in height (and less than 3 cm thick), although plants in cultivation are usually shorter than this. Its most conspicuous feature is the nearly 2 m long pinnate leaves which have numerous narrow leaflets that are covered by a powdery bloom on both surfaces, especially obvious when young. These leaflets are borne in clusters and arise in several ranks along the rachis. The fronds spread in a loose crown from the top of the slender trunk which is ridged at each node. Fruit are relatively small (8 mm long), globose, and black when ripe. The species grows best in tropical and subtropical regions and requires shady conditions. It is well suited to close planting in groups.

Chamaedorea graminifolia

(with grass-like leaves)

CENTRAL AMERICA

A wonderful clumping palm which is valued for its very dense habit and fine, grassy leaflets. The slender, cane-like stems grow to about 3 m tall, generally arise close together and have arching leaves to about 1.5 m long. The narrow leaflets are regularly arranged along the rachis. Male and female inflorescences are branched at the end of a slender peduncle. The black, globose fruit are about 1 cm across. Native to Mexico, Guatemala, Belize and perhaps Costa Rica, this species grows at moderate altitudes in moist forest on limestone. It is a rewarding palm for cooler tropical and subtropical regions. Horticultural hybrids (wrongly called *C. schippii*) have been made between *C. graminifolia* and *C. pochutlensis*.

Chamaedorea hooperiana

(after Louis Hooper, who grew the type plant)

MEXICO

This vigorous, clumping species has excellent prospects as an indoor palm since plants are tolerant of reduced light, low humidity and are relatively free of pests, especially mites. Plants also grow well in outdoor situations. New growths arise adjacently from long-lived basal sheaths in a

Chamaedorea glaucifolia infructescence.

Chamaedorea graminifolia.

unique clustering arrangement and the pinnate leaves are very thick-textured with numerous spreading leaflets. A native of dense, shady, wet forests of Mexico at 1000–1500 m altitude, this palm deserves to become more widely grown. It should succeed in warm subtropical regions.

Chamaedorea klotzschiana

(after J. Friedrich Klotzsch, German botanist)

MEXICO

A neat, small palm which makes a decorative container subject and is also ideal for group planting in a shady garden. A native of Mexico growing at altitudes of 500–1200 m, it has a slender, solitary trunk enlarged at the nodes and a small, extended crown of arching pinnate leaves. The leaves are very distinctive in the genus, having bright green, broad leaflets arranged in sporadic clusters along the rachis. The only other species with such a leaflet arrangement is *C. glaucifolia* and this species is readily distinguished by its very narrow leaflets covered with a powdery bloom and arranged in several ranks.

Chamaedorea linearis

(linear, in reference to the leaflets)

SOUTH AMERICA

A palm of shady slopes in the Andes where it is widely distributed from Venezuela to Bolivia, extending from the lowlands to high altitudes. A variable species, especially in leaflet number, shape and size. Plants are robust with a pale green trunk to 10 m tall surmounted by a crown of spreading, dark green pinnate leaves with pleasant spreading to drooping leaflets. White, much-branched inflorescences are attractive, especially in shady conditions and on female plants the flowers are followed by bright red, jaffa-like fruit which may be up to 2.5 cm across. An excellent palm for tropical and subtropical regions which lends itself well to group planting. *C. polyclada*, *C. poeppigiana* and *C. megaphylla* are synonyms.

Chamaedorea metallica

(with a metallic sheen)

Metallic Palm

MEXICO

The dark blue-green leaves of this small Mexican palm have darker veins and an interesting metallic sheen that particularly shows to advantage when wet. They are usually held obliquely erect and vary from being simple and more or less heart-shaped

Chamaedorea metallica in fruit.

Chamaedorea metallica male inflorescence.

with a deep apical notch, to variously pinnately divided. The slender, solitary trunk, which has prominent nodes and white spots, grows to about 1 m tall and bears aerial roots at intervals along its length. The flowers are purplish to orange and the globose, black fruit, about 1 cm long, contrast with the swollen, orange rachis. This palm likes a cool, shady situation with plenty of water and is suited to tropical, subtropical and warm–temperate regions. Young plants are highly ornamental and look most attractive if closely planted in groups or crowded in pots. Male and female plants have a slightly different and distinctive appearance. This species has been confused with *C. tenella* which has non-metallic leaves with small, blunt teeth along the margins.

Chamaedorea microspadix

(with a small inflorescence)

Bamboo Palm

MEXICO

A wonderful, clumping palm that makes an admirable plant for either indoor decoration or for a shady spot in the garden. Under good conditions it is quite a fast grower and forms a dense clump, spreading by suckers. The stems, which are bamboo-like and grow to about 3 m tall, have pinnate, dull green leaves scattered along their length. The terminal pair of pinnae are united and fishtail-like. The stems bear clusters of bright orange-red fruit at intervals through the warmer months of the year. This palm is quite hardy once established but for best appearance needs plenty of water during dry periods. It is equally at home in a shady situation in the tropics or in a protected position in temperate regions. Large clumps transplant easily and can also be divided into new plants. It is native to Mexico where it grows at moderate elevations, often on limestone. Plants of this species are readily available from nurseries but are often mistakenly sold as *C. erumpens* which is a variant of *C. seifrizii*.

Chamaedorea minima

(the smallest)

COSTA RICA

This species is possibly extinct in the wild (Costa Rica) due to clearing of its habitat. It has, however, been introduced into cultivation but is still mainly a collector's item. It is a true palmlet which often appears stemless, but which may develop a creeping stem to about 25 cm tall. The simple, stiff, dark green leaves are deeply notched and the outer margins are coarsely toothed. The female infructescences have orange rachillae in contrast to the round, black fruit.

Chamaedorea microspadix infructescence.

Chamaedorea neurochlamys

(with covered veins)

CENTRAL AMERICA

A small solitary palm which is spectacular in fruit, with the infructescences carrying lobed or kidney-shaped, yellow to orange or brown fruit attached to red rachises and with a sturdy red peduncle. These fruit clusters are well displayed beneath a spreading crown of dark green pinnate leaves which have long-pointed leaflets. Well suited to tropical and subtropical regions, this species occurs in lowland rainforests of Mexico, Guatemala, Belize and Honduras.

Chamaedorea nubium

(clouds, probably referring to cloud forest)

CENTRAL AMERICA

A splendid clumping palm which has thin stems (1 cm across) which may grow in excess of 4 m tall. Deeply notched, fishtail-like leaves are spread along the stems. These have a slender, pale petiole (and midrib) and the blade has a dark green upper surface and is light green or greyish-green to whitish beneath. The ovoid, black fruit have a powdery bloom. This robust species can form clumps up to 15 m across. Stems root on contact with the ground and this can be a useful technique for propagation. The species is native to Guatemala and Mexico, growing at high elevations in moist forests. Plants from Guatemala have a lanky growth habit with leaves prominently glaucous beneath, whereas those from Mexico are more compact and with leaves green on both surfaces.

Chamaedorea oblongata

(with oblong leaflets)

CENTRAL AMERICA

A slender solitary palm that is widespread from Mexico to Nicaragua, growing in shady, lowland forests, often on limestone. The trunk, which may grow to 3 m tall, is about 2 cm thick and prominently ridged at each node. The arching pinnate leaves with rather broad, bright, shiny green, somewhat dimpled leaflets are borne in a small, sparse crown. Flowers are greenish-yellow and are followed by fruit which are about 1 cm long and black when ripe, contrasting with the orange

Chamaedorea oblongata.

rachis. This is an attractive small palm suitable for mingling with ferns in a shaded shrubbery or group planting. It will succeed in warm–temperate, subtropical and tropical regions. *C. lunata*, *C. fusca* and *C. corallina* are synonyms.

Chamaedorea oblongata infructescence.

175

Chamaedorea oreophila

(mountain-dwelling)

MEXICO

A slender solitary palm growing to 2 m tall and with a crown of finely divided, dark green, shiny pinnate leaves. Female plants bear clusters of bright orange-red fruit on long stalks held well clear of the foliage. These fruit ripen through greenish to yellow and red and make a colourful splash. An excellent pot subject or one for group planting in a sheltered situation. Plants resent a dry atmosphere and may suffer badly from mite attack in such conditions. The species is native to Mexico growing at altitudes up to 1500 m.

Chamaedorea pinnatifrons

(with pinnately divided fronds)

CENTRAL AND SOUTH AMERICA

A widely distributed species which extends from Mexico to Bolivia growing in dense, moist forests at moderate to high elevations. A slender solitary palm which may grow to 3.5 m tall and has a pale green trunk, 4–5 cm thick which, in the young stages, may have attractive cream to white tonings. Pinnate leaves may reach 50 cm long and have numerous thin-textured, pale green, lanceolate leaflets. Inflorescences are ivory white and later, as they mature, become pinkish to orange and bear small, ovoid, red to black fruit. Young plants are fast growing and often flower within two years of germination. A very easy and rewarding palm to grow, this species favours moist, shady conditions. It is well suited to group planting. This species is extremely variable and recent studies show that *C. brevifrons*, *C. concinna*, *C. concolor*, *C. gracilis*, *C. holgrenii*, *C. kalbreyeriana*, *C. macroloba*, *C. micrantha* and *C. minor* are all synonyms. Many plants, believed to be this species, have been sold in Queensland (Australia) nurseries in recent years.

Chamaedorea pochutlensis

(from the area of Pochutl)

MEXICO

A collector's palm that originates in the shady forests of Mexico ranging from near sea-level to about 2000 m altitude. It is a slender clumping species with bamboo-like stems which grow 2–4 m tall and bear pinnate leaves, either scattered along their length, or, on older stems, restricted to the upper section. Leaves are 60–130 cm long and have 40–60 pairs of crowded leaflets, the terminal ones of which are hardly differentiated from the others. These leaflets are bright, shiny green on both surfaces, lanceolate in outline and arise on the rachis at an oblique angle. In general appearance this species is similar to many of the other clumping Chamaedoreas, particularly *C. costaricana* which has prominent ligules on the apex of the leaf sheath (absent in *C. pochutlensis*). Like them it prefers shady, moist conditions in the tropics or subtropics. *C. karwinskyana* is a synonym.

Chamaedorea pumila

(dwarf)

COSTA RICA

This palmlet is similar in general appearance to *C. minima*, but its leaves are dark greyish-green with an iridescent sheen and may have a slight mottling. Plants appear stemless, but may actually have as creeping stem and the leathery, deeply

Chamaedorea radicalis.

notched, simple leaves spread in an attractive rosette. The small, black, round fruit contrast with the swollen, orange infructescence. This attractive little palm (a fantastic container plant) has been cultivated in European glasshouses since the early 1900s. *C. nana* is a synonym.

Chamaedorea quezalteca

(from Quetzaltenango, Guatemala, the type locality)

CENTRAL AMERICA

The dense clustering habit of the slender cane-like stems and the spreading, dark green pinnate leaves of this handsome species impart a pleasing effect to any garden. It is similar in general appearance to *C. costaricana* (but with fibrous, persistent ligules on the leaf sheath), although it is much rarer in cultivation. It is, however, reported to be widely grown in Guatemala. It occurs naturally in Guatemala, El Salvador and Honduras.

Chamaedorea radicalis

(arising from the root)

MEXICO

An unusual habit of this small palm is the erect to arching inflorescences often arising from below or just near ground level. Native to Mexico, this species grows in cool, shady rainforests at moderate to high elevations (about 1000 m) where it forms a sparse clump with the trunk either remaining short and congested or elongating normally above ground. The pinnate leaves are dark green and are usually erect and spreading. Fruit, which are about 1 cm across and orange to red when ripe, are borne in large, showy clusters. This cold-hardy species is easily grown in a shady, moist position and will succeed in temperate and subtropical regions. Plants can flower when very young and are best grown in groups. *C. pringlei* is a synonym.

Chamaedorea robertii

(after Robert Hodel)

CENTRAL AMERICA

A stemless or apparently stemless small palm from Costa Rica and Panama with deeply notched, fish-tail-like leaves and simple, spike-like inflorescences in both male and female plants. An excellent

Chamaedorea radicalis infructescence.

species for containers or a shady position in subtropical regions. Although this species has been introduced into cultivation it remains essentially a collector's item.

Chamaedorea sartorii

(after Carl Sartorius, 19th century German botanist)

CENTRAL AMERICA

Although only moderately ornamental, this palm has become widely grown in many countries. Plants have a solitary, slender, prominently ringed

Chamaedorea sartorii.

Chamaedoria stolonifera (*Botanical Magazine*, Vol. XLVIII of the 3rd series, plate 7265)

trunk to 2 m or more tall and a spreading crown of pinnate leaves. Fruiting inflorescences become arching to pendulous and are bright red with contrasting ovoid, black fruit, each about 1 cm long. A native to Mexico and Honduras, this palm has proved adaptable in subtropical and temperate regions. *C. aurantiaca* is a synonym.

Chamaedorea schiedeana

(after Dr Wilhelm Schiede, original collector)

MEXICO

In general appearance this species resembles *C. oblongata* but with more numerous leaflets on each leaf (20–26 in *C. schiedeana*, 12–16 in *C. oblongata*). It is a tall species (to 4 m) with a slender stem and has spreading inflorescences which branch towards the apex. These are colourful in fruit with the red rachillae contrasting with the small, black, round fruit. The species occurs naturally in Mexico, growing in shady forests at about 1500 m altitude.

Chamaedorea seifrizii

(after William Seifriz, original collector)

Bamboo Palm

CENTRAL AMERICA

This elegant palm has been something of a collector's item but in recent years, due to its exploding popularity, has become more commonly propagated by nurseries. The species can be readily recognised by its stiff, erect, clumping habit and the dark green, finely pinnate leaves scattered up the stems (the best forms have leaflets less than 2 cm wide). The clumps are generally narrow and the stems are a pale bluish-green which provides a pleasant contrast to the dark leaves. Yellow flowers are followed by greenish-black to black, globose fruit about 0.8 cm long. Native to Mexico, Belize, Guatemala and Honduras, this is indeed an attractive palm and one well worthy of garden culture or as an indoor plant. Clumps will grow to 3 m tall and prefer shady conditions with plenty of water during dry periods. Propagation is from seed which germinates readily or by division of the clumps. This species hybridises freely with other members of the genus. *C. erumpens* is a synonym.

Chamaedorea stenocarpa

(with narrow fruit)

Elfin Palm

CENTRAL AMERICA

A dwarf palm from Guatemala, Costa Rica and Panama which is truly a collector's delight. It is small in every respect, being trunkless and having a crown of fronds which are usually no more than 40 cm long. These arch from the ground in a pleasant rosette and impart a strong, fern-like appearance to the plant. They are symmetrically pinnate, a bright, glossy green with darker veins and have an attractive, winged rachis. Any doubt about the plant being a palm is dispelled upon the appearance of the inflorescences which are a simple, fleshy spike or sparsely branched panicle. At fruiting the female panicles are brilliant orange and carry shiny black, rounded fruit that are quite large for the size of the plant. In nature this species inhabits shady forest floors. It grows easily in cultivation and deserves to become widely grown. Excellent for massed planting as a groundcover or for mingling with ferns. Well suited to regions with a cool, moist climate. At present its popularity is limited by a lack of propagating material.

Chamaedorea stolonifera

(spreading by stolons)

MEXICO

An excellent clumping palm with plants spreading to 8 m or more across in suitable situations. Clumps consist of numerous, erect, very slender (about 6 mm across), cane-like stems to 1.5 m tall, each having a crown of dark green, fishtail-like leaves deeply notched at the apex and with toothed margins. Spread is by means of stolons which commonly originate from the basal nodes of the stems (sometimes also from above-ground nodes) and emerge some distance from the parent stem. Unusually, most cultivated plants of this palm are female with males being rarely grown and hence seed production is extremely limited. Plants available from nurseries are usually progagated vegetatively from division of stolons. *C. stolonifera* has been reduced to great rarity in the wild state through overexploitation by collectors. It is native to Mexico where it grows in wet, shady forests, usually on limestone.

Chamaedorea sullivaniorum

(after Joe and Pauleen Sullivan)

CENTRAL AMERICA—COLOMBIA

A truly dwarf palm which appears stemless but which in fact has a creeping stem buried in leaf litter. Thick, leathery, dark green leaves have a velvety, often mottled appearance and the prominent raised veins give a corrugated impression to the leaf surface. In some of its native haunts this species formed a pure dense groundcover but these patches have now mostly been wiped out by avaricious collectors. It grows at low to moderate elevations in Costa Rica, Panama and Colombia.

Chamaedorea tenella

(delicate, tender, soft)

CENTRAL AMERICA

A small, slow-growing palm from Mexico and Costa Rica which rarely reaches more than 80 cm tall in cultivation, although in nature plants can have erect or decumbent trunks to 2 m long. It has a solitary trunk and a crown of broad, entire, thick leaves which are deeply notched apically and with blunt, coarse teeth scattered along the margins. In cultivation it requires shady, moist conditions and is best suited to cool–tropical and subtropical regions. This species has been confused with

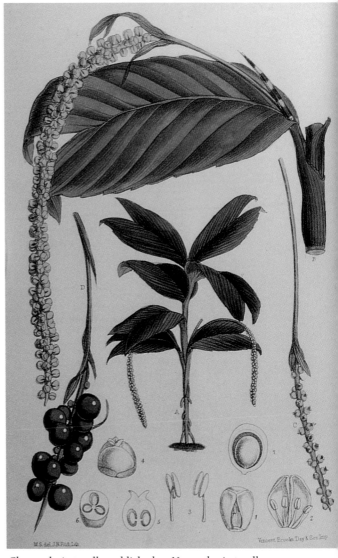

Chamaedoria tenella published as *Nunnezharia tenella* (*Botanical Magazine*, Vol. XXXVII of the 3rd series, plate 6584)

Chamaedorea tepejilote.

C. metallica which has metallic-coloured leaves lacking any marginal teeth. It appears to be rare in cultivation.

Chamaedorea tepejilote

(from a native name)

Pacaya Palm

CENTRAL AMERICA—COLOMBIA

A very slender but robust palm which is widely distributed from Mexico to northern Colombia. Plants, which are usually solitary—although clumping forms are known—have an interesting trunk which has swollen nodes and prominent annular rings with a similar appearance to bamboo and masses of basal prop roots. In very humid situations these roots eventually contact the soil and grow normally. The trunks can grow to 3–7 m tall and at the apex have a loose crown of dark green pinnate fronds which may be 1.5 m long and nearly 1 m broad. Flowers are greenish yellow and fragrant and the ovoid fruit (about 1.5 cm long) are bluish-green to black when ripe, contrasting with the bright orange to red rachis. A shady position is best and the species can be quite fast growing.

Tropical and subtropical regions are most suitable and plants should be grouped in clumps. The unopened male inflorescences are used as an important vegetable (raw or cooked) in countries where it occurs naturally. Selection for the best tasting variants has occurred over centuries and in some areas large plantations have been established to provide the inflorescences which are sold in local markets. *C. casperiana*, *C. exorrhiza*, *C. anomospadix*, *C. sphaerocarpa* and *C. columbica* are synonyms.

Chamaedorea tuerckheimii

(after H.B. von Tuerckheim)

Potato Chip Palm, Ruffles Palm

CENTRAL AMERICA

The unusual common names for this palm arise from a general similarity of its leaves to certain potato crisps. The similarity is superficial, however, for this is one of the most handsome of all palms. With its dwarf habit and stiff, leathery, corrugated, bluish-green or mottled leaves with prominently toothed (often whitish) margins, it is a gem for the avid collector. Unfortunately the species has now been reduced to rarity in the wild because of the greed of collectors. It is native to Guatemala and Mexico where it grows in wet forests at moderate elevations (900–1500 m altitude).

Chamaedorea undulatifolia

(with undulate margins on the leaflets)

COSTA RICA

At first glance the leaves of this species could be mistaken for those of a fern or a ginger. Each plant has only a few leaves in the crown (3–5) and these arise from a very short stem. Characteristically, they have widely spreading dark green leaflets with prominently crisped or wavy margins. This handsome palm is unfortunately difficult to grow to perfection, since plants greatly resent a dry atmosphere and prefer cooler, humid conditions. Its natural occurrence is in the mountains of Costa Rica.

Chamaedorea warscewiczii

(after Joseph Warscewicz, original collector)

CENTRAL AMERICA

Plants of this species can be readily identified by a prominent cream ridge which runs from the base of the petiole down the leaf sheath. They also often have their arching pinnate fronds arranged in an extended crown. A native of Costa Rica and

Chamaedorea tepejilote infructescence.

Chamaedorea tuerckheimii. G. Cuffe

Panama, this solitary species grows well in a shady location in subtropical and temperate regions. *C. latipinna* is a synonym.

Chamaedorea whitelockiana

(after Loran W. Whitelock, noted palm and cycad enthusiast)

CENTRAL AMERICA

This species passes through a growth phase during which where the stem is short and subterranean with closely spaced nodes and the pinnate leaves have narrow leaflets. Plants may remain in this growth phase for a number of years (and even flower) before they produce normal aerial stems which have widely spaced nodes and larger leaves with larger leaflets. This interesting slender palm, native to Mexico and Guatemala, is grown on a limited scale by enthusiasts.

Chamaedorea woodsoniana

(after R.L. Woodson, plant collector in Panama)

CENTRAL AMERICA

The sheathing base of the lowermost leaf of this palm is swollen and at first glance could be mistaken for a crownshaft. The palm, which is widespread in Central America from Mexico to Panama has a stout, solitary trunk to 10 m tall and a spreading crown of pinnate leaves to 2 m long, each with numerous, long, narrow, dark green leaflets. A very handsome species which has been confused horticulturally with *C. costaricana*, a robust clumping palm. It adapts readily to cultivation in tropical, subtropical and warm–temperate regions. *C. vistae* is a synonym.

Chamaedorea zamorae

(after Nelson Zamora, contemporary botanist in Costa Rica)

COSTA RICA

An interesting species which was described from plants cultivated in Hawaii and grown from seeds collected in Costa Rica. At this stage it is still a collector's item. Young plants may have simple notched leaves but the leaves of older plants have up to six pairs of narrow leaflets with a broad, prominently notched apical flabellum. The flowers and fruit are crowded in the inflorescences which are undivided or sparsely branched. Plants eventually reach about 3 m tall.

CHAMAEROPS

(from the Greek, *chamai*, on the ground, *rhops*, a bush)

A monotypic genus found in Europe and Africa. Plants grow in coastal and near-coastal situations, often on rocky headlands and low hills. It is a solitary or clumping fan-leaved palm with thorny petioles. The inflorescences arise among the leaves and bear unisexual or bisexual flowers. Fibre is sometimes obtained from plants in the wild.

Cultivation: A widely cultivated palm popular for its appearance, adaptability and hardiness. Grows very well in temperate and subtropical regions. Generally male and female trees are needed for fertile seed production, however, some trees produce bisexual flowers and are capable of fruiting on their own.

Propagation: Divisions can be successful providing they have sufficient root development. Fresh seed germinates readily, although it may sometimes be slow and erratic. Fruit usually contain a single seed but may have up to three.

Chamaerops humilis

(of low growth)

European Fan Palm

SOUTHERN EUROPE—NORTHERN AFRICA—MALTA

This handsome palm is widespread around the Mediterranean coast of southern Europe and northern Africa extending east to Malta. It is the most widespread palm native to Europe and is a widely cultivated and very hardy species. It is extremely

Above: *Chamaerops humilis* dense clumping variant. Right: *Chamaerops humilis* inflorescence.

Chamaerops humilis sparse clumping variant.

Chamaerops humilis 'Green Mound'.

variable in growth habit and plants may have one to many trunks, be dwarf or tall and with green or glaucous leaves. The petioles are strongly spined and rather unfriendly. Flowers are cream to yellow-ish and the fruit are brown. This palm is very hardy and best suited to temperate regions, being slow and somewhat difficult in the tropics. It is cold-tolerant and remains unaffected by heavy frosts. Plants prefer a sunny position and are adaptable to a variety of soils but will not tolerate poor drainage. An excellent, hardy, tub plant for the patio or verandah. Several variants of European Fan Palm are grown. The most popular variant forms a low-spreading, dense clump with several trunks and has pruinose to glaucous green leaves. Another variant, sometimes called *C. humilis* var. *elatior*, has a single, tall trunk. Variants having short petioles and long petioles are known and a variegated cultivar is sometimes grown. The cultivar 'Green Mound' branches freely, has a very dense, compact habit and bright green foliage. It is grown on a limited scale in California.

CHAMBEYRONIA

(after Captain Charles Chambeyron, 19th century French naval commander)

A small genus of two species both endemic to New Caledonia where they grow in moist rainforests at low to moderate elevations. They are solitary, unarmed palms with a well-developed crownshaft and a spreading crown of pinnate leaves. The inflorescence arises below the crownshaft and bears unisexual flowers of both sexes.

Cultivation: Highly ornamental palms with much to offer horticulture. One species has proved to be adaptable to cultivation whereas the other seems to have stringent requirements. They need warm, humid conditions, well-drained loamy soil and an abundance of water. Single plants are capable of producing fertile seeds.

Propagation: Fresh seed germinates readily. Each fruit contains a single seed.

Chambeyronia lepidota

(with small, scurfy scales)

NEW CALEDONIA

This palm remains a collector's item and limited experience in cultivation suggests that it may be difficult to grow. In New Caledonia it is restricted to the north-east growing on schistose soils at low to moderate elevations. Plants have stiffer leaves than *C. macrocarpa*, narrower leaflets, with the lowest ones often reduced to ribbons and smaller fruit (3 cm x 2 cm).

Chambeyronia macrocarpa

(with large fruit)

NEW CALEDONIA

This handsome New Caledonian palm has proved to be adaptable and easy to cultivate and is gaining in popularity. It is a striking species with a slender trunk and a crown of dark green, curved pinnate fronds. New fronds are a brilliant red to reddish-orange and provide a spectacular contrast to the

Chambeyronia macrocarpa.

A large clump of *Chrysalidocarpus lutescens.*

mature crown. Plants are very slow growing, with each new frond taking up to five months to expand and many years before flowering occurs. Flowers are pink to cream and the large fruit (4.5 cm x 2.5 cm) are crimson when ripe. Its neat habit makes it a very desirable garden subject. The plants like well-drained, organically rich soil and need a shady position at least for the first few years. Plants have been successfully grown in tropical, subtropical and warm–temperate regions.

CHRYSALIDOCARPUS

(from the Greek *chrysos*, gold, *carpos* a fruit)

A genus of twenty species of palm restricted to the Indian Ocean islands of Madagascar, Comoros and Pemba north of Zanzibar. They grow in a range of habitats from coastal forests to montane forests and open situations. They are solitary or clustered, unarmed feather-leaved palms with a crownshaft. The inflorescence arises among the leaves and bears unisexual flowers of both sexes.

Cultivation: Highly ornamental palms with only a couple of species being commonly grown. These have proved to be excellent horticultural subjects and more species should be tried. They grow readily in a range of situations. Solitary plants are capable of producing fertile seed.
Propagation: Clumping species can be divided but the divisions are often slow to establish. Fresh seed germinates readily. Each fruit contains a single seed.

Chrysalidocarpus cabadae

(after Dr Cabada from Cienfuegos, Cuba, who first grew the palm)

Cabada Palm

CUBA

A suckering palm which forms large, spreading clumps to more than 7 m tall and 5 m across. It is reputedly a fast-growing species and could be an ideal palm for parks, but at present is mainly to be found in botanical collections. The crown on each trunk is quite long and drawn out by comparison with *C. lutescens* and the leaves are dark green and the leaf sheaths are covered with a glaucous bloom. The trunks are also green to bluish green, much thicker and prominently swollen at the base. The much-branched inflorescences carry greenish-yellow fragrant flowers which are followed by ovoid, scarlet fruit, each about 11 mm x 5 mm. This palm was described from plants cultivated in Cuba and it is unknown in the wild. It is an attractive palm which seems most suitable for tropical regions.

Chrysalidocarpus lutescens

(becoming yellow)

Golden Cane Palm, Butterfly Palm, Areca Palm.

MADAGASCAR

The first common name of this palm arises from the clumps of slender, golden stems and leaf stalks that each plant produces. The stems grow up to 10 m tall and each is topped with a crown of curved, yellowish-green pinnate fronds. New canes are produced from the base of the clump as the older canes mature. This spineless and very decorative palm is native to Madagascar where it grows in large thickets along streams, in littoral forests and in sand dunes. The golden colouration is especially pronounced in starved specimens or those grown in a very hot, sunny position. Golden Cane Palm is very popular in the tropics and subtropics but can even be grown in temperate regions. It is, however, rather cold-sensitive when young. Plants prefer a sunny position in rich, well-drained soils. Large plants are often used for indoor decoration but need very bright light. The species is often wrongly sold in the nursery trade as *Areca lutescens*. Seed germinates easily but it may take four to five months for seedlings to appear.

Chrysalidocarpus madagascariensis

(from Madagascar)

MADAGASCAR

As the specific name indicates this species is native to the island of Madagascar where it is widespread

Above: *Chrysalidocarpus lutescens* inflorescence.
Right: *Chrysalidocarpus madagascariensis*.

and locally common. It is a clumping palm (occasionally solitary) with fairly thick, prominently ringed, grey stems which may grow to 8 m or more tall. Each stem has a fairly sparse crown of plumose fronds. Although this species and the previous one are both clumping palms, the habit of each is very different. *C. lutescens* forms a dense, crowded clump while that of *C. madagascariensis* is open and sparse by comparison. The latter species is still uncommonly grown but plants will succeed in tropical, subtropical and perhaps warm–temperate regions.

A variant, previously well known as *C. lucubensis*, was introduced into cultivation many years ago and is now widely grown and esteemed as a horticultural subject. It thrives best in the tropics but can also be successful in the subtropics, although many plants suffer from cold damage in their early years. In the tropics it is a very fast-growing palm and appreciates deep, rich soils and plenty of moisture. Most plants have a solitary trunk but occasional specimens branch to form a pair. Mature plants have a stately, graceful appearance and the palm is ideal for home gardens as well as for parks and avenues. They are also very useful in coastal districts tolerating a fair degree of exposure to salt-laden winds. In nature, this variant is restricted to the island of Nossi Be which is off the coast of Madagascar.

Chrysalidocarpus pembanus
(from Pemba Island)

TANZANIA

A rare palm from Pemba, off the coast of Tanzania. Natural populations have been seriously reduced by clearing of its habitat and the palm is officially classified as being vulnerable. It is cultivated as an ornamental in Zanzibar and seeds have recently been distributed in the USA. It is a handsome palm with a smooth, strongly ringed, green trunk and an attractive crown of arching, dark green fronds. The narrow leaflets are held obliquely erect. Clusters of crimson fruit provide a splash of colour. It is interesting to note that plants of this palm may have either a solitary trunk or produce basal suckers and form a spreading clump.

CHUNIOPHOENIX
(after W.Y. Chun and *phoenix*, a palm)

A small genus of two species occurring in southern China, Hainan and Vietnam, where they grow as understorey plants in humid forests. They are small, clumping, unarmed fan palms. The leaves lack a hastula and the blades are irregularly divided into groups of folded segments. The simple or simply branched, spike-like inflorescences arise among the leaves and bear bisexual flowers or a mixture of unisexual and bisexual flowers. Clusters of scarlet fruit are highly ornamental.
Cultivation: Excellent ornamentals which are becoming more widely grown. Decorative features include modest dimensions, a clumping habit, attractive foliage and colourful fruit. Plants are easily grown in tropical and subtropical regions. Single plants are capable of producing fertile seed.
Propagation: Fresh seeds germinate readily two

Chuniophoenix nana.

to four months after sowing. Seedlings establish quite quickly. Each fruit contains a single seed.

Chuniophoenix hainanensis

(from Hainan Island)

SOUTHERN CHINA

A delightful small palm which forms spreading clumps to about 3 m tall. The slender stems are prominently ringed and the crown of leaves relatively open. The blades are dull green on both surfaces and the segments are attractively folded. The inflorescences have unusual, slender, spidery rachillae which bear creamy flowers followed by round fruit (2–2.5 cm across) which ripen from green to scarlet. This palm which is native to Hainan Island, is becoming popular for its neat habit and adaptability in cultivation. Plants have best appearance in a sheltered location but will also tolerate some exposure to sun.

Chuniophoenix nana

(dwarf, small)

SOUTHERN CHINA—VIETNAM

A novelty palm which is popular with enthusiasts and deserves to become more widely grown for its ornamental qualities. Plants form a compact clump with numerous, very slender, greenish stems (about 1 cm thick) which are prominently ringed and the leaf blades are borne on very thin (3–5 mm wide), arching petioles about 25 cm long. The blades themselves are nearly round in outline and are divided into about six segments. Clusters of scarlet fruit (about 1.5 cm across) are very showy. A native of Vietnam and southern China which grows well in tropical and subtropical regions. *C. humilis* is synonymous with this species.

CLINOSTIGMA

(with an inclined or drooping stigma)

A genus of thirteen species of palm distributed in various Pacific islands, including the Caroline Islands, Samoa, Fiji, New Hebrides, New Ireland and the Solomon Islands. They grow as understorey palms in dense rainforests, often at moderate to high elevations. They are solitary, unarmed feather-leaved palms with a prominent crownshaft. The inflorescence arises below the crownshaft and carries unisexual flowers of both sexes. A few species, previously placed in the genus *Exorrhiza*, have prominent stilt roots at the base of the trunk.
Cultivation: Palms of this genus are rare in cultivation. It seems they require moist to wet humid conditions in a shady position. Because of the high altitude influence, these palms may be reasonably cold-tolerant and may succeed in warm–temperate and cooler subtropical zones. Single plants are capable of producing fertile seeds.
Propagation: Fresh seed germinates readily two to four months from sowing. Each fruit contains a single seed.

Clinostigma exorrhizum

(with uncovered roots)

FIJI

In Fiji this palm is common on ridges and slopes in wet forests which are often covered by fogs and cloud. Young plants grow in the understorey but older specimens commonly emerge through the forest canopy. Plants are very reminiscent of a species of *Archontophoenix* but are more difficult to grow, being demanding of moisture and shelter. *C. seemannii, C. thurstonii* and *Clinostigmopsis seemannii* are synonyms.

Clinostigma gronophyllum

(resembling *Gronophyllum chaunostachys*)

GUADALCANAL

A massive stilt-rooted palm which is common on the Pacific island of Guadalcanal, where it grows as an emergent in rainforest. Plants have a crown of arching leaves with stiffly erect leaflets and a green crownshaft. Emerging leaves may have purplish tonings. Although apparently rare in cultivation, this palm would make an excellent novelty item for the tropics.

Clinostigma ponapensis

(from Ponape in the Caroline Islands)

CAROLINE ISLANDS

A tall palm (to 20 m), with a whitish trunk and a crown of long (3–5 m), arching leaves. These have arching to spreading leaflets, to 80 cm x 4.5 cm,

Clinostigma ponapense in cloud forest. A. Bowden-Kerby

which usually fray at the apex. The much-branched inflorescences are whitish and the oblong fruit, to 2.2 cm x 1.5 cm, are laterally compressed. Although restricted to the island of Ponape, this species is locally common and dominates wet forests from low to moderate altitudes.

Clinostigma samoense

(from Samoa)

SAMOA

An incredibly beautiful palm from Samoa where it grows in protected situations in high rainfall forests. The most notable feature of the species is its crown of very finely divided fronds. These spread gracefully and have strongly drooping, long-pointed leaflets. This graceful crown tops a prominent, pale green crownshaft and a slender, prominently ringed, grey trunk which grows to about 12 m tall. Inflorescences are intricately and finely branched and the fruit are small and unusually humped. The performance of this lovely palm in cultivation is largely unknown. A few plants

are in the collections of enthusiasts and it is to be hoped that they succeed with its culture. In nature it grows in well-drained soils in areas with an extremely high and evenly spread annual rainfall. *C. onchorhynchum* is a synonym.

Clinostigma savoryanum

(after Nathaniel Savory, chief founder of the settlement of Bonin Islands)

BONIN ISLANDS

This species, from the Bonin Islands, was widespread and common prior to the Second World War, but its populations were decimated by Japanese soldiers and the species was reduced to great rarity on some of the Islands. Plants have a slender, prominently ringed, greyish-green trunk to about 15 m tall, a whitish to glaucous crownshaft and a large crown (about twelve leaves) of erect to arching fronds (to 3 m long) which have strongly pendulous leaflets. The inflorescences are large and much branched and the slightly wrinkled fruit (12 mm x 9.5 mm) are black when ripe. This highly desirable palm has been introduced into cultivation but is grown on a very limited scale by collectors. It has been successsfully grown in subtropical and warm–temperate regions.

COCCOTHRINAX

(from the Greek *coccos*, grain, berry and *Thrinax*, another genus of palm)

Broom Palms

An interesting genus of forty-nine species of palm predominately developed in the West Indies, with the major centre of evolution in Cuba. They grow in exposed, rocky situations on limestone or magnesium-rich serpentinite formations. All are unarmed, solitary or clustered fan palms with numerous fibres on the young parts of the trunk. The inflorescence arises among the leaves and bears bisexual flowers. Clusters of fruit ripen through white and mauve shades before maturity. Species of this genus are often confused with those of *Thrinax* but can be readily distinguished by the solid, non-splitting leaf-bases and the fleshy, purple-black fruit (white in *Thrinax*).

Cultivation: Interesting and ornamental palms which deserve to become more widely grown. Most species are of moderate size and are ideally suited to

Coccothrinax alta.

garden culture. They are not particularly demanding if given a sunny location and unimpeded drainage. Many species grow well in coastal districts. Single plants are capable of producing fertile seed.

Propagation: From fresh seed which germinates readily two to six months after sowing but seedlings may be slow growing. Each fruit contains a single seed.

Coccothrinax alta

(tall, lofty)

Puerto Rican Thatch Palm, Tyre Palm

PUERTO RICO—VIRGIN ISLANDS

Like *Coccothrinax argentea*, the leaf segments of this species have numerous transverse veinlets noticeable on the undersurface. Plants also have a sparse, open crown but the leaf segments of *C. alta* are drooping, whereas those of *C. argentea* are stiff and rigid. These are dark green above and silvery beneath. The trunk (to 10 m tall) is mostly naked except for loose, coarse, webbed fibres near the

crown. Clusters of white flowers are followed by purplish-black fruit. An attractive, relatively fast-growing member of the genus which can be grown in tropical and subtropical zones. It is native to Puerto Rico and the Virgin Islands where it grows on low hills of calcareous soil.

Coccothrinax argentata

(silvered, silvery)

Silver Palm, Silver Thatch Palm, Silvertop, Florida Silver Palm

BAHAMAS—FLORIDA

Although mainly grown in enthusiast's collections, this fan palm is deserving of much wider cultivation. It has many interesting features, including a slender trunk covered with woven fibres and a sparse crown of deeply divided leaves which have narrow, drooping segments. These are dark green and glossy on the uppersurface and stark silvery white beneath. Both surfaces contrast attractively when the leaves move in the wind. White flowers are followed by purplish-black fruit. Silver Palms are fairly slow growing and this is their main drawback, although they may flower when quite small. They will tolerate sun when quite small and need an open, sunny position in very well-drained soil. Plants will withstand almost as much coastal exposure as the Coconut. Once established, the plants

Coccothrinax argentata inflorescence.

are quite hardy but for best appearance should be watered during dry periods. This palm is native to Florida and the Bahamas and succeeds best in tropical and subtropical conditions.

Coccothrinax argentea

(of silvery tints)

Hispaniolan Silver Thatch

HISPANIOLA

Although similar in name and general appearance to *C. argentata* this species is quite distinct. It is endemic to the island of Hispaniola (both Haiti and Santo Domingo) where it grows in colonies in open situations, often in dry, rocky areas. Plants have a sparse, open crown and the leaves are deeply divided into rigid segments which are dull green above and silvery beneath. The leaflets have transverse veinlets (absent in *C. argentea*). The trunks may reach 10 m tall and have a covering of leaf-sheath fibres which appear as if woven. White flowers are followed by brownish to black seeds about 9 mm across. An interesting, if somewhat sparse, palm

which has proved to be quite adaptable and will grow in temperate as well as tropical regions. Plants need well-drained soil in a sunny position.

Coccothrinax barbadensis

(from Barbados)

Lesser Antilles Silver Thatch

LESSER ANTILLES

A graceful fan palm distributed in the Lesser Antilles, south to Barbados and Trinidad, growing in coastal scrubs, often in calcareous soil. Plants have a slender trunk to 10 m or more tall and a crown of large, fan-shaped leaves which are green above and shiny, silvery-white beneath. Most of the trunk is naked but thick fibre adheres to the upper part just below the leaves. The leaves are not deeply divided into segments and are most attractive as their surfaces alternate in windy weather. Clusters of white flowers are followed by purple-black fruit. This palm succeeds best in tropical regions and needs a sunny position in well-drained soil. *C. dussiana* is a synonym.

Coccothrinax argentea.

Coccothrinax crinita.

Coccothrinax crinita

(long-haired, with a mane)

Old Man Palm, Palma Petate

CUBA

The trunk of this palm is truly remarkable, being completely covered with a thick layer of white to brown, woolly fibre and what is more, this fibre is quite conspicuous when the plants are only 10 cm tall. The palm's crown is rather nondescript by comparison with the hairy trunk, consisting of drooping fan leaves which are greyish beneath and shiny green above. This palm is a collector's dream and is uncommon to rare in cultivation. It seems to succeed best in tropical regions but may adapt to the subtropics. Young plants are quite tolerant of sunshine and need well-drained soil. The species is native to Cuba and consists of two subspecies.

Coccothrinax crinita ssp. *crinita*

This subspecies has long, persistent, flexuose hairs on the trunk and the leaves are shiny beneath, with prominent rusty glands. It is native to western Cuba and grows in low-lying areas which are seasonally flooded.

Coccothrinax crinita ssp. *brevicrinis*

(with short hair)

Short-haired Old Man Palm

This subspecies has short hairs on the trunk and the leaves are dull and rusty beneath. It is native to southern Cuba, growing in montane areas on serpentinic soils.

Coccothrinax ekmanii

(after Dr E.L. Ekman, Swedish collector)

Gouane Palm

HAITI

This species was first described in the genus *Cocothrinax* in 1929 and later transferred to the monotypic genus *Haitiella* in 1947. Although this is a very unusual palm, there is no significant character to support its retention in a distinct genus and it is best accommodated in *Coccothrinax*. The species grows in a single locality, in Haiti, occupying limestone

Characteristic fibrous trunk of *Coccothrinax crinita*.

outcrops which lack any trace of surface soil. Plants have an unusual stiff, dry appearance with the grey to silvery leaves (glossy when young) held, more or less, stiffly erect. The slender trunk reaches to about 5 m tall and the young parts are covered with interwoven fibrous sheaths. The mature fruit are rounded, greenish-yellow to tawny and have a rough, warty surface. This is a novelty palm, which would appear to have excellent prospects for cultivation.

Coccothrinax fragrans

(sweet-smelling)

Fragrant Cuban Thatch, Yuraguana

CUBA

Flower clusters of this palm are very noticeable, being bright yellow and highly fragrant. The species is native to eastern Cuba and is cultivated on a limited scale. Plants have a slender trunk to about 8 m tall and a sparse crown of fronds which are glossy above and dull grey below with a covering of scaly hairs. This species has been confused with *C. jamaicensis* which has white, fragrant flowers.

Coccothrinax inaguensis

(from Great Inagua Island, Bahamas)

BAHAMAS

This species is variable in the attitude of its leaf segments, with those on some plants moderately stiff and others drooping as if the plant is wilting. A useful guide to identification can be found in

Coccothrinax miraguama.

the bowl-like depression on the upper surface of the leaf blade where the segments are united. The species is a slender palm (trunk about 7.5 cm across) native to the southernmost island of the Bahamas where it grows in coastal thickets on limestone or sand. Plants grow to about 5 m tall and the leaf blades are green on both surfaces. Creamy-white, fragrant flowers are borne in noticeable clusters and the mature fruit is purple. The species is grown on a limited scale but has potential for greater use in coastal districts of the tropics and subtropics.

Coccothrinax jamaicensis

(from Jamaica)

Jamaican Silver Thatch

JAMAICA

Endemic to the island of Jamaica, this palm grows in coastal districts at altitudes ranging from sea-level to about 450 m, often occurring on stabilised dunes and limestone formations. The leaves are woven locally into attractive baskets, hats and bags. A highly ornamental palm which performs well in

coastal localities, growing successfully on limestone or sand and tolerating exposure to salt-laden winds. Plants have a slender, naked trunk to about 8 m tall and circular leaves which are glossy green on the upper surface and silvery white beneath from a dense covering of scales. Emerging leaves are covered with similar white scales. Attractive clusters of white, fragrant flowers are borne several times a year and are followed by juicy, purplish-black fruit. This species is highly variable in factors such as trunk thickness, leaf blade size, crown size and the density of white scales on the leaf undersurface.

Coccothrinax miraguama

(a Cuban name)

Miraguama Palm

CUBA

Cuba is the home of this elegant palm, the leaves of which make a graceful silhouette against the sky. These leaves are relatively small but very rigid and, with the segments widely spaced, allow plenty of light to pass through, thus imparting a characteristic appearance. They are a shiny, dark green above and greyish and hairy beneath. Most of the trunk is covered with a closely woven, woollen fibre which adds greatly to the palm's decorative appeal. Although slow growing, the species is easy to grow and is suited to tropical and subtropical regions. It is primarily a collector's palm but deserves to be more widely grown.

Four subspecies have been described, all from various parts of Cuba.

Coccothrinax miraguama ssp. *miraguama*

The typical subspecies has two layers of fibres in the leaf sheath, 8–10 stamens with the filaments fused only at the base and fruit 7–9 mm across. It is widespread in Cuba, growing in savanna on rocky hills of serpentinic origin.

Coccothrinax miraguama ssp. *arenicola*

(growing on sandhills)

Wiry Miraguama Palm

This subspecies has very wiry fibres in the leaf sheath, more numerous leaf segments, orbicular leaf blades, staminal filaments fused only at the base and fruit 7–9 mm across. It grows in sandy savanna in western Cuba and Isla de Pinos.

Coccothrinax miraguama ssp. havanensis

(from Havana)

Havana Miraguama Palm

This subspecies has woody leaf sheaths, twelve stamens with the filaments fused in the basal third and fruit 8–12 mm across. It grows around Havana on calcareous coastal sands and rocky hills of serpentinic origin.

Coccothrinax miraguama ssp. roseocarpa

(with rose-coloured fruit)

Rose-fruited Miraguama Palm

This subspecies has woody leaf sheaths, 8–10 stamens with the filaments fused only at the base and rose-purple fruit (all other subspecies are purple-black). It grows on siliceous hills in northern Cuba.

Coccothrinax munizii

(after O. Muniz, contemporary Cuban botanist)

CUBA

The crown of this slender palm is very sparse with the few fronds (about ten) being held more or less erect to obliquely erect on long, slender petioles.

Coccothrinax miraguama fibrous trunk.

Coccothrinax miraguama ssp. *roseocarpa*.

The segments of the blades do not spread widely, adding to the wispy appearance of the species. The crown tops a slender, fibrous trunk which can grow in excess of 10 m tall. Plants have a nodding inflorescence about 60 cm long and the round to nearly globose fruit are roughened with small prickles. This rare species is native to eastern Cuba where it grows on rocky, limestone hills.

Coccothrinax proctori

(after George R. Proctor, botanist specialising in the Caribbean region)

Cayman Thatch Palm, Proctor's Silver Palm

CAYMAN ISLANDS

A solitary palm which is endemic to the Cayman Islands where it is common in open situations growing on limestone. Plants have a smooth, grey trunk to 8 m tall and nearly circular leaves which are dark, glossy green on the upper surface and silvery or golden beneath with a dense covering of scales. Young expanding leaves are similarly clad.

Clusters of creamy-white, fragrant flowers are followed by purplish-black fruit. This attractive palm grows readily in subtropical regions in an open, sunny location. It is tolerant of coastal exposure and limestone. The leaves are excellent for making rope and were once widely employed for this purpose in its native country.

Coccothrinax readii

(after R.W. Read, 20th century palm botanist)

Mexican Silver Palm, Knacas

MEXICO

Described as recently as 1980, this species hails from the Yucatan Peninsula of Mexico where it is common in near-coastal rainforests and also extends to exposed situations on sand dunes. It is a typical *Coccothrinax* with a slender trunk (3–8.5 cm across) to 4 m tall and a sparse crown of deeply segmented fan leaves carried on very slender, arching petioles. These are dark green above and silvery beneath. Clusters of fragrant white flowers are followed by juicy, purplish-black fruit. A ready means of identification of this species is that the narrow, triangular hastula is frequently notched at the apex. Although generally rare in cultivation, seeds of this species have been distributed to enthusiasts. Plants would probably require similar cultural conditions to those of *C. barbadensis*.

Coccothrinax spissa

(sticky)

Guano, Swollen Silver Thatch

DOMINICAN REPUBLIC

This species can be readily recognised by its thick trunk (20–30 cm across) which is usually stoutly swollen above the middle. Plants grow to about 8 m tall with the fibres adherent only around the lower leaves. The crown is rather sparse, with the finely divided blades held on long, slender petioles which tend to arch with age. The drooping leaf segments are deeply divided into narrow, whip-like lobes and are green above and greyish beneath. The bright purple fruit are flattened on the end. An interesting palm which is native to the Dominican Republic where it grows along the margins of deciduous broadleaf forests. Plants have performed well in tropical and subtropical regions.

Above: *Coccothrinax proctori.*

Above: *Coccothrinax readii.*
Below: *Coccothrinax readii* infructescence.

Coccothrinax spissa.

Cocos

(from the Portuguese for monkey; an apparent reference to facial markings on one end of the nut)

A monotypic genus which probably originated on islands of the western Pacific but now so ubiquitous that its origin is uncertain. Plants typically line the coast in tropical regions and are responsible for much of the atmosphere associated with beaches in the tropics. The species is an unarmed solitary

palm with a large, spreading crown of pinnate fronds. The inflorescence arises among the leaves and bears unisexual flowers of both sexes. The Coconut is of tremendous significance to humans not only because of its large, edible nuts but also because of the other numerous uses to which its parts can be put (see page 47).

Other palms have previously been included in the genus *Cocos* but have now been transferred to other genera—for example, *Acrocomia*, *Arecastrum*, *Butia*, *Lytocaryum* and *Syagrus*. Nursery growers may still sell plants of *Syagrus romanzoffiana* as *Cocos plumosus*.

Cultivation: Widely planted as an ornamental and for its commercial products. Plants must have a warm, sunny position and grow best where they can tap underground water.

Propagation: Seeds take five or six months to germinate. They are often planted direct into the soil where they are to grow, or are sown singly in large containers.

Cocos nucifera

(bearing nuts)

Coconut Palm

WESTERN PACIFIC

Although most familiar in its natural habitat lining tropical, sandy beaches, the Coconut will also grow in warm inland areas and on near-coastal tropical tablelands up to about 1000 m altitude. The secret of its success seems to be a warm to hot, humid

Cocos nucifera.

Copernicia alba young plant.

climate and access to underground water. Coconuts are widely planted in coastal districts from the tropics to warm–temperate regions but they rarely fruit in the subtropics or further south. There are numerous varieties of Coconuts suited to different climatic zones, bearing nuts of different sizes and being tall or dwarf-growing. Coconut Palms make an excellent street tree (although the plants tend to lean) and are widely planted in gardens of the tropics. Their ability to withstand severe coastal conditions is unparalleled in palms. Coconuts germinate readily in warm temperatures and their subsequent growth is quite fast. They respond strongly to the use of nitrogenous fertilisers.

COPERNICIA

(after Copernicus, famed Polish astronomer)

Wax Palms

A genus of about twenty-five species of palm distributed mainly in Cuba with outliers in Hispaniola and South America. They are solitary, spiny palms with stiff, deeply divided, fan-shaped leaves and the stems densely covered with a persistent petticoat of brown leaf sheaths. The inflorescence arises among the leaves and bears bisexual flowers. In some species the smaller bracts of the inflorescence are completely tubular.

Cultivation: Impressive or even spectacular palms which deserve to be more widely grown, but their popularity is limited by their very slow growth. Essentially tropical palms, they demand a sunny position and unimpeded drainage. All species hybridise freely and care should be taken when planning plantings, or collecting seed in mixed collections. Single plants are capable of producing fertile seeds.

Propagation: Fresh seeds germinate readily three to six months from sowing but seedlings are very slow growing. Each fruit contains a single seed.

Copernicia alba

(white)

Caranday Palm

SOUTH AMERICA

A palm which is widely distributed in South America where it forms extensive colonies in savanna. Countries where it grows include Brazil, Bolivia, Paraguay and Argentina. The total population of wild palms has been estimated at one billion plants. An attractive palm with a slender trunk to about 25 m tall and a rounded crown of stiff leaves. These have an orbicular blade and the surfaces are

Copernicia baileyana.

Copernicia curtissii.

densely covered in wax. The species can be recognised by the smaller bracts on the inflorescence being tubular.

Copernicia baileyana

(after L.H. Bailey, noted American botanist and horticulturist)

CUBA

The most spectacular feature of this palm is its huge, deeply segmented, bright green fan leaves which are at first stiffly erect, then as they age are held at greater angles, the tips of the segments eventually becoming lax and drooping. The crown of the palm is large and crowded, with the leaves almost overlapping. Plants about ten years old look most impressive, then as they begin to develop a trunk, they take on a different appearance. Native to Cuba, the plants prefer a sunny position, although they will grow in partial shade. Well-drained soil is essential for success.

Copernicia berteroana

(after Carl Guiseppe Bertero, 18th & 19th century Italian physician and botanical collector)

DOMINICAN REPUBLIC

The trunks of this palm provide a tough, durable wood which is used for construction purposes and the leaves are used to make hats. Its leaves, which

are finely divided into about 100 segments, have a green, waxless upper surface and the lower surface is only slightly waxy. A handsome palm which occurs naturally in the Dominican Republic (Haiti and Hispaniola).

Copernicia cowellii

(after J.F. Cowell, original collector)

CUBA

The smallest of the genus, this Cuban species is of restricted distribution and grows in savanna on serpentinic soils. Plants only grow to about 2 m tall and begin flowering when only half the size. The slender trunk is mostly hidden beneath a dense crown of stiff, overlapping fronds with a skirt of dead fronds beneath. The fronds are green above and waxy white beneath. An interesting but very slow-growing palm for tropical and warm subtropical regions.

Copernicia curtissii

(after A.H. Curtiss, original collector)

CUBA

An impressive palm with a slender (25 cm across), columnar, smooth trunk to 7 m tall and a rounded crown of large leaves which have a coarsely spiny petiole and orbicular blade. Long, arching inflorescences hang well clear of the crown and persist

197

Copernicia ekmanii.

Copernicia fallaense.

long after fruiting is finished. The species is endemic to Cuba where it grows in coastal savanna and woodland. Some plants—a minority—produce basal suckers which can be used for propagation. The resulting plants retain the ability to sucker.

Copernicia ekmanii

(after Dr E.L. Ekman, Swedish collector)

HAITI

A distinctive species readily recognised by the dense, waxy covering on the leaves which makes them appear white. This is obvious on young leaves and less so on old leaves. Because of this waxy covering, plants have an unusual appeal. They are also of relatively modest dimensions with large specimens only being about 4 m tall. The species is restricted to Haiti where it grows on coral rocks and sand along the coast.

Copernicia fallaense

(from Falla, the original area of collection)

CUBA

Possibly the most impressive member of the genus and certainly the species with the largest leaves. Plants develop a cylindrical trunk to 20 m tall and

80 cm across. This is topped by a massive crown of large leaves, each growing to more than 3.5 m long with an orbicular blade about 2 m long. A magnificent palm which must be given ample room to develop. Ideal for avenue planting or planting in groups. The species is native to Cuba where it grows in savanna and woodland.

Copernicia gigas

(giant)

CUBA

A truly majestic palm which grows to about 15 m tall and with a wide (50 cm across), smooth, grey, columnar trunk topped by an impressive rounded crown of green to greyish fronds. The blade of each frond tapers to the petiole in a distinct wedge and is deeply and stiffly pleated. The conspicuous arching inflorescences may grow to 3 m long. Immature trees have a characteristic featherduster or shuttlecock shape and appear to erupt out of the ground. This species is native to Cuba where it forms large stands, often close to mangroves.

Copernicia gigas.

Copernicia glabrescens.

Copernicia glabrescens

(becoming glabrous)

CUBA

A rather untidy palm which has a slender trunk, mostly covered with coarse leaf-bases and leaves which have a rounded blade with stiff to lax segments. Arching inflorescences hang well clear of the crown and persist for a number of years, adding to the general untidiness of the plant. This palm, which suckers to form sparse clumps, occurs naturally in Cuba where it grows in coastal savanna.

Copernicia hospita

(hospitable—usually to parasites)

CUBA

A common Cuban palm which forms colonies on hills and slopes in savanna and woodland. Its leaves, which are greyish from a thick coating of wax on both surfaces, are used locally for making hats and for thatching. A handsome ornamental of moderate dimensions for tropical regions.

Copernicia hospita young plants.

Copernicia macroglossa inflorescence.

Copernicia macroglossa with its dense petticoat of dead leaves.

Copernicia macroglossa

(with a large tongue)

Cuban Petticoat Palm

CUBA

This must be one of the most spectacular of all palms and it deserves to be much more widely grown. The plants bear a spiral crown of closely packed, almost stalkless, glossy green leaves, but the most remarkable feature is the dense, brown petticoat formed by the dead leaves. This petticoat is a solid mass of closely packed, dead leaves and, in plants up to about twenty-five years old, extends right to the ground. Individual leaves are about 2 m across and are deeply divided into stiff, pointed segments. The Cuban Petticoat Palm likes a sunny aspect in well-drained soil. It should be grown in subtropical and tropical regions. Seeds are reported to germinate readily but plants are very slow growing. This species was previously known as *C. torreana*.

Copernicia prunifera

(bearing plums, a reference to the fruit)

Carnauba Wax Palm

BRAZIL

In Brazil where this palm is native, it is widely grown in commercial plantations for the production of carnauba wax. This wax is harvested from the leaves and used for a wide variety of purposes. The export is worth millions of dollars to Brazil and research programs are underway to produce high-yielding cultivars. The palms grow to about 10 m tall and have a large, rounded crown of deeply divided, fan-shaped leaves held on long petioles. The trunk is very hard and patterned with the bases of fallen leaves. Although quite ornamental, this palm seems to be rarely grown outside its natural habitat. It is well suited to tropical and subtropical conditions. Seed takes eight to ten months to germinate. This species was previously known as *C. cerifera*.

Copernicia rigida

(stiff, rigid)

CUBA

The juvenile and early adult growth phases of this novel palm (especially when still acaulescent), resemble a narrow, dense shuttlecock. This happens because the young leaves emerge from the crown in an erect manner and press against the stiff older leaves to form a dense tuft of overlapping fronds. Older plants loose this appearance, especially when the outer parts of the older fronds fold downwards and the trunk is covered with the decaying remains of the frond bases. Mature plants change again, consisting of a slender, bare trunk topped with a rounded pompom of crowded fronds, with the stiff narrow segments radiating in a ball. At all stages this Cuban palm is a splendid conversation piece.

Copernicia yarey
(a Cuban name)

CUBA

A common palm in the western parts of Cuba which grows in savanna and woodland. Plants have a slender trunk to 8 m tall and finely divided orbicular blades which are densely waxy, especially on the undersurface. Long inflorescences arch well clear of the leaves and hang below the crown when heavy with fruit. This species is grown on a limited scale in Florida.

Copernicia prunifera.

CORYPHA
(from the Greek *coryphe*, summit or hilltop, probably a a reference to the spectacular terminal inflorescence)

A genus of eight species of huge palms distributed in India, Sri Lanka, Vietnam, Laos, Malaysia, the Philippines, Indonesia, New Guinea and Australia. They grow in open situations, particularly along flood plains. They are solitary fan-leaved palms which grow to massive proportions and die after flowering and fruiting but once. The inflorescence, arising terminally, is the largest produced by any flowering plant and carries millions of bisexual flowers. Various parts of these palms are used for a wide variety of purposes.
Cultivation: Because of their huge size, these palms need plenty of room to grow and develop. They are best suited to acreage planting or in parks and large gardens. Interesting effects can be obtained by planting in groups, particularly using plants of mixed age or adding new plants to a colony at regular intervals. Single plants are capable of producing fertile seeds.

Copernicia yarey.

Propagation: Fresh seeds germinate readily two to six months from sowing. Each fruit contains a single seed.

Corypha talliera

(from Tallier in Bengal)

Tallier Palm

INDIA

A native of northern India which has similar appearance and features to those of *C. umbraculifera*, but with crowded, overlapping leaf segments and larger (to 3.5 cm across) dark green fruit. Once commonly cultivated in India, this interesting palm is now rarely grown, having been replaced by more ornamental species which are less demanding of space. Its cultural requirements are similar to those of *C. umbraculifera*.

Corypha umbraculifera young plant.

Corypha umbraculifera

(in the form of an open umbrella)

Talipot Palm

INDIA—SRI LANKA

This giant palm is suitable only for very large gardens, parks and acreage planting since it develops into an immense plant before eventually flowering, fruiting and dying. Individual leaves may be more than 5 m across (the broadest of all palms) and hence one can imagine the space occupied by a single plant. The large, leathery, bright green, costapalmate leaves are carried on a stout petiole about 4 m long. This petiole has numerous small teeth along the margin. Younger parts of the trunk are covered with persistent leaf-bases. Plants grow for thirty to eighty years, achieving heights of 12–25 m before flowering. Flowering plants are an impressive sight indeed, with a terminal panicle more than 6 m high and bearing millions of tiny, cream flowers. These are followed by dull green, rounded fruit which mature about twelve months after flowering (after which the palm dies). Talipot palms are quite hardy in cultivation but have limited appeal because of their awesome size. They are also very slow growing. Once established, plants are very hardy to dryness and other adverse conditions. The species is native to India and Sri Lanka and grows best in tropical zones.

Corypha utan

(a native name)

Gebang Palm, Buri Palm

INDIA—SOUTH-EAST ASIA—NEW GUINEA—AUSTRALIA

A widely distributed palm which extends from India and Burma through northern Malaysia, the Philippines, Indonesia and New Guinea to northern Australia, where it is distributed sporadically across the top end of the Northern Territory and Cape York Peninsula in northern Queensland. It favours tropical climates which are seasonally dry and often grows on wet sites, usually forming extensive colonies. In the Philippines it forms huge palm forests covering many square kilometres. In Australia, this palm grows along the margins of watercourses subject to flooding. It is common along some of the rivers flowing west from the high country of Cape York Peninsula into the Gulf of Carpentaria and in some situations forms extensive

Corypha umbraculifera in flower. J. Dransfield.

erate elevations. They are solitary, spiny fan-leaved palms with large, branched spines present on the roots. The inflorescences arise among the leaves and bear bisexual flowers.

Cultivation: Interesting palms which are uncommon in cultivation. Plants grow easily and have proved to be adaptable to a range of soils and situations. Single plants are capable of producing fertile seeds.

Propagation: Fresh seeds germinate erratically. Each fruit contains a single seed.

Cryosophila warscewiczii

(after Joseph Warscewicz, 19th century collector in South America)

Rootspine Palm

PANAMA

The trunk of this palm is its most distinctive feature, being covered with short spines which are in fact aerial roots. These are called rootspines and are frequently branched. If in contact with the soil the lowest ones may take root. The palm is native to Panama and is a tall, slender species with a crown of large fan-shaped leaves. These may be 2 m across and are dark green above and greyish beneath. Plants are rather cold-sensitive and are best suited to the tropics but will also succeed in a warm position in subtropical regions.

thickets. On a windy day a colony is very noisy, with the large leaves creaking and scraping with each movement. The colonies consist of plants in all stages of growth from seedlings to flowering and fruiting specimens and even those in the advanced stages of collapse and decay. This palm is easily grown in tropical areas, particularly where the plants have access to ground water. Plants are moderately common in cultivation and can be recognised by the conspicuous spiral markings on the trunk and the leaves breaking in the middle when old, with the leaf-base retained on the trunk. *C. elata* is a synonym.

CRYOSOPHILA

(from the Greek *cryos*, frost, *philos*, loving)

A genus of eight species of palm native to Mexico, Central America and northern Colombia. They grow in drier forests and rainforests at low to mod-

CYPHOPHOENIX

(from the Greek *cyphos*, a tumour and *phoenix*, a palm)

A small genus of two species of palm endemic to New Caledonia. They grow in diverse habitats: one in moist forests on acidic soils and the other in littoral rainforest on decomposed coral. They are solitary, unarmed feather-leaved palms with a prominent, bulging crownshaft. The inflorescence is borne below the crownshaft and bears unisexual flowers of both sexes.

Cultivation: Although highly ornamental these palms are rarely cultivated. They require excellent drainage and a protected position at least when young. Single plants are capable of producing fertile seeds.

Propagation: Results with seed are variable. Each fruit contains a single seed.

Cyphophoenix elegans

(elegant)

NEW CALEDONIA

A rather tough-looking, but elegant, palm which grows as an emergent in moist forests near streams on acidic soils. It is rarely grown but is deserving of wider recognition. Its slender trunk is topped by a prominent, swollen, grey-green crownshaft and a relatively small, neat crown of arching to recurved, greyish-green fronds with stiffly erect leaflets. These fronds are relatively stiff and tough, apparently withstanding wind quite well. The plants appreciate a well-drained, organically rich soil and need a shady position, at least for the first few years. They are apparently moderately adaptable and will grow in warm–temperate and subtropical regions. Seeds can be slow and difficult to germinate and seedlings are very slow growing.

Cyphophoenix nucele

(with a nut-like seed)

NEW CALEDONIA

This species is a component of littoral rainforests developed on decomposed coral. It is restricted to Isle Lifou in the Loyalty Islands off the coast of New Caledonia. It has been introduced into cultivation and seeds are reported to germinate readily but seedling growth is slow. Plants of this species would probably be well suited to coastal districts in the subtropics on alkaline soils. The species can be readily distinguished from *C. elegans* by its spreading (not arched) leaves with more numerous dark green, glossy leaflets.

CYPHOSPERMA

(from the Greek *cyphos*, a tumour and *sperma* a seed)

A small genus of four species restricted to Fiji, Vanuatu and New Caledonia. They are solitary, unarmed feather-leaved palms which may or may not have a crownshaft. The inflorescence is borne among the leaves or below them and bears unisexual flowers of both sexes. The genus *Taveunia* is synonymous.

Cultivation: Novelty palms which are rare in cultivation. Plants are generally slow growing and require unimpeded drainage in a sheltered position

in tropical and subtropical regions. Single plants are capable of producing fertile seed.

Propagation: Seeds may be slow and erratic to germinate. Each fruit contains a single seed.

Cyphosperma balansae

(after Benedict Balansa, 19th century French botanical collector)

NEW CALEDONIA

This is an uncommon and little-known collector's palm from New Caledonia where it grows on acid soils in dense rainforest at moderate to high elevations. It is a handsome, feather-leaved palm with a solitary trunk lacking a crownshaft and a crown of dark-green fronds which have long, pointed pinnae. Little is known about the cultural requirements of this palm except that plants are generally slow growing and seed may be difficult to germinate. It probably requires shade, certainly when young, and a well-drained, organic soil with extra water during dry periods.

Cyphosperma tanga

(a Fijian name)

FIJI

Endemic to the island of Viti Levu, Fiji, where it grows as an understorey palm in rainforest on basaltic soils. Plants have large leaves which may be undivided or split at irregular intervals towards the apex and a sparsely branched inflorescence. A very attractive palm which appears to be virtually unknown in cultivation.

Cyphosperma trichospadix

(with a hairy spadix or inflorescence)

FIJI

Unlike other members of the genus, this species has a prominent crownshaft and carries its inflorescences below the leaves. Plants often have a shiny trunk and spreading pinnate leaves. The fruit are larger than those of other species of the genus, being about 2 cm long. A native of the islands of Vanua Levu and Taveuni in Fiji, this collector's palm grows in rainforest at moderate to high elevations.

Cyphosperma voutmelense

(from Voutméle, a peak near the type locality)

VANUATU

A recently described (1993) species which occurs in Vanuatu on the island of Espirito Santo. It is similar to *C. balansae* but with a smaller inflorescence and smaller fruit. Plants have a closely ringed stem to about 6 m tall (lacking a crownshaft) and an open crown of fronds each to 1.5 m long, with the falcate leaflets (to 45 cm x 3 cm) widely spaced on the rachis. Arching to pendulous inflorescences are about 90 cm long and after flowering carry clusters of nearly globose, red fruit (each to 1 cm x 8 mm). This species is known only from a small population in the wild, growing in moist forest on volcanic soils at about 1000 m altitude. It would appear to be an attractive palm for tropical and subtropical regions.

CYRTOSTACHYS

(from the Greek *cyrtos,* arched, *stachys,* a spike, in reference to the curved inflorescence).

A small genus of eight species, with one widespread from Thailand to Indonesia and the others in New Guinea and the Solomon Islands. They are solitary or clustered, unarmed feather-leaved palms with a prominent crownshaft. The inflorescence arises below the leaves and bears unisexual flowers of both sexes.
Cultivation: One species, *C. renda*, is widely cultivated in the tropics, the other species are ornamental but are rarely grown. All species appear to be very sensitive to cold and are therefore strictly tropical in their requirements. Single plants are capable of producing fertile seeds.
Propagation: Fresh seed germinates readily two to four months from sowing. Each fruit contains a single seed.

Cyrtostachys elegans

(elegant)

NEW GUINEA

A New Guinean species which shares many features with *C. glauca*, but which may be immediately distinguished by the green crownshaft. The leaves also have very short petioles, prominently drooping leaflets and thicker inflorescence branches which bear prominent, dark brown flowers. An elegant palm which should be excellent in tropical gardens.

Cyrtostachys glauca

(bluish-green)

NEW GUINEA

A clumping palm from the rainforests of New Guinea which should be an attractive acquisition to tropical gardens. Plants normally have one or two dominant stems and a few basal suckers. Like other members of the genus, this palm has a prominent crownshaft (more than 1 m long), but in this case it is markedly bluish-green rather than being brightly coloured. The leaves have spreading to drooping, bright green pinnae (to 1 m x 2.8 cm) and the small (11 mm x 6 mm), ellipsoid, black fruit are sunken in a yellowish to reddish perianth, with the branches of the infructescence being yellow. This rarely grown species probably requires similar cultural conditions to others in the genus.

Cyrtostachys peekeliana

(after Miss Peekel, a missionary in New Guinea)

NEW GUINEA

Although this species appears to be grown on a limited scale, it has excellent prospects in tropical areas. Native to the island of New Britain in New Guinea, it is a solitary palm with a slender, prominently ringed trunk and a conspicuous crownshaft below an upright crown of dark green fronds. This combination gives the species a very elegant and tropical appearance. The palms bear a massive, intricately branched inflorescence which at fruiting time carries masses of small, slender black fruit.

Cyrtostachys renda

(a native name)
Sealing Wax Palm, Lipstick Palm, Maharajah Palm, Pinang Rajah

SOUTH-EAST ASIA

The brilliant, glossy, scarlet leaf-bases and petioles which characterise this tropical species make it one of the most colourful and ornamental of all palms. With this colour contrasting with its dark green, erect leaflets and added to the neat, clumping habit,

Cyrtostachys renda colourful crownshaft and petioles.

it is perhaps surprising that the species is not more widely grown than it is at present. There is some suggestion that it may not be an easy plant to establish and it is certainly very tropical in its requirements. Plants will grow in shade or full sun and require plenty of water at all times. The slender stems grow to about 6 m tall. Also excellent in large containers. The palm is native to Thailand, Malaysia, Sumatra and Borneo and grows in near-coastal swamps. Seed should be sown soon after collection as it loses its viability rapidly. Germination usually takes place within two months of sowing. The species was previously very well known as *C. lakka*.

DAEMONOROPS

(from the Greek *daimon*, *denion*, evil spirit, *rhops*, a shrub, in reference to the spiny nature of these palms)

A large genus of 115 species of spiny palms occurring in southern China, India, Malaysia, the Philippines, Indonesia and New Guinea where there is a single species. The greatest diversity in the genus occurs in Malaysia and Indonesia. Most species occur in rainforests, with a few growing on forest margins. They are mostly climbing palms (some are dwarf, others have erect non-climbing stems) with long, slender, spiny stems and spread-

A large clump of *Cyrtostachys renda*. C. Goudey.

ing pinnate leaves, each usually with a thorn-bearing cirrus at the tip. A small number of species are solitary, others have a clumping habit. In a few species the stems die after fruiting. The inflorescences arise in the leaf axils and bear unisexual flowers. The fruit are scaly and in some species have a layer of edible flesh around the seed. In some countries the slender, flexible stems of these palms are harvested for rattan cane, which is used widely in furniture manufacture.

Cultivation: Few species of *Daemonorops* are grown, mainly because of their spiny nature and climbing habit. Many species have highly ornamental leaves and would make an interesting addition to a shady garden or understorey. Some also make excellent container plants. As a general rule, these palms require well-drained loamy soils, but some are very restricted in their distribution and may have very specific requirements. Male and female plants are essential for fertile seed production.

Propagation: Clumping species can be divided, although the divisions are slow and difficult to establish. Fresh seed generally germinates readily, although some species are reported to be slow and erratic. Each fruit contains a single seed.

Daemonorops angustifolia

(with narrow leaves)

MALAYSIA

A vicious climbing palm from Malaysia where it is common on stream banks, frequently forming tangled thickets. Its slender, climbing stems are prickly and the spreading, bright green leaves are divided into numerous narrow, crowded segments. Each leaf ends in a slender cirrus which is effective at entangling in surrounding vegetation. When damaged, the stems of this species bleed white sap. Because of its decidedly unfriendly nature, this palm has limited appeal for cultivation. Enthusiasts may find some attraction in the graceful fronds. The species succeeds best in tropical regions.

Daemonorops calicarpa

(with cup-like fruit)

MALAYSIA—SUMATRA

A dwarf clumping palm from Malaysia and Sumatra where it is very common in hilly country and frequently forms colonies in rainforest. The plants are actually trunkless, with the stems branching beneath the soil, each bearing a rosette of erect, prickly leaves to 5 m long. The inflorescences are unusual, with each terminating an axis which dies after fruit has ripened and is replaced by new suckers arising from the base. The flowering axis is a compact, crowded structure borne among the leaf-bases and contains reduced leaves, the axils of which bear the dumpy inflorescence. Plants may be male or female. Although unknown in cultivation, this species would be a very interesting palm for a shady position in the tropics.

Daemonorops draco

(a dragon)

Dragon's Blood Palm

INDONESIA

A reddish resin obtained from this Indonesian palm is used for dyeing, varnishing and medicinal purposes. The general colour and appearance of

this material gives rise to the unusual common name of the palm. The species is a clustering palm with slender stems to about 10 m long and spreading pinnate leaves. It is an interesting palm for tropical regions.

Daemonorops jenkinsiana

(after Father Jenkins, a collector in India)

INDIA

A native of the mountains of northern India, this climbing palm has prickly stems about 3.5 cm across and large pinnate leaves with prominently hairy leaflets. All parts of the leaf are armed with hooked, often lobed thorns. The rounded fruit, about 1.5 cm across, are covered with overlapping yellowish-brown scales. Although rarely grown, this species will apparently succeed in subtropical as well as tropical regions.

DECKENIA

(after Baron K.C. von Decken, 19th century explorer)

A monotypic genus of palms which is endemic to Seychelles. It forms large colonies, with a dense litter-layer over the ground, on ridges and slopes at low altitudes. The species is a huge, sturdy, spiny, solitary feather-leaved palm with a prominent, spiny crownshaft. The inflorescence arises below the crownshaft and bears unisexual flowers of both sexes.

Cultivation: Cold-sensitive and strictly tropical in its requirements. Plants have proved to be difficult to establish in some localities. Must have excellent drainage. Single plants are capable of producing fertile seeds.

Propagation: Fresh seed germinates sporadically. Each fruit contains a single seed.

Deckenia nobilis

(noble, tall, lofty)

SEYCHELLES

A tall palm which has a slender trunk which may reach more than 35 m in height. This is topped by a spiny, whitish crownshaft and a graceful crown of dark green pinnate leaves. Individual leaflets are slender with a long, pointed tip and are hairy

Deckenia nobilis spiny trunk and petioles.

beneath. Sheathing bases and petioles are prickly but this is mainly obvious on young plants. Ovoid fruit are dark purple to black when ripe. In its native habitat of the Seychelles this species is becoming rare, partly through clearing of its habitat and also because the palms are harvested for their edible cabbage. The leaf sheaths are also used as feed troughs for animals and may be made into sandals. Plants sold as this species are often wrongly identified.

DICTYOSPERMA

(from the Greek *dictyon*, a net, *sperma*, a seed)

A monotypic genus of palms which is endemic to the Mascarene Islands of Rodrigues, Mauritius, Round Island and Réunion, where it grows in coastal and near-coastal forests at low elevations. The species is a solitary, unarmed feather-leaved palm with a prominent crownshaft. The inflorescence arises below the crownshaft and bears unisexual flowers of both sexes. This palm has become very rare in its natural habitat, mainly due to collection of its cabbage.

Cultivation: Popular in cultivation and excellent in coastal districts. Plants need good drainage but grow well where their roots can tap ground water. Single plants are capable of producing fertile seed.

Propagation: Fresh seed germinates readily within two to four months if sowing. Each fruit contains a single seed.

Dictyosperma album

(white)

Princess Palm, Hurricane Palm

MASCARENE ISLANDS

A rather cold-sensitive palm which is well suited to tropical conditions but will also succeed in warm subtropical and even warm–temperate areas, especially in near-coastal districts. It is an excellent palm for the coast, tolerating salt-laden winds quite happily without the burning usually suffered by palms in such areas. In fact, it is frequently called Hurricane Palm because of its ability to survive strong blows. It is a solitary palm with a dark grey, ringed trunk which grows to more than 10 m tall and has a prominent, bright green, white or reddish crownshaft. The pinnate fronds spread in a graceful crown and when young are frequently reddish. This trait is usually obvious on young plants. The flowers of this palm are large, fragrant and quite showy, being reddish and are followed by purplish-black fruits in large clusters. One characteristic feature which often serves to identify this species is the habit of the youngest developing frond to stand erect, like a sentinel. In cultivation this species needs well-drained, rich soil and a warm, sunny position. Plants respond to nitrogenous fertilisers and water during dry periods. When young, they make excellent specimens for indoor decoration. *D. album* is somewhat variable and studies have shown that these variants may

Dictyosperma album showing characteristic spear frond.

have been originally associated with specific islands and habitats. Natural populations are now, however, reduced to great rarity and cultivated plants of this species hybridise readily, hence it is difficult to be certain if the following varieties are in cultivation.

D. album var. *album* occurs on the islands of Réunion and Mauritius. It is a tall-growing palm with distinctly green petioles and leaf rachises and the leaflets have one or two prominent veins between the midrib and margin.

D. album var. *aureum* occurs on the island of Rodrigues. It too is a tall-growing palm, with the petiole and leaf rachis having a distinct yellow to orange stripe on the underside and the leaflets lacking any prominent veins other than the midrib.

D. aureum var. *conjugatum* occurs on Round Island. It is a short-growing palm with a stout trunk and the leaflet tips united by nearly persistent reins.

DRYMOPHLOEUS

(from the Greek *drymos,* wood or forest, *phloios,* bark)

A small genus of about fifteen species of palm from Indonesia, Samoa, the Solomon Islands and New Guinea. They grow as understorey plants in dense rainforest. They are solitary, unarmed, rather slender feather-leaved palms with a crownshaft and broad leaflets which are irregularly cut and notched at the apex. The inflorescence arises below the crownshaft and bears unisexual flowers of both sexes. This genus includes those species previously placed in *Coleospadix*.
Cultivation: Being highly ornamental, these palms have excellent prospects for cultivation in the tropics. They require unimpeded drainage, shade and an abundance of water. Single plants are capable of producing fertile seed.
Propagation: Fresh seed germinates readily within two or three months of sowing. Each fruit contains a single seed. The fleshy layer of the fruit contains stinging crystals and should be handled with care.

Drymophloeus beguini

(after V.M.A. Beguin, Dutch collector in Indonesia)
INDONESIA

A slender Indonesian palm which is uncommonly grown. The plants may reach about 5 m tall and carry a small but elegant crown of dark green,

glossy fronds. These have widely spaced, broad segments, the apexes of which appear as if they have been roughly trimmed. In cultivation this palm needs a shady position and liberal applications of mulch and water. Plants are reported to be cold-sensitive and should be grown in the tropics and perhaps also in warm subtropical zones. The juice from the fruit is reported to be an irritant.

Drymophloeus ceramensis

(from Ceram)
CERAM

As the specific name suggests, this elegant palm is native to the island of Ceram where it grows in wet, shady forests. Plants develop a slender, woody trunk to 4 m tall topped by a slightly bulging crownshaft and with a delicate crown of arching fronds. The dark green leaflets are narrowly wedge-shaped, with the outer margins appearing as if irregularly chewed or bitten. This species is very close to *D. oliviformis* and may not be distinct.

Drymophloeus oliviformis

(in the shape of an olive)
AMBON

A native of the island of Ambon, this neat, small, elegant palm is well suited to tropical and subtropical regions. Although rarely grown, it has all the hallmarks of a decorative pot or garden subject. Plants are smaller and more slender than the previous species and the leaflets impart a ruffled appearance to the fronds. The leaves are very sensitive to sunburn and plants require a shady, moist position with plenty of water during dry periods and protection from strong winds.

Drymophloeus pachychladus

(with thick stems)
SOLOMON ISLANDS

This tall, slender palm is native to the Solomon Islands. In rainforest, where they grow naturally, plants may reach 12 m tall and have an elegant, if somewhat untidy, crown of arching or even twisted fronds which have distinctively broad segments. These are shiny green above and dull grey beneath. The crownshaft is prominent and bright green.

Mature fruit, about 2 cm long, are elliptical in shape and red when ripe. A few plants are in the collections of enthusiasts in Australia but at this stage the cultural requirements of the species are hardly known. Plants would probably grow best in the warm, moist, tropical lowlands.

DYPSIS

(an obscure term, perhaps from the Greek *dyptein*, to dip or dive)

A genus of twenty-one species of palms endemic to Madagascar where they grow as small understorey plants in rainforest. They are slender, solitary or clumping, unarmed palms with entire or pinnate leaves and slender, cane-like stems, often with a crownshaft. The sparsely branched inflorescence arises among the leaves and bears unisexual flowers of both sexes. The small fruit are bright red when ripe.

Cultivation: As a group these are exceptionally ornamental palms which have much to offer horticulture but unfortunately they are rarely grown, probably because of a lack of propagating material. Plants need warm, moist conditions in a shady location and must have excellent drainage. Single plants are capable of producing fertile seeds.

Propagation: Fresh seeds germinate readily and clumping species can be divided. Each fruit contains a single seed.

Dypsis louvelii in fruit. J. Dransfield

Dypsis procera.

Dypsis louvelii

(after M. Louvel, original collector)

MADAGASCAR

A truly dwarf palm which develops a solitary, slender stem to about 20 cm tall and with a small crown of deeply notched, pleated entire leaves which are a soft dark green (described as resembling velvet). Small red fruit are borne on a cream-coloured infructescence. This delightful palmlet is common in moist, shady forests on the eastern side of Madagascar. It would make a delightful addition to any tropical garden.

Dypsis pinnatifrons

(pinnately divided fronds)

MADAGASCAR

A delightful Madagascan palm with a sturdy, erect trunk which may exceed 5 m in height. The trunk starts out very slender (about 1.5 cm thick) but remarkably can increase in diameter with age, eventually reaching about 10 cm thick. The species has pinnate fronds with broad, attractively clustered or whorled leaflets which are dark, shiny, lustrous green. The fronds are scattered up the stem and may sometimes be produced at irregular intervals. The youngest frond is usually pointed and erect, forming the perfect tip to the spear-like stem. Young plants of *D. pinnatifrons* make very attractive pot plants and are also excellent for garden culture in the tropics. Being very sensitive to sun damage, they should be grown in a shady, protected position and provided with plenty of moisture. The species was previously known as *D. gracilis*.

Dypsis procera
(tall, lofty)

MADAGASCAR

A dwarf palm with a clumping habit and slender stems rarely exceeding 30 cm in length. The relatively short leaves are divided into a few narrow, stiffly spreading, dark green leaflets which are prominently pleated and toothed on the end. Plants would be excellent as a container subject and also for a shady position.

ELAEIS

(from the Greek *elaia*, for the olive tree)

Oil Palms

A small genus of two species of palm, one occurring in Africa and the other in Central and South America. Both species grow in open situations, such as along streams, in swamps and savanna. They are stout, armed, solitary feather-leaved palms which lack a crownshaft. Separate male and female inflorescences arise among the leaves and the infructescence is large and densely packed with fruit. The seeds and pulp of the fruit are very rich in oil. One species is widely grown in plantations in the tropics for this product. The genus *Corozo* is included in *Elaeis*.

Cultivation: Oil palms are generally easy to grow and adaptable as to soil type. They are sun-lovers and being large should be given sufficient room to develop. Single plants are capable of producing fertile seed.

Propagation: Seeds take three to six months to germinate, although some may take longer. The seeds have a thick shell and cracking prior to sowing may be an advantage. Similarly, pre-soaking in hot water and sowing over bottom heat may be beneficial practices. Most fruit contain a single seed, sometimes two.

Elaeis guineensis.

Elaeis guineensis
(from Guinea)

African Oil Palm

AFRICA

This palm, native to numerous countries in tropical Africa, has certainly become more widely distributed as a result of cultivation activities by humans. The palm is of major importance to the economies of countries such as Nigeria because high-quality oils are extracted from both the fruit pulp and kernel of the seed. Here it is estimated that more than 250 million mature African Oil Palms exist. These oils are exported to industrialised countries where they are used in manufacture and as a lubricant. Plantations of this palm have also been established in many tropical countries, particularly in South-East Asia. A small industry was established in north-eastern Queensland, Australia, before the Second World War but did not persist because of the slow growth of the palms. African Oil Palms are tough, distinctive plants and a large, healthy specimen is quite ornamental. The large, solid trunk is very rough, with persistent leaf-bases and bears a large crown of spreading, graceful, shiny, green pinnate fronds. Male and female flowers are borne on separate inflorescences on the same tree and each type is very distinctive. The male inflorescence is a cluster of furry, simple spikes which resembles a hand; while the female inflorescence is a dense, compound head. The shiny fruit are black when ripe. African Oil Palms thrive best in the tropics but are sufficiently hardy to grow in cool, subtropical areas where their rate of growth is, however, much slower. Once established they are a hardy palm, but for best appearance should be watered during dry spells. Epiphytes such as ferns and orchids can be successfully grown on the rough trunks and often naturalise on the upper trunk. Plants do very well in coastal districts and will withstand some salt spray. Well-drained soil is essential and the palms grow rapidly where their roots can tap ground water.

African Oil Palms are variable in some features. Most plants produce black seeds but some have green seeds and the flesh of the fruit is usually orange but in some is whitish. Occasional plants have all the leaflets fused to form an entire leaf.

Elaeis oleifera

(oil-bearing)

American Oil Palm

CENTRAL AND SOUTH AMERICA

This palm is native to Central America and northern South America where it grows in moist to swampy, sandy soils often close to the coast and in savanna. Plants have an unusual decumbent habit, with the young part of the trunk erect and the older parts prostrate on the ground. The stout leaf-bases are clustered towards the apex of the trunk and the leaves spread in a graceful crown. Separate male and female inflorescences and large clusters of black fruit are similar to those of *E. guineensis*. This species grows readily in tropical regions but is more cold-sensitive than *E. guineensis* and is unsuitable for the subtropics. Plants require an abundance of water.

ELEIODOXA

(from the Greek *eleio*, wet, *doxa*, glory, in reference to its wet habitat)

A monotypic genus of palms which occurs in south Thailand, Malaysia and Indonesia where it grows in

Clump of *Eugeissona minor* showing stilt roots.
J. Dransfield

freshwater swamps, forming colonies and thickets. The species is a spiny, clumping feather-leaved palm with subterranean stems. The inflorescences form a compound head which terminates a stem. Plants are unisexual and the fruit are covered in scales. This genus was formerly included in *Salacca*.

Cultivation: An interesting palm which is rarely grown. Plants need an abundance of water and should be planted in a low-lying site or around the margins of ponds and pools. Plants are very cold-sensitive and are best suited to tropical regions. Male and female plants are essential for fertile seed production.

Propagation: Seed may be slow and erratic to germinate. Each fruit contains a single seed. Clumps can be successfully divided.

Eleiodoxa conferta

(crowded)

Kelubi Palm

SOUTH-EAST ASIA

E. conferta is a prickly palm which forms thickets in swampy lowland areas, usually in or near rainforest. The trunks are subterranean and branch freely to form spreading clumps. Each trunk, at its apex, bears a tuft of tall, erect, willowy fronds which have coarse, broad leaflets which are silvery

Large clump of *Eugeissona utilis*. G. Cuffe

beneath. The petioles, rachises and even the midribs of the leaflets are liberally coated with long, sharp, black spines imparting a very formidable appearance. The inflorescence is a compound structure terminal on the trunk. The fruit are about 3 cm long, somewhat pear-shaped and covered with overlapping scales which, when ripe, are dull yellow. The sour flesh of the fruit is widely used in cooking. In cultivation this palm needs a hot tropical climate and must be provided with an abundance of water. The species was previously called *Salacca conferta*.

EUGEISSONA

(from the Greek *eu*, good, *geisson*, cornice of a roof a reference to its use as thatch)

A small genus of six species of palm, two occurring in Malaysia and the remainder in Borneo. They mainly grow on hills and ridges in forests, usually forming large colonies and thickets. They are clumping palms, with the stems being either subterranean or supported on strong stilt roots. The leaves are pinnate and have black, flattened spines on the petiole and rachis. Each stem bears a long, complex, terminal inflorescence which has short branches, each with paired flowers—a solitary male flower subtended by a solitary bisexual flower. Each trunk dies after flowering. The fruit is covered by fringed scales.

Cultivation: Interesting but not highly ornamental palms with an unfriendly nature. They are rarely encountered in cultivation and experience with them appears to be limited. They appear to be tropical in their requirements and must be given ample room to spread. Male and female plants are necessary for fertile seed production.

Propagation: Plants can be propagated by division of basal offsets. Fresh seed germinates in four to six months from sowing. Each fruit contains a single seed.

Eugeissona brachystachys

(with a short spike)

MALAYSIA

This Malaysian palm is more attractive and less rampant than its counterpart, *E. tristis*. Plants form smaller clumps with attractive, arching leaves and interesting leaflets which have a long, drawn-out drip tip. They are also less spiny than *E. tristis*. Damaged parts exude a white sap. This palm is grown on a limited scale in Malaysia and is appreciated for its ornamental appearance. It has good prospects for other tropical regions.

Eugeissona minor

(smaller, lesser)

BORNEO

This palm is similar to *E. utilis* with the trunks supported by prop roots, however, it is smaller and more slender in all its parts. It occurs naturally in Borneo where it grows in thickets on ridges and moist to wet heathland. Because of its relatively smaller size, this species would seem to have good prospects for cultivation. It would be suitable for tropical and subtropical regions.

Eugeissona utilis

(useful)

BORNEO

A native of Borneo, this species grows on forested ridges and low-lying heathy sites, often in colonies. Leaves are used by the local people for thatch; petioles and roots for structural purposes in buildings; and sago is extracted from the pith. An interesting palm which has the spiny trunks supported above the ground on long prop roots. Plants form dense clumps to about 20 m tall.

Eugeissona tristis

(dull-coloured)

Bertam Palm

MALAYSIA

A rather untidy palm from Malaysia where it is widespread from the lowlands to about 800 m altitude. It usually grows in timbered areas and in favourable sites may dominate the vegetation forming extensive thickets. On logged sites the growth may be so dense as to disrupt revegetation of commercially important trees. Each clump of this palm is large, dense and usually contains numerous dead leaves imparting an untidy appearance. Individual leaves are held stiffly erect, are crowded and may grow to 8 m tall. The leaf stalks are spiny. Inflorescences may rise to 3 m tall and carry brownish flowers (with purple pollen) and large fruits (to 10 cm x 5 cm). Immature fruits are edible and the leaves are prized locally for thatching huts. *E. tristis* is a very hardy and interesting palm which will grow in a sunny or shady position in tropical regions.

EUTERPE

(a mythological name for one of nine goddesses of the liberal arts)

A genus of twenty-eight species of palm distributed in the Lesser Antilles, Central America and South America south to Peru. They grow in moist habitats, including swamps, streamside vegetation and rainforest and range from near sea-level to high elevations. They are solitary or clustered, unarmed palms with spreading pinnate leaves and a prominent crownshaft. The inflorescence arises below the crownshaft and bears unisexual flowers of both sexes. These palms have sweet, tasty, edible cabbage and some species have been exploited heavily for this product.

Cultivation: Highly ornamental palms which deserve to be more widely grown. Most species require shelter from excess sun and wind when young. They can be fast growing and are highly responsive to fertilisers and regular watering. Some species grow well in subtropical regions; others are strictly tropical in their requirements. Single plants are capable of producing fertile seed.

Propagation: Fresh seed of some species germinate rapidly within a month of sowing, others are much slower. For the difficult species, seed germination may be improved by leaching in water held at 30°C for seventy-two hours. Each fruit contains a single seed.

Euterpe edulis

(edible)

Jucara Palm

BRAZIL

Native to Brazil where it is locally common, this palm is now widely grown around the world. It is a very useful species for indoor decoration, tolerating dark conditions and neglect. Larger plants develop a slender, tall trunk that is topped by a prominent crownshaft and graceful, dark green pinnate fronds which have crowded, drooping leaflets. After flowering, the inflorescence bears masses of small, round, brown to black fruit on drooping clusters. Plants of this species are easy to grow and seem best suited to subtropical and temperate regions. They like rich, well-drained soil and young plants need protection from direct sun for the first few years. Seed germination can be markedly improved by leaching in water held at 30°C for seventy-two hours (eighty per cent germination compared with thirty per cent germination untreated after twenty-four days).

Euterpe oleracea

(used as a vegetable)

Assai Palm, Acai Palm

BRAZIL

The common name of this palm is derived from a thick, refreshing drink which is concocted from the ripe fruit in its native country of Brazil. This material can also be used to flavour ice cream and is a very popular product. As well, the cabbage is highly regarded, so this palm is a very useful member of the community. It is a solitary species with a tall, slender trunk and a graceful crown of finely cut, drooping leaves. Plants can be quite fast growing and revel in deep, rich soils and will also tolerate poor drainage. They will grow very successfully in subtropical and tropical regions and look most attractive when planted in groups.

GASTROCOCOS

(from the Greek *gaster*, belly and *Cocos*, another genus of palm)

A monotypic genus of palms endemic to Cuba where it grows in forests and open habitats on calcareous soils. The species is a solitary feather-leaved palm with a remarkable, bulging, spiny trunk. The inflorescence arises among the leaves and bears unisexual flowers of both sexes.
Cultivation: An intriguing palm which is hardy and adaptable, but very slow growing. Plants need free drainage and are tolerant of calcareous soils. Single plants are capable of producing fertile seeds.
Propagation: Seed may be slow and erratic in its germination. Each fruit contains a single seed. Seedling growth is very slow.

Gastrococos crispa

(finely crinkled, crisped)

Corojo Palm, Cuban Belly Palm

CUBA

A distinctive palm with a grotesquely swollen trunk which is supported on a slender base. The trunk itself is very woody and armed with rows of flat spines; a row on each annular ring. The crown of arching fronds is quite dense towards the centre. Individual fronds have numerous leaflets which are dark, glossy green above and prominently glaucous beneath. Each inflorescence is subtended by a large, persistent, woody spathe which is covered with brownish fur and spines. The flowers are bright yellow and showy and the orange fruit have an edible, oily flesh. *G. crispa* grows well in tropical and subtropical regions. Plants are generally slow but may grow in spurts. They tolerate sun from an early age and like well-drained soil and plenty of water during the warm months of the year. This species was previously known as *Acrocomia armentalis* and *A. crispa*.

GAUSSIA

(after Caroli Frederici Gauss, 19th century astronomer)

A small genus of four species of palm, two occurring in Mexico and the others in Cuba, Puerto Rico, Belize and Guatemala. They are hardy palms from harsh, open, rocky habitats, often in skeletal,

Gastrococos crispa.

limestone soils on cliffs, crevices and steep slopes. They are solitary, unarmed feather-leaved palms which lack a crownshaft. Small, prickly lateral roots arise from the main roots and form a mass at the base of the trunk. The inflorescence arises among the leaves and bears unisexual flowers of both sexes. Species of the genus *Opsiandra* are now included in *Gaussia*.
Cultivation: Hardy palms of moderate ornamental appeal. They grow well in subtropical regions and prefer an open, sunny situation. Plants are generally slow growing and are tolerant of calcareous soils. Single plants are capable of producing fertile seeds.
Propagation: Fresh seed germinates readily within two to four months of sowing. Each fruit usually contains a single seed but may carry two or three. Seedlings are slow growing. The fruit of some species contain stinging crystals and should be handled with care.

Gaussia attenuata

(drawn out, tapered)

Llume Palm

PUERTO RICO

The trunk of this palm is widest at the base (sometimes bulging prominently) and then tapers evenly throughout until it is quite slender just below the crown. It is also unusual because its roots wander prominently across the surface of the soil and have whorls of spine-like rootlets. Native to Puerto Rico, it may reach 20 m tall in nature. Its crown is usually rather depauperate and disappointing, as if the plant could do with a good dose of fertiliser. The fruit are ovoid to pear-shaped, orange to red, about 1.5 cm long and carried in large clusters. In cultivation this palm has proved to be very hardy, succeeding in tropical, subtropical and perhaps even warm–temperate areas. Young plants will tolerate exposure to full sun. An annual dose of lime may be beneficial since in nature this species is restricted to limestone soils.

Gaussia gomez-pompae

(after Dr Arturo Gomez-Pompa, original collector)

MEXICO

This species was originally described in the genus *Opsiandra* and later transferred to *Gaussia*. It is native to Mexico where it is a component of forests on rugged limestone hills. Plants have a columnar trunk to about 12 m tall and a spreading crown of fronds in which the narrow, crowded leaflets are arranged in four rows and have a prominent yellow midrib. Globular, red fruit are borne in dense clusters. An interesting collector's palm for subtropical regions.

Gaussia maya

(a native name)

Maya Palm

CENTRAL AMERICA

This palm grows in the rainforests of Mexico, Guatemala and Belize. It has a tall, slender, columnar trunk and the unusual characteristic of delayed floral development—that is, the inflorescences grow to a certain stage but develop no further until several others form above them. After some triggering mechanism, flowering begins in order of maturity and a plant may have fifteen or more inflorescences present at the one time, all in different stages of development. This odd characteristic is further complicated by the persistence of viable male flowers for long periods on an inflorescence and also the ability of sections of the inflorescence to flower much later than others and produce succeeding crops of red fruit. The unusual flowering habit adds to the ornamental appeal of this palm, although it is mainly one for the collector. It will grow in tropical and subtropical zones and there are also instances of it succeeding in warm–temperate areas. It likes shade, plenty of water and mulch. Previously this palm was well known as *Opsiandra maya*.

Gaussia princeps

(chief, foremost)

CUBA

The trunk of this Cuban palm is broad and swollen at the base and then narrows suddenly upwards towards the crown, which is rather sparse, consisting of four to six short fronds. Dense clusters of pear-shaped fruit may be orange or purple when ripe. An interesting palm which lends itself to planting in groups. Best suited to tropical and subtropical zones in a sunny location.

GRONOPHYLLUM

(from the Greek, *gronos*, a cavern, *phyllon*, a leaf)

A genus of thirty-three species of palm occurring in Sulawesi, the Moluccas, New Guinea, islands in the Bismarck Archipelago and northern Australia. Many species are understorey palms or emergents in dense forests; others occur in more open habitats. They are solitary or clustered, small or tall, unarmed palms with pinnate leaves with regular or unevenly arranged leaflets and a well-defined scaly or hairy crownshaft. The inflorescence arises below the crownshaft and bears unisexual flowers of both sexes. Recently all species of the genus *Nengella* were transferred to *Gronophyllum*.

Cultivation: This genus includes many attractive palms, especially the smaller-growing, shade-loving species. Very few have been introduced into cultivation and little is known of their requirements. The Australian species has proved to be difficult to grow. Single plants are capable of producing fertile seeds.

Propagation: Response from seed is variable, with some species germinating within two to four months and others appearing sporadically over twelve months or more. Each fruit contains a single seed.

Gronophyllum apricum

(sun-loving)

NEW GUINEA

An interesting and highly ornamental small palm from the upper Sepik River Basin in Papua New Guinea where it grows on exposed limestone ridges at about 300 m altitude. Plants have a solitary, slender trunk to 5 m tall and a sparse crown of short, arching to lax fronds which have narrowly wedge-shaped leaflets grouped along the rachis in irregular clusters. Male flowers are cream with purple tips, female flowers are dark purple and the globose fruit are bright red. This rare novelty is unknown in cultivation.

Gronophyllum brassii

(after L.J. Brass, leader of Archbold expeditions)

NEW GUINEA

This solitary palm may develop a trunk in excess of 15 m tall but no thicker than 10 cm across. Plants have a prominent crownshaft and a small

Gronophyllum microcarpum.

crown of leaves which have wedge-shaped leaflets with ragged tips arranged irregularly along the rachis. An interesting species from the rainforests of southern New Guinea.

Gronophyllum chaunostachys

(with loose spikes)

NEW GUINEA

This tall, elegant palm is to be found in the highlands of New Guinea where it juts above the forest canopy. It is a very prominent, stately palm with a columnar trunk and a crown of attractively arching fronds which have erect and spreading slender leaflets. The trunk is slender (20–39 cm across), ringed and grey and the crownshaft covered with woolly, brown hairs. Bright red, smallish fruit are carried in dense clusters about 1 m long. This palm has many features to offer the enthusiast and should be tried in tropical and subtropical regions.

Gronophyllum flabellatum

(fan-shaped)

NEW GUINEA

This prized collector's palm from New Guinea grows in shady, moist situations under dense rainforest. It is a very slender, clumping palm reminiscent of both *Linospadix* and *Chamaedorea* and the bamboo-like stems may reach more than 5 m tall. The leaves are scattered up the stem and are divided into a few leaflets which are usually irregularly cut and toothed on the ends. The end pair are usually united and fishtail-like. Flowers and fruit are carried on simple, unbranched spikes and may be quite colourful. This delicate, highly ornamental palm likes deep shade, plenty of moisture and tropical or subtropical conditions. Seeds take four to five months to germinate and seedlings are very slow growing. Previously known as *Nengella flabellata*.

Gronophyllum microcarpum

(with small fruit)

MOLUCCAS

Native to the Maluku Islands and Ceram off the coast of the Indonesian island of Sulawesi, this palm grows in shady forests. A solitary species with a trunk to 8 m tall and a spreading crown of long fronds, with the broad leaflets being notched apically and with thickened margins. The basal and distal leaflets are regularly spaced but those in the centre are often clustered. The small red fruit are used locally as a substitute for betel nut. An elegant palm suitable for tropical regions.

Gronophyllum microspadix

(with a small spadix or inflorescence)

SULAWESI

The slender, cane-like, ringed trunks of this species may be no more than 2 cm across but reach more than 2 m tall. Short, arching fronds have wedge-shaped, dark green leaflets irregularly arranged along the rachis. Each leaflet has a lacerated apex as if bitten off by some ragged-toothed herbivore. A collector's palm from the shady tropical forests of Sulawesi.

Gronophyllum oxypetalum

(with sharp petals)

MOLUCCAS

An Indonesian palm which has been cultivated in the Bogor Botanic Gardens. Plants have a robust

Guihaia argyrata. J. Dransfield

habit similar to *G. chaunostachys* but with a distinctly rounded crown of broad fronds. These have long, almost scimitar-shaped leaflets narrowed to the base and irregularly cut off at the apex. The apical pair of leaflets is united in a fishtail pattern. Plants will tolerate full sun in the tropics. It is native to the island of Pulau Mangole.

Gronophyllum pinangoides

(like a *Pinanga*)

NEW GUINEA

This palm from north-western New Guinea has had a chequered taxonomic history with no less than twenty synonyms being listed in recent studies. It is a small, clumping species with very slender, cane-like stems (2–3 cm across) and arching leaves which have broad, wedge-shaped, lustrous leaflets with a ruffled appearance. An excellent small palm for a shady position in the tropics.

Gronophyllum ramsayi

(after P. Ramsay)

Northern Kentia Palm

AUSTRALIA

An Australian endemic which is restricted to northern parts of the Northern Territory where it forms colonies in sparse forest on sandy soils. An attractive, tall, sun-loving palm which unfortunately has proved to be very slow growing and difficult to establish. Plants are very sensitive to cold, need good drainage, plenty of water during summer and appear to resent strong fertiliser applications.

GUIHAIA

(based on the ancient name for Guilin and Guangxi, China)

A small genus of two species of palm endemic to southern China and northern Vietnam where they grow in crevices and niches of limestone hills at low elevations. They are dwarf, clumping, unarmed fan-leaved palms which appear trunkless but actually have a short, decumbent or erect stem which is clothed with persistent leaf-bases and coarse fibres. Inflorescences arise among the crownshaft and bear unisexual flowers.

Cultivation: Attractive small palms with excellent

Gulubia costata in habitat.

ornamental appeal. Their behaviour in cultivation is largely unknown but they are certainly worth trialling. They should be suited to warm–temperate and subtropical climates and would more than likely be intolerant of bad drainage. Male and female plants are essential for fertile seed production.
Propagation: Fresh seed of *G. argyrata* germinates readily two to four months after sowing. Seedlings are slow growing. Each fruit contains a single seed.

Guihaia argyrata

(silvery)

SOUTHERN CHINA

This species was described in 1982 from Guangxi, China, where it grows in sparse forest on hills of limestone origin. It is a small clumping species with short trunks 0.5–1 m tall covered with interlaced brown fibres. The leaf blade, which is about 90 cm across, is deeply divided into about twenty narrow segments and supported on a slender petiole about 1.8 m long. The undersurface of the segments and the inflorescence are covered densely with closely appressed silvery to brown hairs. The small, rounded fruit are about 5 mm across. This species was inadvertently introduced to cultivation when thousands (or even millions) of seeds were imported into the USA and Australia in the mistaken belief that the seeds were *Rhapis excelsa*.

Seedlings have grass-like leaves with silvery undersides. *Trachycarpus argyratus* is a synonym.

Guihaia grossefibrosa

(with coarse fibres)

SOUTHERN CHINA—VIETNAM

This species was originally described as a *Rhapis* in 1937 but its unique features were only recognised in 1982 after *Trachycarpus argyratus* (now *Guihaia argyrata*) was described. The species, which occurs in northern Vietnam and southern China, is similar to *G. argyrata* but with an erect trunk to about 1 m tall bearing old leaf sheaths at the apex only, the lower leaf surface being similar to the upper surface (not silvery) and with ellipsoid, blue-black fruit, 6–8 mm long. This species grows in a similar habitat to that of *G. argyrata* but is not known to be in cultivation.

GULUBIA

(apparently from *gulubi*, an Indonesian name)

A small genus of nine species of palm occurring in Fiji, New Guinea, Palau, New Hebrides, Solomon Islands and north-eastern Australia. They commonly grow in rainforest, often in colonies, with the crowns emerging above the surrounding forest. They are solitary, unarmed feather-leaved palms with a tall trunk, long crownshaft and large, rounded crown of fronds. The leaflets may be drooping or erect. The inflorescence arises below the crownshaft and bears unisexual flowers of both sexes. The genus *Gulubiopsis* is synonymous.
Cultivation: These palms have potential for group planting in tropical regions, but are rarely encountered in cultivation. Some species may be fast growing and single plants are capable of producing fertile seed.
Propagation: Each fruit contains a single seed and fresh seed germinates readily.

Gulubia costata

(ribbed, in reference to the fruit)

AUSTRALIA—NEW GUINEA

A tall, graceful palm which has a slender, grey trunk, long crownshaft and a large crown of spreading fronds with drooping leaflets. A fast-growing palm which is well-suited to planting in groups. Plants are very sensitive to cold and are suitable only for the tropics. They respond strongly to an abundance of water and regular fertilising. The species is very common in lowland areas of New Guinea and also extends to north-eastern Queensland, Australia.

Gulubia cylindrocarpa

(with cylindrical fruit)

NEW HEBRIDES

This species is from the New Hebrides, where it is widespread on a number of islands and may also extend to the Solomon Islands. It is a tall palm which grows in colonies, the crowns commonly being emergent above the forest canopy. Young plants have prominent stilt roots and entire or sparsely divided leaves, whereas mature plants have a large crown of arching pinnate fronds with the light, grey-green leaflets held erect. The unusual, cylindrical fruit are yellow when ripe. An attractive palm for the tropics.

Gulubia hombronii

(after M. Hombron, 20th century collector in Polynesia)

SOLOMON ISLANDS

This distinctive palm is common in the Solomon Islands where it grows on ultrabasic soils. Young plants have undivided leaves, whereas mature plants have pinnate leaves with stiffly erect, dark green, sword-shaped leaflets. The inflorescences are sparsely branched and after flowering carry oblong to nearly club-shaped fruit, about 2 cm x 6.5 mm. This is an attractive palm which is grown on a limited scale.

Gulubia longispatha

(with a long sheath)

NEW GUINEA

A tall palm from the highlands of New Guinea where it grows in small groups on steep ridges. Plants have a slender, grey trunk and a large crown of arching fronds, with the leaflets spreading widely but drooping at the tips. Although apparently not cultivated, this species would be worth trying in subtropical zones.

Hedyscepe canterburyana, top of Mount Gower, Lord Howe Island.

Gulubia macrospadix

(with a large spadix or inflorescence)

SOLOMON ISLANDS

The leaves of this palm ascend in a large crown (20–25 leaves in the crown) and the tips arch in an interesting manner. Each leaf has about sixty leaflets (green above, bluish-green beneath) which are held obliquely erect and have drooping tips. The crown is subtended by a prominent greenish yellow crownshaft. Small clusters of bright crimson fruit (to 1.5 cm x 8 mm) make a colourful splash below the crownshaft. This interesting palm is common in the Solomon Islands (including Bougainville) where it grows in the lowlands to about 1000 m altitude.

Gulubia microcarpa

(with small fruit)

FIJI

The very small seeds of this palm can be used as a diagnostic feature and gave rise to its specific name. It is native to Viti Levu, Fiji, where it grows in colonies on slopes at low elevations. It is a tall palm with a large, arching crown and erect leaflets. Experience in cultivation is very limited and the species would seem to be best suited to warm, tropical climates.

Gulubia palauensis

(from the Palau Islands, Micronesia)

PALAU ISLANDS

Populations of this species suffered drastically during the Second World War when thousands of trees were cut down for their edible cabbage. It occurs naturally on a number of islands in the Palau Group and grows on coral rock and limestone. It is a tall palm (to 20 m) with a rounded crown of leaves, each about 2 m long and a prominent crownshaft. The leaflets, to 23 cm x 1.2 cm, are crowded towards the base of a leaf and usually have frayed tips. Although rarely grown, this species would make an interesting palm for the tropics.

HEDYSCEPE

(from the Greek, *hedys*, sweet, *scepe*, a covering)

A monotypic genus of palm which is endemic to Lord Howe Island where it grows in mountainous forests and cliffs above the sea. The species is a solitary, unarmed feather-leaved palm with a crownshaft and bears unisexual flowers of both sexes.

Cultivation: Highly ornamental but only moderately popular in cultivation and slow growing. Single plants are capable of producing fertile seeds.

Propagation: Fresh seed germinates erratically, with seedlings appearing sporadically from five to eighteen months after sowing. Fruit take about four years to mature and is difficult to judge when it is ripe. Each fruit contains a single seed.

Hedyscepe canterburyana

(after Count J.H.T. de Canterbury, former Governor of Victoria)

Umbrella Palm, Big Mountain Palm

LORD HOWE ISLAND

A slow-growing, but highly ornamental palm which is gaining favour with enthusiasts in cooler subtropical and temperate climates. It is an excel-

lent palm for a container and plants perform very well indoors. They also grow well close to the coast. Ornamental features include a slender, ringed trunk, prominent silvery crownshaft and compact crown of dense, dark green, stiffly arching leaves with erect leaflets. Plants need a protected situation in well-drained soil and must be sheltered from excessively hot sun for the first five years or so. The species grows in forests and on ridges on the mountains of Lord Howe Island, Australia, between 400 and 750 m altitude.

HETEROSPATHE

(from the Greek *heteros*, different, of another kind and *spathe*, bract enclosing an inflorescence)

A genus of about thirty-two species of palm occurring in the Philippines, Mariana Islands, Indonesia and the Solomon Islands, with the centre of diversity in New Guinea where there are about sixteen species. Most species are understorey palms which grow in rainforest. They are slender, solitary or clumping feather-leaved palms which lack a crownshaft. The inflorescence arises either among the leaves or below the leaves and bears unisexual flowers of both sexes.

Cultivation: Attractive small palms best suited to tropical zones but worth trying elsewhere. Young plants seem to need some protection but older plants may relish sun. Single plants are capable of producing fertile seed.

Propagation: Fresh seed germinates two to four months from sowing. Each fruit contains a single seed.

Heterospathe elata

(tall)

Sagisi Palm

PHILIPPINES

This tall, slender palm from the Philippines and adjacent islands is rather cold-sensitive and succeeds best in lowland tropical regions, although plants are known to survive in the subtropics. It has a large crown of gracefully curving, dark green fronds and long, tapering leaflets. The young fronds are an unusual pale pink or brownish colour. Small, white fruits are borne in dense, hanging clusters among the leaves. In good conditions this palm can be fast growing and young

Heterospathe elata naturalised on Guam. A. Rinehart

plants may then have an extended crown of graceful fronds. It is mainly of interest for the collector living in the tropics. This species was introduced to Guam sometime between 1900 and 1920 and it has since become a significant weed, spreading and crowding out ravine species on the island.

Heterospathe humilis

(low, dwarf)

NEW GUINEA

An attractive New Guinean palm which deserves to be more widely cultivated in tropical gardens. It is a dwarf clumping palm with thin, cane-like stems and small leaves which may be either pinnate or simple. Pinnate and simple-leaved forms may grow side by side in a population and it seems uncertain whether these are juvenile and mature leaves or merely different growth forms. The species grows in shady positions in rainforest.

Heterospathe negrosensis

(from the island of Negros, the Philippines)

PHILIPPINES

A slender Philippine palm which occurs naturally in shady rainforests. Its trunk can grow to 3 m tall and reach a diameter of about 5 cm. Pinnae of the

leaves are 35–40 cm long and 2.5 cm broad and are generally dark green and shiny. Unlike the other species, its inflorescence is sparse, usually with only two branches. An attractive palm for culture in the tropics.

Heterospathe philippinensis

(from the Philippines)

PHILIPPINES

The slender trunk of this palm is 2–3 cm thick and can reach 3 m tall. It is a graceful species with the leaves divided into numerous leaflets which are 25–30 cm long. New leaves are a pinkish-bronze colour. The fruit are ovoid and suddenly contracted at the apex into a small point. As the specific name suggests, the palm is native to the Philippines where it grows in rainforest at moderate altitudes. Like the other species, it needs a shady position in a tropical or subtropical area.

Heterospathe sibuyanensis

(from the island of Sibuyan, the Philippines)

PHILIPPINES

This is one of the tallest of the Philippine *Heterospathe* species, with the trunks growing to 10 m with a thickness of 12 cm. It is characterised by having small, brown scales on the underside of the leaflets and a branched inflorescence which after flowering bears pointed, conical fruit. Plants require a shady aspect in moist soil and are best suited to the tropics.

Heterospathe woodfordiana

(after C.M. Woodford, original collector)

SOLOMON ISLANDS

New leaves of this slender palm are a colourful, deep red and similarly coloured to those of the delightful *Chambeyronia macrocarpa*. As each matures, it becomes deep green and a new frond provides a colourful contrast to the old. The palm has a slender brown trunk which grows about 4 m tall and the leaf sheaths have distinctive dark blotches. Pinkish flowers are followed by brilliant red fruit, each about 1 cm long. Native to the Solomon Islands, this attractive species is suitable only for the tropics.

HOWEA

(from Lord Howe Island)

Sentry Palms

A small genus of two species of palm which is endemic to Lord Howe Island. Both species grow in extensive colonies at low to moderate altitudes and form an interesting habitat on the island. They are medium-sized, slender, solitary feather-leaved palms which lack a crownshaft. The simple, unbranched inflorescence arises among the leaves and bears unisexual flowers of both sexes.

Cultivation: These are among the most sought after palms for horticulture in the world. Large quantities of sprouted seeds are exported from Lord Howe Island to nurseries in many countries. They are among the most successful of all palms for indoor use and are also excellent for coastal districts in the subtropics and temperate regions. Both species lend themselves very well to group planting. Single plants are capable of producing fertile seed.

Propagation: Fresh seed germinates erratically, with seedlings appearing sporadically over one to three years. Bottom heat and fungicide treatment greatly improves germination. Commercial growers commonly sow large beds of seed from which they remove sprouted seed at regular intervals. Seeds take three or four years to mature and it is difficult to judge when they are ripe as they colour slowly.

Howea belmoreana (*Botanical Magazine*, Vol. XIV of the 4th series, plate 8760)

Howea belmoreana

(after the Earl of Belmore, former Governor of New South Wales)

Curly Palm, Sentry Palm

LORD HOWE ISLAND

This palm can be readily recognised by its distinctive crown of strongly arching fronds in which the dark green leaflets are held stiffly erect. It is a slender species native to Lord Howe Island where it grows in colonies at intermediate levels. It is a popular palm in cultivation but is not as commonly grown as *H. forsteriana*. It grows well in subtropical and warm–temperate regions and looks particularly appealing when planted in groups. Plants do very well in coastal districts. A highly successful palm for indoor decoration.

Howea forsteriana in habitat, Lord Howe Island.

Howea forsteriana

(after William Forster, New South Wales Senator)

Kentia Palm, Sentry Palm, Thatch Palm

LORD HOWE ISLAND

Although restricted to Lord Howe Island, this palm is one of the most commonly grown species in the world, being prized for its graceful fronds and the tremendous ability of potted specimens to withstand neglect. It is undoubtedly one of the best plants for indoor decoration and has been used to beautify hallways, ballrooms, offices and houses since the 1850s. Although very tolerant of neglect, it is best to rest the palms at regular intervals by moving them outside to a shady, humid situation where accumulated dust and dirt can be washed from the leaves and the plants refreshed. Outdoors this species lends itself well to group planting. With its slender trunk and graceful crown of spreading fronds with drooping, dark green leaflets, this palm has become a firm favourite in the landscape. Plants grow very well in coastal districts withstanding considerable exposure to buffeting, salt-laden winds. They will tolerate direct sun from about five years of age but need protection when small. Mild frosts are tolerated without setback and plants grow best in subtropical and warm–temperate regions.

HYDRIASTELE

(from the Greek *hydor*, water, *stele*, a column)

A small genus of eight species of palm restricted to New Guinea, islands of the Bismarck Archipelago and northern Australia. They commonly grow in swamp forests or along streams in rainforest, both in lowland and highland situations. They are solitary or clumping, unarmed palms with a prominent crownshaft and pinnate leaves which have broad leaflets either regularly arranged or in clusters along the rachis. The stems, although slender, may grow very tall. The inflorescence arises below the crownshaft and bears unisexual flowers of both sexes.

Cultivation: Attractive clumping palms suitable for tropical and, to a lesser extent, subtropical regions. Young plants make ornamental specimens for large containers. There is some confusion as to the correct names of cultivated plants because most species are of similar general appearance. Single plants are capable of producing fertile seeds.

Propagation: Clumping species can be propagated

by division. Fresh seed germinates within two to four months of sowing. Each fruit contains a single seed.

Hydriastele beccariana

(after Odoardo Beccari, noted 19th century German palm botanist)

NEW GUINEA

A tall, slender palm found in hot, steamy, lowland jungles of New Guinea. The stems are about 8 cm thick and are topped with a small but dense crown of six to eight fronds. These fronds are dark green, a little over a metre long and more than 65 cm wide. The apical pinnae are conjoined and fishtail-like. The trunk also has quite a long, greyish crownshaft and dense, drooping inflorescences arise at its base. Clusters of fruit are bright red. This is a solitary palm and is not commonly seen in cultivation. It deserves to be more widely grown and is best suited to lowland tropical regions.

Hydriastele microspadix

(with a small spadix or inflorescence)

NEW GUINEA

A New Guinean palm which is widespread and common in moist to wet forested areas. It grows in clumps, develops tall, slender stems and the short leaves have oddly clustered, sharply truncate pinnae. White flowers are produced in masses from short, branched inflorescences and are followed by round, bright red fruit. This ornamental species deserves to be widely grown but unfortunately it is hardly known. It is best suited to tropical districts.

Hydriastele rostrata

(beaked)

NEW GUINEA

A clumping palm from New Guinea where it grows in lowland swamps in hot, humid conditions. Its slender trunks may grow to more than 10 m tall and have a dull green crownshaft which has a roughened texture and a small crown of relatively few fronds. The fronds may be 2 m long and have well-scattered, irregularly placed leaflets which are dull green and have irregularly cut tips. This is a rather tall but impressive palm that should be more

Howea forsteriana is one of the most successful palms for indoor decoration. C. Goudey

frequently grown. It would appear to be somewhat cold-sensitive and suitable only for planting in tropical regions. The plants like wet conditions or plenty of water.

Hydriastele wendlandiana

(after Hermann Wendland, 20th century palm botanist)

AUSTRALIA

An Australian species which occurs in near-coastal areas of the top end of the Northern Territory and north-eastern Queensland. It grows in soaks and swamps usually in, or close to, rainforest. It is a clumping palm with very tall stems (to 25 m or more), a whitish crownshaft and fronds about 2 m long in a sparse crown. Leaflets are irregularly spaced, with the terminal ones confluent and fish-tail-like. Slender strings of bright red fruit are highly decorative. Although attractive, this palm is uncommonly grown and is mainly found in the collections of enthusiasts. It succeeds best in the tropics in a sheltered situation. Plants respond to regular watering.

HYOPHORBE

(from the Greek *hys*, a pig, *phorbe*, food, in reference to the fruit being eaten by pigs)

A small genus of five species of palm endemic to the Mascarene Islands. They once grew in great abundance in forests and savanna but all are now reduced to great rarity, with two verging on extinction in the wild and one extinct in the wild and reduced to a solitary cultivated individual. They are solitary unarmed palms, sometimes with a grotesquely swollen trunk, a prominent crownshaft and a small crown of stiff pinnate leaves Developing inflorescences are held stiffly erect from the base of the crownshaft and resemble horns. Some species were previously included in the genus *Mascarena*.

Cultivation: These are attractive palms which should be grown for conservation purposes. Two species have become well established in cultivation and are popular accent plants because of their swollen trunks and stiff fronds. All species are sun-loving palms of slow, but steady, growth. Single plants are capable of producing fertile seed.

Propagation: Fresh seed germinates within three months of sowing. Each fruit contains one or two seeds. Seedlings are slow growing.

Group of *Hyophorbe lagenicaulis*.

Hyophorbe amaricaulis
(with a bitter-tasting stem)

MAURITIUS

Although once abundant in the mountainous forests of Mauritius, this palm is now reduced to the verge of extinction, being known from a single individual surviving in forest remnants in a botanic garden. This plant rarely produces fertile seed and the survival of the species is precarious indeed.

Hyophorbe indica
(from India)

MASCARENE ISLANDS

Endemic to the Macarene Island of Réunion where it grows on basalt soils at low to moderate elevations. This species is known from several populations in the wild and is relatively secure. It is a handsome palm with a slender grey trunk and a crown of five spreading leaves with numerous widely spreading narrow leaflets. The flowers are white and the elongated bright red fruit are beaked or hooked at the apex.

Hyophorbe lagenicaulis
(with a flask-shaped stem)

Bottle Palm

MASCARENE ISLANDS

An intriguing palm from the Mascarene Islands which obtains its common name from the unusual bloated trunk which in some specimens resembles a bottle. Also distinctive are the dark green pinnate fronds which have a characteristic, prominent twist and the crown which consists of a small number of expanded fronds (usually four to six) at any one time. Bottle Palms are rather cold-sensitive and are best suited to tropical regions, although they can succeed in a warm position in the subtropics. They grow very well in coastal districts and will tolerate considerable exposure to salt-laden winds. Plants like a sunny aspect and are very slow growing. They are readily propagated from seeds which usually germinate within six to eight months of sowing. This species is native to the Mascarene Island of Round Island and may also once have occurred on Mauritius. Although once common, it is now greatly endangered, being reduced to a single wild population.

Group of *Hyophorbe verschaffeltii*.

Hyophorbe verschaffeltii

(After Ambrose Colletto Alexandre Verschaffelt, 19th century Belgian nurseryman)

Spindle Palm

MASCARENE ISLANDS

As the common name suggests, the trunk of this palm is fusiform or spindle-shaped, being narrow at either end and with a prominent bulge in the middle. The trunk is usually grey and topped with a bright green crownshaft which may be expanded at the base. The crown consists of six to ten arching, feathery, dark green fronds which are most attractive. The inflorescences, when young, are carried in unusual, erect, curved, horn-like spathes and usually several at a time appear from near the base of the crownshaft. The flowers are bright orange and intensely fragrant. This palm is very popular in cultivation, especially in tropical areas, but plants will withstand some cold in the subtropics and also coastal exposure. A sunny position is essential for success. Young plants make excellent tub specimens and they can be held in the same container for many years. Plants are generally slow growing and, although hardy, they are best watered during dry periods. Seeds germinate readily within six months of sowing. This species is native to the Mascarene Island of Rodriguez where it grows on calcareous soils at low elevations. Once numbering thousands it is now reduced to great rarity, with perhaps fewer than fifty individuals left in the wild.

Hyophorbe vaughanii

(after Dr R.E. Vaughan, former curator, Mauritius Herbarium)

MAURITIUS

This species, native to Mauritius where it grows at moderate altitudes, is now reduced to rarity because of clearing operations. An ornamental species which has a slender prominently ringed trunk and a crown of stiffly arching fronds. The flowers are orange and the globose fruit are red when ripe. Cultivation will play a major role in the future survival of this distinctive palm.

HYPHAENE

(from the Greek, *hyphaino*, to entwine, weave, in reference to the fibres in the fruit)

Doum Palms

A genus of nine or ten species—although some authorities record as many as forty. They are widely distributed in coastal and near-coastal parts of Africa north from Natal, as well as Madagascar, Arabia and the west coast of India. They grow in open habitats, sometimes dominating the vegetation to form 'palm veld', otherwise along streams and on forest margins. They are solitary or clumping fan palms and are unique because their trunks have the ability to branch by forking. The degree of branching varies with the species. The large, stiff leaves are costapalmate and the petioles have stout thorns. The inflorescences arise among the leaves and bear unisexual flowers. The large, brown fruit have a layer of edible flesh. All parts of these palms are used by local people for a wide variety of purposes.

Cultivation: Hardy, sun-loving palms which must be given plenty of room to develop. Plants are generally slow growing and are well suited to planting in groups. They grow strongly if the roots can tap ground water. Male and female trees are essential for fertile seed production.

Propagation: Seed germination is erratic and may take ten months or longer. Removing the outer fleshy layer prior to sowing is an advantage. The seeds should be sown direct into the ground or in

Hyphaene coriacea.

Hyphaene coriacea fruit.

a deep container, because a long vertical shoot develops before the leaves emerge and seedlings are very sensitive to handling and disturbance. Each fruit commonly contains one seed but may contain up to three.

Hyphaene coriacea

(leathery)

Ilala Palm

AFRICA

Widely distributed in Africa from Natal into the tropics and growing in a range of habitats, usually in low-lying areas or near water. In the open areas of Zululand it dominates the vegetation to form palm veld or 'ilala'. Plants can either sucker from the base to form a clump or develop a single, sturdy trunk which is deeply ringed. The grey-green, finely divided fan-shaped leaves are crowded at the apex of the trunk. Female trees carry clusters of dark brown, shiny, pear-shaped fruit, each 5–6 cm across and with a thin, sweet layer of edible flesh. These fruit take two or three years to ripen and are favoured by elephants, baboons and fruit bats. This palm grows well in tropical regions but provenances from southern habitats should be suited to temperate zones. *H. natalensis* is a synonym.

Hyphaene indica

(from India)

Indian Doum Palm

INDIA

A native of the west coast of India, this species branches in a similar way to *H. thebaica*. Its leaves, however, are much larger with the lamina more than 1 m long and are divided into more numerous segments (about forty) compared with about twenty in *H. thebaica*. Female trees bear clusters of pear-shaped, brown fruit each about 7 cm across. These are edible and are eaten locally. The leaves are also made into hats, mats, bags and baskets and the seeds into buttons and beads. *H. indica* is grown to a limited extent in tropical and subtropical regions. This interesting palm is now becoming rare in its natural habitat due to clearing.

Hyphaene indica inflorescence.

Top: *Hyphaene indica*. Above: *Hyphaene indica* fruit.

Hyphaene petersiana

(after W.C.H. Peters, original collector)

Vegetable Ivory Palm

AFRICA

The trunk of this palm often has a characteristic bulge near the middle or towards the crown. Its large, grey-green, fan-shaped leaves are folded centrally and deeply divided into fine, stiff segments. They top a woody trunk which may grow to 15 m tall. Huge clusters of cricket ball-sized, shiny brown fruit are borne on the female trees. These

Hyphaene petersiana in habitat, Zimbabwe. C. Goudey

have a layer of sweet, edible flesh around a hard, bony kernel which has been used for vegetable ivory. This hardy palm is distributed from south-west Africa and Botswana northward into tropical Africa. *H. ventricosa* is a synonym.

Hyphaene thebaica

(of Thebes)

African Doum Palm, Gingerbread Palm

AFRICA

The Doum Palm was revered by the early Egyptians, with large quantities of fruit being recovered from the tombs of the pharaohs and outlines of the trees being commonly depicted at that time. It is even recorded that plants of this palm were planted in gardens as early as 1800 BC. Perhaps the most distinctive of all palms, this species is native to coastal districts of northern and eastern Africa. The Doum Palm is renowned for its regularly forking trunk, with an old plant consisting of many branches, each ending in a small crown of leaves. The leaves are stiffly fan-shaped, grey-green to glaucous and strongly costapalmate. They are carried on the end of strong petioles which have thick, black teeth along the margins. The oblong to pear-shaped fruit are orange to brown when ripe (about 8 cm long) and have a mealy flesh which is edible and reported to taste like gingerbread. Doum Palms may reach 15 m tall and may be nearly as far across. They are a botanical curiosity because of their branching habit. They will grow in warm–temperate, subtropical and tropical regions and need an open, sunny position.

IGUANURA

(after a reptile with a tail resembling the inflorescence of this palm)

A genus of about eighteen species of palm native to south Thailand, Malaysia, the Philippines and Indonesia. They are slender, understorey palms of rainforests, often occurring in scattered colonies. They are solitary or clumping, unarmed palms with or without a crownshaft and the leaves variable—from entire with or without an apical notch; to pinnate with regularly spaced or unevenly arranged leaflets. The inflorescence arises either among the leaves or, in those species with a crownshaft, below the leaves and carries unisexual flowers of both sexes. (For a recent treatment of the genus see R. Kiew, 'The genus *Iguanura* (Palmae)', *Gardens Bulletin*, Singapore 28 (2): 191–226.)

Cultivation: Lovely small palms with much to offer tropical horticulture, however, few species are cultivated. Some species have proved to be difficult to maintain in cultivation and they may have specific requirements. They relish shelter, loamy soils and an abundance of moisture. Single plants are capable of producing fertile seed.

Propagation: Suckers can be successfully removed from some of the clumping species. Experience is limited but seed of some species germinates readily within three months of sowing. Each fruit contains a single seed.

Iguanura bicornis

(with two horns, in reference to the fruit)

MALAYSIA—THAILAND

A small clumping palm which would make an excellent groundcover for a sheltered position in

Hyphaene thebaica.

the tropics. Plants form suckering clumps to about 2 m across with very slender, sinuous stems to about 2 m tall. The leaves are divided irregularly into broad segments and the unusual fruit have two prominent humps at the apex. The species is native to southern Thailand and northern Malaysia.

Iguanura borneensis

(from Borneo)

BORNEO

A delightful small palm which has paddle-shaped leaves to about 40 cm x 20 cm, each with a prominent, deep apical notch, the surface with prominent veins and the margins with numerous blunt, irregular teeth. Sometimes the leaves are divided into a few broad segments. The very slender stem (1–1.5 cm across) grows to about 1 m tall. All in all, a delightful small palm from Borneo which would appear to have great potential for cultivation.

Iguanura geonomiformis

(resembling a *Geonoma*)

MALAYSIA

A common, small, Malaysian palm which forms congested clumps up to 3 m tall (occasional plants have solitary stems). The stems are very slender and sometimes have clusters of stilt roots at the base. The leaves, which are about 1 m long, arise at a steep angle to the stem and are divided into irregularly shaped, folded leaflets which are green above and grey beneath. The inflorescence is a simple, pendulous spike which, after flowering, carries small, red fruit. In nature the fronds of this palm trap litter, which builds up untidily around the crown. This is a delightful small palm suitable for tropical gardens where it needs a sheltered, shady position. It should also make an attractive pot plant.

Iguanura macrostachya

(with large spikes)

BORNEO

A dwarf clustering species with slender stems about 40 cm long and pinnate leaves which have broad leaflets unevenly clustered along the rachis.

Iguanura wallichiana.

Bright red fruit are carried on long slender, unbranched spikes. An attractive species from Borneo which would make an excellent small palm for a shady garden in the tropics.

Iguanura polymorpha

(in many forms)

BORNEO

This species occurs in Malaysia and Borneo, growing in shady forests. Plants grow to 2 m tall and the pinnate leaves (rarely simple and entire) have a variable number of segments (four to twelve) with the main veins running from the midrib to the margin. An attractive palm for tropical gardens.

Iguanura wallichiana

(after Nathaniel Wallich, 19th century Danish botanist)

MALAYSIA

This palm is similar in many respects to *I. geonomiformis* but can be distinguished by the inflorescence which has up to ten branches. It is a small, suckering palm native to the Malaysian rainforests and has excellent potential as a garden plant for the tropics. Its leaves are variable, usually being divided, but with either coarse or fine leaflets. Occasionally however, the leaves are entire.

231

Johannesteijsmannia altifrons.

JOHANNESTEIJSMANNIA

(after Johannes E. Teijsmann, 19th century Dutch botanist)

A genus of four species of majestic palms which have been described by enthusiasts as the most beautiful palms in the world. They occur in south Thailand, Malaysia and Indonesia (Sumatra and Borneo). One species is very widely distributed, the other three are localised. They are strictly understorey palms which occur in undisturbed rainforest, perhaps surviving in selectively logged areas but quickly disappearing from badly disturbed sites. They are solitary palms either acaulescent or with a short trunk and very large, simple and entire leaves which are mostly diamond-shaped but may be elongate. These leaves form a large, erect to spreading rosette and have unusual hooked teeth on the lower margins of the blade. The narrow petioles bear small, sharp teeth. The inflorescences, which arise among the leaves, have unusual tubular, inflated bracts and bear bisexual flowers. The rounded, knobbly fruit have numerous, unusual, corky protuberances. Seedlings are rare in nature. The leaves of these palms are commonly used for thatch. (For a recent treatment of the genus see J. Dransfield, 'The genus *Johannesteijsmannia* H.E. Moore', *Gardens Bulletin, Singapore* 26 : 63–83.)

Cultivation: Magnificent palms which are in demand by horticulturists but are still extremely rare in cultivation. Plants are generally best suited to the tropics but are proving to be adaptable and success has been achieved in the subtropics. They require a sheltered position protected from excess sun and wind. Well-drained, loamy soils rich in organic matter, mulches and regular watering are all useful for these lovely palms. Single plants are capable of producing fertile seed. The fruit are borne low down in the crown and it is reported that the seeds are often not viable.

Propagation: Fresh seed takes about three months to germinate but may be erratic. Seedlings are slow growing and must be given ample room to develop since the radicle buries about 10 cm deep in the soil before the first leaf is formed. Each fruit can contain up to three seeds.

Johannesteijsmannia altifrons

(with tall fronds)

Joey Palm, Diamond Joey

SOUTH-EAST ASIA

The most widespread of the genus, this species occurs in disjunct populations in southern Thailand, Malaysia, Sumatra and Borneo. Plants grow in undisturbed rainforest on ridge tops and slopes from near sea-level to about 1200 m altitude. Plants are variable in robustness and leaf size

Johannesteijsmannia lanceolata in habitat, Malaysia. G. Cuffe

between populations but more research is needed to determine if this variation is genetically controlled. This palm has a creeping, subterranean trunk and a cluster of up to thirty leaves which form an impressive erect tussock. Each leaf, to 6 m tall, has a slender, thorny petiole to 2.5 m long and a large, dark green, diamond-shaped blade which is prominently ribbed and with coarsely toothed margins. The white flowers have a sour smell and the corky fruit, about 4 cm across, are brown when ripe. This palm is well known in the area where it grows because the large fronds provide an excellent shelter in a sudden downpour. They are also ideal for thatching huts. Cultivated on a limited scale but becoming more widely grown.

Johannesteijsmannia lanceolata

(lance-shaped)

Slender Joey

MALAYSIA

A distinctive species with a subterranean creeping stem and narrow leaves with the blade lanceolate in outline. The leaves may grow to 3.5 m tall but the blade is 30 cm wide or less and has soft, brown scales on the undersurface. As in *J. magnifica*, the petiole of this species has two conspicuous yellow stripes. A native of central-west Malaysia, *J. lanceo-*

Johannesteijsmannia magnifica in habitat, Malaysia. G. Cuffe

Johannesteijsmannia perakensis in habitat, Malaysia. G. Cuffe

lata grows in rainforest and at one locality mingles with *J. magnifica*, but hybrids are unknown.

Johannesteijsmannia magnifica

(magnificent, beautiful)

Silver Joey

MALAYSIA

The leaf blade of this magnificent, apparently trunkless palm is the broadest in the genus (to 3 m long x 2 m wide) and the undersurface is covered with white hairs to give a silvery impression. Also, the slender petiole is ornamented with two conspicuous yellow lines. This species grows on steep slopes and ridges in undisturbed rainforest of central-western Malaysia and in some areas is locally common. It is also reported to survive in forest that has been selectively logged.

Johannesteijsmannia perakensis

(from the type locality in Perak, Malaysia)

MALAYSIA

In general form, this species resembles *J. altifrons* but plants develop a distinct trunk to 4 m tall, the leaves tend to be more widely spreading, the inflorescence has more branches and the fruits are larger. Its flowers are also sweetly scented. This majestic palm is restricted to parts of Perak State in north-western Malaysia where it is locally common and forms loose colonies. Altitudinal range is about 175–850 m. A delightful, shade-loving palm for the tropics.

JUBAEA

(after King Juba of Numidia, north Africa)

A monotypic genus of palm endemic to Chile. The species grows in woodland and on rocky ridges and is now reduced to great rarity in the wild because the plants are cut down for wine and sugar. The species is a solitary palm with a massive, woody trunk and a spreading crown of pinnate leaves. The inflorescence arises among the leaves and bears unisexual flowers of both sexes. The fruit resemble miniature coconuts. Rare hybrids occur with *Butia capitata*.

Cultivation: A popular species prized for its massive trunk. An excellent accent plant often seen in parks and also sometimes seen lining driveways.

Jubaea chilensis.

Well suited to temperate regions, particularly those with a warm, dry climate. Drainage must be free and unimpeded. Single plants are capable of producing fertile seeds.

Propagation: Seeds are large and must be sown fairly deeply. They are very difficult to germinate, with often a very low success rate (less than 10 per cent). Seedlings appear erratically, taking six to fifteen months to germinate. Each fruit contains a single seed.

Jubaea chilensis

(from Chile)

Chilean Wine Palm, Coquito Palm, Honey Palm

CHILE

This distinctive palm is readily recognised by its stout, grey trunk crowned with feathery, spreading fronds. It is native to Chile, where it has apparently become rare because mature plants are cut down for their sugary sap which is distilled to make the delicacy known as palm honey. It is also used for the production of palm wine and sugar; the yield from each trunk being substantial. Palms are now protected in their native state and more recently have been propagated for reafforestation purposes. This palm grows well in temperate climates but is difficult in the tropics. It is very tolerant of cold and can withstand severe frosts even while small. Plants are slow growing when young but accelerate after a trunk is produced. They require a sunny position in well-drained soil and respond to nitrogen-rich fertilisers. Mature plants grow to 25 m tall with a trunk 1–2 m across. The species was previously known as *J. spectabilis*.

JUBAEOPSIS

(resembling the genus *Jubaea*)

A monotypic genus of palm endemic to South Africa where it is restricted to the lower reaches of three rivers (the Mtentu, Msikaba and Mzintlava) in Pondoland. It forms colonies on rocky bluffs on the north banks of the rivers close to the sea. The species is a clumping, unarmed feather-leaved palm with the fronds arranged in five ranks. The inflorescence arises among the leaves and bears unisexual flowers of both sexes. The fruit resemble miniature coconuts.

Cultivation: An interesting palm suitable for

Jubaeopsis caffra.

are brown when ripe and have an oily flesh which is eaten by local people. Although not highly ornamental, this species is a popular item with collectors and is often represented in botanical gardens.

KENTIOPSIS

(resembling a *Kentia*)

A monotypic genus of palm which is endemic to New Caledonia where it is restricted to acidic soils in the central south-west of the island. It is a solitary, tall, unarmed feather-leaved palm with a prominent crownshaft. The inflorescence arises below the leaves and bears unisexual flowers of both sexes.
Cultivation: Rarely grown and mainly a collector's item. Suitable for temperate and subtropical regions. A slow-growing species which needs shelter when small.
Propagation: Seed can be erratic to germinate. Each fruit contains a single seed.

Kentiopsis oliviformis

(in the shape of an olive)

NEW CALEDONIA

At first glance this palm could easily be mistaken for *Howea forsteriana* but a mature specimen can be distinguished readily by the vastly different inflorescence, which resembles a straw broom. The trunk, which can grow to 30 m tall, is prominently ringed and swollen at the base. It is surmounted by a long, pale green crownshaft and the spreading fronds have a short petiole and lustrous, dark green leaflets. The elongated fruit are red when ripe. This palm is not as easy as *Howea* to grow and requires moist, shady conditions. There are several plants growing in the Royal Botanic Gardens, Sydney, Australia.

KERRIODOXA

(after A.F.G. Kerr and the Greek *doxa*, glory)

A monotypic genus of palm endemic to western Thailand where it grows as an understorey plant in coastal forests. It is a solitary, slender, unarmed fan-leaved palm with the circular leaf blades held more or less in the one plane. The inflorescence arises among the leaves and bears unisexual

temperate and subtropical regions. Plants are slow growing and favour sandy soils where ground water is available.
Propagation: Seeds are slow and erratic to germinate and the percentage germination is often very low. Improved germination has been noted by sowing seeds on the surface of propagating mix and covering lightly with sphagnum moss. Each fruit contains a single seed.

Jubaeopsis caffra

(of Caffraria, a region in eastern Cape Province)

Pondoland Palm, Kaffir Palm

SOUTH AFRICA

This palm is distantly related to the genus *Jubaea*, hence the generic name. It is also related to the Coconut and its survival in South Africa as a relict species is remarkable. Plants have a rather untidy habit with numerous dead fronds clothing the main trunks. The live leaves have a slight twist in the rachis with the stiff, dark green leaflets spreading widely. The fruit, which are about 3 cm across,

flowers. The large fruit have small surface pustules and a spongy mesocarp.

Cultivation: A delightful palm which will surely become very popular with growers. Excellent for a warm, humid, shady position. Male and female plants are essential for fertile seed production.

Propagation: Fresh seed apparently germinates readily but may be slow. Each fruit contains one or two seeds.

Kerriodoxa elegans

(neat, elegant)

THAILAND

An attractive fan palm of modest dimensions, with a grey trunk to about 5 m tall and 20 cm across. The leaves are carried on slender, dark green, shiny petioles which have very sharp margins. The circular blades, to 2 m across, are dark green and shiny on the upper surface and chalky white beneath. The lamina is not deeply divided and the segments do not droop markedly. Female inflorescences are more robust than the male, both carrying either creamy yellow or yellow flowers. The fruit are large (to 4.5 cm x 3 cm) and unusual, being orange-yellow and covered in short pustules. This species is native to the island of Phuket off the west coast of Thailand, where it grows on sheltered slopes in dryish evergreen forest. It has been introduced into cultivation and is an excellent small palm for the tropics, requiring a sheltered location. Plants also make very decorative specimens for containers.

LACCOSPADIX

(from the Greek, *laccos*, a pit, *spadix*, a palm inflorescence, the flowers are immersed in pits)

A genus of one or possibly two species which is endemic to north-eastern Australia, growing in rainforest at high altitude. They are small to medium-sized, solitary or clumping, unarmed feather-leaved palms which lack a crownshaft. The simple, unbranched inflorescence arises among the leaves and bears unisexual flowers of both sexes. Red fruit hang in long strings.

Cultivation: Moderately popular, slow-growing palms which are highly ornamental, especially when in fruit. Single plants are capable of producing fertile seeds.

Propagation: Established suckers can be removed

The Blue Latan Palm *Latania loddigesii.*

fairly readily. Fresh seeds germinate readily three to five months from sowing. Each fruit contains a single seed.

Laccospadix australasica

(Australian)

Atherton Palm

AUSTRALIA

An Australian palm which has become moderately popular in its own country, but is uncommonly grown elsewhere. Fruiting plants are readily distinguished by long (to 1.5 m), pendulous strings of bright red fruit. These hang from the leaf-bases, with the finely divided pinnate fronds forming a graceful crown. Plants grow to about 8 m tall and are common in some highland areas of north-eastern Queensland between 800 m and 1600 m altitude. Some populations are solitary whereas others are clumping, indicating that possibly two distinct taxa are involved. An easy and rewarding palm to grow in highland tropical, subtropical and warm–temperate regions. Plants require a sheltered situation and respond to mulches, fertilisers and watering during dry periods. They are also successful as an indoor plant, tolerating some neglect and low light intensities.

LATANIA

(from a native name for one of these palms)

A genus of three species of interesting palms which are endemic to the Mascarene Islands. They were once common components of open habitats but have now been reduced to great rarity by clearing. They are robust, solitary fan palms with the leaf-bases prominently split. The leaf lamina are large and striking. The inflorescence arises among the leaves and bears unisexual flowers. All species are basically similar in appearance and can usually be distinguished by colouration in the leaves and petioles and the size and shape of the hastula.

Cultivation: Handsome, sun-loving palms which have proved to be hardy and adaptable. They must have unimpeded soil drainage but are generally not demanding. Male and female trees are essential for fertile seed production. Hybridisation is frequent in cultivated plants of this genus.

Propagation: Seeds germinate readily within two

Latania loddigesii fruit.

to four months of sowing and the seedlings are surprisingly fast growing. Each fruit contains one to three seeds.

Latania loddigesii

(after Conrad Loddiges, founder of a famous English nursery)

Blue Latan Palm

MASCARENE ISLANDS

The large, very glaucous fan leaves of this palm spread stiffly from the crown and in mature plants the petioles and basal parts of the leaves are covered with a thick, white wool which adheres strongly. The plants develop a slender trunk to more than 8 m tall and this is topped by a rounded, dense crown of the fronds, each of which may be over 4 m long. Large, shiny brown, plum-like fruit are carried on inflorescences nearly 2 m long and ripen slowly. This very attractive palm, which is native to Round Island in the Mascarenes, is deserving of much wider cultivation since it has proved to be adaptable to a range of climates. It grows well in warm–temperate areas as well as in the tropics. Plants will tolerate sun from a very early age.

Underside of leaf-lamina of *Latania loddigesii*.

Young plant of *Latania lontaroides*.

Latania lontaroides

(like the genus *Lontarus*)

Red Latan Palm

MASCARENE ISLANDS

Young plants of this palm have a striking red colouration in the leaves and often also have bright red petioles. Older plants have greyish-green leaves with streaks of red (actually red veins) in the leaves and along the petioles. The leaves also have a prominently pointed hastula which is raised above the leaf surface. In mature plants the trunk is grey and naked throughout. This is a hardy palm which needs a sunny position in well-drained soils. Plants will succeed in warm–temperate regions and can also be grown in the tropics. It is native to the Mascarene Island of Réunion where it is now reduced to a few isolated individuals. This species has also been known as *L. borbonica* and *L. commersonii*.

Latania verschaffeltii

(after Ambrose Colletto Alexandre Verschaffelt, 19th century Belgian nurseryman)

Yellow Latan Palm

MASCARENE ISLANDS

A robust fan palm native to the Mascarene Island of Rodriguez and similar in many respects to *L. loddigesii*. Its petioles and leaf-bases are also cov-

ered with a dense, white wool at least on mature trees but the leaves are green rather than glaucous. In young plants the leaf veins and petioles are an attractive bright orange colour with this colouration even extending to the leaf lamina of some plants. The leaves also have a short, blunt hastula which is quite different from that of the other species.

LEPIDORRACHIS

(from the Greek *lepis*, *lepidos*, a scale and *rachis*, the stem of a leaf or inflorescence)

A monotypic genus of palm endemic to Lord Howe Island where it grows in stunted moss forests at about 750 m altitude. The species is a solitary, small to medium-sized, unarmed feather-leaved palm with a bulging, incomplete crownshaft. The inflorescence arises below the leaves and bears unisexual flowers of both sexes.

Cultivation: A handsome palm which is rarely grown. Plants are extremely slow growing and difficult.

Propagation: Fresh seed takes six to twelve months or longer to germinate. Seedlings appear sporadically, are very slow growing and rot readily.

The Yellow Latan Palm *Latania verschaffeltii*.

Young plant of *Latania verschaffeltii*.

Lepidorrachis mooreana

(after C.E. Moore, first Superintendent of Sydney Botanic Gardens)

Little Mountain Palm

LORD HOWE ISLAND

This species is locally common on the upper slopes and summits of Mount Gower and Mount Lidgbird on Lord Howe Island, Australia, growing in low, mossy forests. Most plants grow within the forest canopy but some are exposed to the elements. Clouds frequently cover the mountain tops where this palm grows. Ornamental or interesting features include a short, moderately stout, ringed trunk, bulging leaf-bases and a compact crown of arching or slightly twisted fronds. The small, globular fruit ripen red. Unfortunately this palm has proved to be slow and difficult to maintain. It needs cool, moist conditions in a temperate climate.

LICUALA

(from the Moluccan name, *leko wala*)

A large genus of ornate palms consisting of 108 species occurring in southern China, India, South-East Asia, Malaysia, the Philippines, Indonesia, New Guinea and Australia. The centre of development appears to be in Indonesia and New Guinea. Most species are understorey palms, although some grow in extensive colonies and may dominate local areas of vegetation. They are solitary or clumping (sometimes acaulescent), unarmed or spiny fan palms with the blade entire or variously divided into wedge-shaped segments. The inflorescence arises among the leaves and usually bears bisexual flowers (occasionally unisexual on separate plants).

Cultivation: Highly ornamental palms of manageable dimensions. Some species are well entrenched in cultivation, but there are many with tremendous horticultural potential awaiting introduction to cultivation. New Guinea alone has thirty-six species, some of them dwarf to small palms. As a group these palms are shade-lovers (at least when young) and are generally tropical in their requirements. Plants are usually slow growing but respond well to regular watering and nitrogenous fertilisers. Single plants of most species are capable of producing fertile seed.

Propagation: Clumping species can be propagated from division but these are often slow to establish. Fresh seeds germinate readily but often take up to twelve months and seedlings may appear sporadically. Bottom heat may promote more rapid, even germination. Seedlings are generally slow. Each fruit contains a single seed.

Lepidorrachis moorei top of Mount Gower, Lord Howe Island.

Licuala angustiloba

(with narrow lobes)

NEW GUINEA

A small palm with a slender trunk to about 1.5 m tall and about 12–14 fronds in the crown. The spiny petioles are about 50 cm long and the semi-circular blades are divided into approximately nine narrowly linear segments (to 43 cm x 3.5 cm). These are dark green on the upper surface and paler beneath with numerous brown scales. Branched inflorescences carry white flowers followed by small, globose red fruit. This species, which is native to lowland forests along the Palmer River in the Western Province of Papua New Guinea, appears to be unknown in cultivation.

Licuala anomala

(unusual, abnormal for the genus)

NEW GUINEA

An intriguing species from Irian Jaya which has entire or partially divided leaf blades with the central segment deeply notched. The blades are fan-shaped in outline, tapered to a wedge-shaped base and the main veins on the undersurface are covered with rusty brown scales. Flowers and fruit are crowded into the apical section of a very short spike. A novelty palm which is unknown in cultivation. It grows in southern areas at about 750 m altitude.

Licuala arbuscula

(a small tree)

BORNEO

The leaf blades of this palm are divided into about twelve, more or less equal, narrow segments (to 22 cm x 1.5–4 cm) which radiate in a semi-circle from the top of the petioles. The petioles are thin to 40 cm long, spiny at the base and arise from a slender trunk (25 mm thick) 1–1.5 m tall. Sparsely branched inflorescences longer than the fronds carry glabrous, white flowers which are followed by oblong, red fruit about 15 mm long. This Bornean species is similar in general appearance to *L. pumila*.

Licuala bellatula

(neat and beautiful)

NEW GUINEA

A small palm less than 1 m tall with a very slender stem (5–7 mm thick), slender petioles about 25 cm long with reflexed spines at the base and the leaf blade divided into about six, widely divergent linear segments to 22 cm long. The central segment is wider than the rest and with a deeply notched apex. The sparsely branched inflorescence is shorter than the leaves and bears small, white flowers about 3 mm long. By all accounts this is a beautiful, small palm which awaits introduction to cultivation. It is native to Irian Jaya, growing in forests by streams.

Licuala bidentata

(with two teeth)

SARAWAK

A palmlet which has a subterranean or very shortly emergent trunk and long, slender petioles with the leaf blades divided into numerous (about twenty-four) very narrow segments. These spread widely from the petiole, each having two tooth-like lobes at the apex. Long inflorescences have short branches and shortly hairy, acute bracts. An attractive

Licuala borneensis in habitat. G. Cuffe

species which is grown on a limited scale mainly by enthusiasts. It is native to Sarawak, growing in humid forests.

Licuala bintulensis

(from the area of Bintulu, Sarawak)

SARAWAK

The leaf segments of this species number about nine, with the outer ones being narrow and the inner ones broadly wedge-shaped. The palm is a dwarf species native to Sarawak where it grows in lowland forests. The inflorescence is sparsely branched and held erect among the leaves. The species is an attractive, small palm suitable for container culture or in a shady garden. Plants will grow in tropical and subtropical regions. The species has been introduced into cultivation and at this stage is mainly to be found in the collections of enthusiasts.

Licuala borneensis

(from Borneo)

BORNEO

A dwarf, bushy species from Borneo where it grows in humus-rich soil in lowland forests near streams. The stem is very short (almost superficial) and the fronds appear to erupt out of the ground in a tuft. The slender (4 mm across) petioles (to 80 cm long) have small spines towards the base and the leaf blades are entire and broadly paddle-shaped, or divided into about five, unequal, narrowly wedge-shaped, dark green, segments. These spread widely and the central segment is broader (to 33 cm x 8 cm) than those on the outside (to 33 cm x 5 cm). The outer margin of all the segments is shallowly toothed. The slender inflorescence (about 60 cm long) is sparsely branched into three spikes towards the apex and bears white flowers about 4 mm long followed by ellipsoid red fruit about 8 mm long. A desirable ornamental which is mainly grown by enthusiasts.

Licuala brevicalyx

(with a short calyx)

NEW GUINEA

A robust species growing to about 7 m tall with a crown of large, semi-circular, bright green fronds. The petioles have prominent, spreading spines at the base (to 5 mm long) and the blades are divided into about twenty wedge-shaped segments. The central segments are larger (to 78 cm x 17 cm) than the outer segments (to 50 cm x 8 cm), with the outer margins of all segments being truncate and bluntly toothed. Large, branched panicles carry small, white flowers (about 3 mm long) followed by clusters of small (7–8 mm across), red globose fruit. An attractive palm from Irian Jaya which awaits introduction to cultivation.

Licuala cabalionii

(after Pierre Cabalion, 20th century French ethnopharmacologist)

VANUATU

A recently described (1993) species which occurs in Vanuatu on the islands of Vanua Lava and Malekula. It grows in volcanic soils in coastal forests and rainforests at low altitudes, often in dense stands with the ground covered in seedlings. It is a slender solitary palm with a trunk to about 5 m x 8 cm and an open crown of about twelve fronds. These have long, thin petioles (to 3 m x 1 cm), spiny near the base and the blade is divided into about twelve spreading or drooping, narrowly wedge-shaped segments which have the outer margins sharply toothed. Pendulous inflorescences to 1.8 m long carry flowers about 12 mm long, followed by globose yellow to orange fruit about 1 cm across. This species would appear to be an attractive palm for tropical and subtropical regions.

Licuala calciphila

(lime-loving)

VIETNAM

An interesting palm which grows on outcrops of limestone near streams in Vietnam. It is a clumping palm with slender stems to about 60 cm tall and a crown of small, segmented fronds. The petioles are very thin (to 30 cm x 2 mm) with spreading spines at the base. The blades are divided into three to five segments (about 16 cm long) with the central segment broadly wedge-shaped and the outer segments linear to narrowly wedge-shaped. Slender, arching to nodding inflorescences shorter than the petioles have two or three scattered branches. This species would be an excellent addition to tropical gardens and should be tolerant of calcareous soils.

Licuala cordata. G. Cuffe

Licuala concinna

(neat, trim)

NEW GUINEA

A small palm growing to about 2 m tall with numerous (20–25) bluish-green, erect fronds in the crown. The petioles are about 1 m long, spiny throughout and purplish-brown on the underside. The semi-circular leaf blades are divided into four to six oblanceolate segments (to 25 cm x 6 cm), with the central segments broader (to 12 cm) and deeply notched. Slender inflorescences about 65 cm long bear white flowers about 4 mm long followed by globose, red fruit about 1 cm across. The species is native to lowland forests along the Palmer River in the Western Province of Papua New Guinea. It appears to be unknown in cultivation.

Licuala cordata

(heart-shaped)

SARAWAK

One of the most delightful of all palms for cultivation, this dwarf species is much sought-after by enthusiasts and commands high prices. Plants grow to about 70 cm tall and present a neat, compact appearance. The stem is short and slender, the petioles (to about 60 cm long) are slender with spiny margins and the blades are undivided and an attractive bright green. Each is orbicular to fan-shaped, prominently cordate at the base and with attractively wavy and lobed margins. On well-grown plants the fronds overlap and present a neat

appearance. The species is very similar to *L. orbicularis* but with a cordate base to the leaf. It is native to Sarawak.

Licuala corneri

(after E.J.H. Corner, 20th century English botanist)

MALAYSIA

A Malaysian species which grows in shady forests at about 150 m altitude. It is a solitary palm, 2–4 m tall, with a stem to 2 m tall and short, spiny petioles to about 1 m long. The blades are divided into twelve to fourteen, broadly wedge-shaped segments, to 35 cm x 4.5 cm, with the outer margins toothed. These are dark green above and paler beneath. Sparsely branched inflorescences about 1 m long bear greenish-white flowers about 8 mm long which are followed by orange-red, ellipsoid fruit about 8 mm long. This species is a collector's item which appears to be rarely cultivated, if at all.

Licuala debilis

(weak, frail, small)

NEW GUINEA

A little-known species which was named from material collected on Mount Resi in Irian Jaya, at about 600 m altitude. It is a small palm growing to about 1.5 m tall with a slender trunk about 2.5 m thick. Slender petioles about 1 m long have a few small spines at the base and the blade is divided into about eight segments, with the central segments much broader (13–17 cm) than the rest (2–6 cm). The sparsely branched, erect inflorescence bears white flowers about 3 mm long. Details of the fruit are lacking.

Licuala densiflora

(densely flowered)

SARAWAK

Seed of this species has been available for a few years, but plants are mainly to be found in the collections of enthusiasts. It is a small palm with slender petioles spiny at the base and the blades divided into about seven wedge-shaped segments to 24 cm x 4 cm. The inflorescences are sparsely branched and the flowers are crowded into a section near the apex of each branch. The species is native to Sarawak.

Licuala dransfieldii

(after John Dransfield, contemporary English botanist specialising in palms)

MALAYSIA

A litter-collecting palm which has a crown of about fifteen, dark glossy green leaves and yellowish-green leaf sheaths. Plants develop a slender trunk to about 3 m tall, with the younger plants usually solitary and the older plants producing basal suckers. The petioles have recurved, black-tipped spines towards the base and the blades are divided into 7–19 narrowly wedge-shaped segments. These are all more or less equal in size, although sometimes the median segment is slightly larger and may be stalked at the base. Short, sparsely branched inflorescences with short, thick rachillae are hidden among the leaves and carry brownish, scattered flowers followed by globose, orange fruit about 12 mm across. This attractive species is endemic to the Ulu Endau area of eastern Johore, Malaysia where it is a common, lowland palm growing in sheltered forests to about 150 m altitude. It has excellent potential as a small ornamental palm for the tropics.

Licuala ferruginea

(rusty brown)

MALAYSIA

Fruiting plants of this palm are very noticeable with their clusters of brilliant red fruit. The large floral bracts and branches of the inflorescence are densely covered with dark brown scales and hairs, a character which gives rise to the specific epithet. The species is a small solitary palm with long, spiny petioles (to 1.3 m long) and leaf blades about 1 m across. Emerging leaves are covered with long, brown hairs and scales. The blades are divided into three to seven, broadly wedge-shaped segments. An attractive Malaysian palm which deserves to become more widely grown.

Licuala ferruginoides

(resembling Licuala ferruginea)

SUMATRA

A little-known species from sheltered, humid forests in eastern Sumatra. It is a tall, robust palm with unarmed petioles and large leaf blades (1–1.5 m across), which are divided into numerous, wedge-shaped segments about 6.5 cm across. Sparsely branched inflorescences about 1 m long have a zig-zagged rachis and are densely covered with brown hairs. White flowers about 7 mm long are followed by sparsely hairy red fruit about 1 cm across.

Licuala flavida

(yellowish)

NEW GUINEA

This species has slender petioles about 15 cm long and the more or less circular leaf blade is irregularly divided, with the two outer lobes being very narrowly linear and the rest more or less wedge-shaped (to 21 cm x 4 cm). The tips of the segments are irregularly divided into narrow lobes 5–10 mm long. Dull yellow flowers are carried on slender inflorescences about 30 cm long. This interesting species is native to Irian Jaya, New Guinea, growing in rainforest at about 800 m altitude.

Licuala furcata

(forked)

BORNEO

A small species from mountainous forests in Borneo which is similar in some respects to L. flabellum but with the central leaf segment being deeply notched. Plants have short or subterranean stems and small fronds (all just 50–70 cm long). The leaf blades are divided into three to seven segments, 35–37 cm long, with the central segment widest (4.5–6.5 cm across). Sparsely branched inflorescences much longer than the fronds bear white flowers about 3 mm long followed by top-shaped fruit. A choice collector's item which appears to be rarely, if at all, cultivated.

Licuala glabra var. glabra

(hairless)

MALAYSIA

This fan palm may be stemless or can develop a trunk that can reach 2 m tall. It has a crown of spreading leaves that have slender petioles up to 3 m long, each topped with a dark green, orbicular blade which may be 60 cm across. The blades are divided into 12–18 narrowly triangular, overlapping segments with the central segment divided,

Licuala glabra in habitat. G. Cuffe

somtimes being split to the base. Long, slender inflorescences (sometimes reaching more than 3 m) arch above the foliage and after flowering carry small, dull orange, ellipsoid fruit. This palm is very decorative and is common in the forests of Malaysia up to about 1300 m altitude. In cultivation it needs a shady position in the tropics or subtropics.

Licuala glabra var. selangorensis

(from Selangore, Malaysia)

MALAYSIA

This variant, from Peninsular Malaysia, has smaller leaf blades than var. *glabra* (about 30 cm across), divided into five to eight broader segments, with arching margins and the main central segment

Licuala glabra var. *selangorensis*. G. Cuffe

divided about halfway to the base. It has similar cultural requirements to var. *glabra*.

Licuala gracilis

(graceful)

JAVA

This clumping palm resembles a miniature L. *rumphii*, forming as it does, a dense, rounded clump to about 1.5 m tall. The stems are fibrous and slender (3 cm thick) and the thin petioles (to 70 cm long) have short marginal spines in the basal half. The leaf-blades are divided into seven to ten, radiating, widely separated segments, with the apical segment being neatly split in two (an uncommon feature in the genus). The segments are broadest at the apex (where truncate and toothed), taper to the base and the upper surface is strongly pleated. Short, sparsely branched inflorescences carry small, white, unisexual flowers and on female plants these are followed by red, spherical fruit about 6 mm across. This West Javan palm (erroneously recorded from Sulawesi) performs well in tropical regions in a partially sheltered position. Seed of this species may have been distributed as L. *flabellum*, which is a synonym.

Licuala grandiflora

(with large flowers)

NEW GUINEA

A species of palm from lowland areas of Irian Jaya which can be recognised by its large, white flowers (5 mm long) densely crowded in thick spikes. The leaves have slender petioles about 80 cm long and the blade is divided into about twenty segments, each to 60 cm x 3 cm, with the blunt tips having two prominent teeth.

Licuala grandis

(large, big)

Vanuatu Fan Palm, Palas Payung

SOLOMON ISLANDS—VANUATU

Some palms are prized horticultural subjects and this species is one of the most sought-after. It originates in Vanuatu where it is widely distributed on many of the islands and is also found in the Solomon Islands (San Cristobal Island and the Santa Cruz Group). It commonly grows as scattered

Licuala grandis (*Botanical Magazine,* Vol. XXXIX of the 3rd series, plate 6704)

individuals in disturbed lowland rainforest, especially in moist soils. In some areas, the species forms extensive colonies. Established plants may tolerate considerable exposure to sunshine. This palm has become popular because of its wonderfully symmetrical, glossy dark green leaves and neat habit of growth. The circular, pleated leaves are up to 1 m across and have notched margins. Young leaves are held erect to obliquely erect whereas older leaves become nearly pendulous. Long spikes of red fruit are a further decorative feature. The species is essentially tropical in its requirements and is rather sensitive to cold, although it can be grown in a warm, protected position in the subtropics. The plants are generally slow growing and make excellent subjects for tub culture. They require shaded conditions and

Licuala grandis in fruit. G. Cuffe

protection from strong winds. A choice variant having variegated leaves is grown in Singapore.

Licuala kemamanensis

(from Kemaman on the Malay Peninsula)

MALAYSIA

A dwarf acaulescent palm from a restricted area in the shady rainforests of Malaysia. Plants have slender petioles to 65 cm long and the leaves are divided into about fourteen narrowly wedge-shaped segments. The outer segments spread widely and are mainly free from the others. The inflorescence, to 60 cm long, is relatively simple, having a couple of thickened branches towards the apex. This species shares some similarities with *L. ferruginoides* but is dwarf, with smaller leaves and the inflorescence has a straight axis.

Licuala kiahii

(after Kiah bin Hadji Modamed Salleh, original collector)

MALAYSIA

A small species with a reduced, subterranean trunk, the leaves appearing to emerge directly from the ground. The thin petioles, to 70 cm long, are sparsely spiny and reddish at the base. The leaf blades are divided into five to seven unequal segments, with the outer two segments smallest and spreading widely away from the others. The

central segment, the largest of all, is broadly wedge-shaped and flanked closely by the intermediate segments. The outer margins of all the segments are bluntly toothed. The erect inflorescence, to 25 cm long, is sparsely branched and has brown scales. The fruit are an unusual shape, being cylindrical to club-shaped and curved (to 25 mm x 4.5 mm). An unusual palm from shady forests in Malaysia which has excellent ornamental qualities.

Licuala klossii

(after C.B. Kloss, 19th century English zoologist)

NEW GUINEA

A compact species with slender, prickly petioles about 50 cm long and nearly round blades which are divided into about twelve narrow segments, shortly toothed on the apex. Slender, arching inflorescences about 40 cm long bear small, white flowers followed by globose, reddish fruit about 5 mm across. This species is native to Irian Jaya, growing at about 200 m altitude.

Licuala kunstleri

(after H.H. Kunstler, 19th century German collector)

MALAYSIA

A small palm which may be nearly trunkless or has a trunk to about 0.5 m tall and about 6 cm thick. The slender petioles, to about 1 m long, have prominent stout thorns on the margin. The large, impressive leaf blades are approximately 1 m across and are divided into about fifteen spreading lobes, with the central lobe much larger than the rest. The outer lobes are about 3 cm wide and have prominent apical teeth to 5 cm long. Short inflorescences (to around 30 cm long) have silvery scales on the bracts and coarse brown hairs on the rachis. An impressive palm which is native to Malaysia where it grows in shaded, humid forests.

Licuala lanata

(woolly or cottony)

BORNEO

A clumping, apparently stemless species which forms a tussock, with each stem supporting a crown of up to ten leaves. Each leaf is about 1.5 m long with the basal parts of the petioles bearing brown spines about 4 mm long. The blade is divided into five to nine (usually five) large, leathery, wedge-shaped segments which are dark shiny green. The central segment is the largest (to 50 cm x 25 cm). All parts of the sparsely branched inflorescence (to 50 cm long) are densely woolly and the densely crowded flowers bear shaggy hairs. This highly desirable ornamental palm is not known to be in cultivation. It is native to Borneo where it grows on forests on alluvial flats near limestone.

Licuala lanuginosa

(woolly)

MALAYSIA

A small palm which grows to about 2 m tall and has a crown of about twelve leaves. These have slender petioles which are spiny towards the base and the blades are divided into about fifteen narrowly wedge-shaped segments. These are dark green and somewhat shiny and the outer margin of each is coarsely toothed. The sparsely branched inflorescences are shorter than the leaves and the short, thick rachillae are covered densely with cream, woolly hairs. The widely scattered flowers bear silky hairs and the globose, orange fruit are about 1.5 cm across. This is an attractive small palm which occurs in the shady forests of central and western Johore, Malaysia.

Licuala lauterbachii

(after C.A.G. Lauterbach, German plant collector)

NEW GUINEA

A small to medium-sized fan palm from New Guinea which carries its leaves on very slender petioles nearly 1.5 m long. The blades themselves are about 1 m across, circular, bright green and are divided to the base into thirty to thirty-five narrowly triangular segments which have pointed tips. These are not symmetrically arranged, often appearing as if grouped or clustered. Bright red, globose fruit 12–13 mm across are carried on erect inflorescences about 1 m long. This attractive fan palm is widespread throughout New Guinea, growing in shady forests. A distinct variety (var. *bougainvillensis*) occurs in the Solomons. Tropical conditions are needed for the culture of both.

Licuala lauterbachii.

Licuala leprosa

(scaly, scurfy)

NEW GUINEA

A small palm with a stem to about 1.5 m tall and very short petioles which are spiny towards the base. The leaf blades are divided into about 12–14, more or less equal, broadly linear segments (to 40 cm x 4 cm). These are dark green on the upper surface, with conspicuous brown scales beneath. The inflorescences are large and much-branched and after flowering bear spherical, reddish fruit about 1 cm across. This species is native to lowland areas of north-eastern New Guinea, along the Fly River.

Licuala linearis

(linear)

NEW GUINEA

A small palm with a slender trunk less than 1 m tall and numerous spreading to drooping semi-circular fronds in a large crown. These are dark green on the upper surface and minutely scaly beneath. Thin petioles about 60 cm long are finely spiny towards the base. The blades are divided into approximately eleven segments, with most being very narrow (7–23 mm wide) but the central segment is much larger (to 38 cm x 13 cm) and wedge-shaped. Globose, red fruit about 9 mm across are carried on branched inflorescences about 1 m long. A choice collector's item which originates in central New Guinea growing in forests at about 500 m altitude.

Licuala longicalycata

(with a long calyx)

MALAYSIA

A moderately robust palm with a trunk to about 3 m tall and an attractive crown of large fronds. Each has a slender petiole to 3 m long and a large blade more than 1 m across which is divided into about sixteen more or less equal segments. These are more or less wedge-shaped with the central segments slightly larger than the rest (to 55 cm x 16 cm); especially the outer segments which are quite narrow. The inflorescences are shorter than the leaves (to 80 cm long) and have four or five major branches. White flowers about 8 mm long have a prominent calyx and are followed by oblong fruit, about 14 mm x 10 mm, which are wrinkled when dry. An interesting species from shady humid forests in Malaysia.

Licuala longipes

(with long stalks)

BURMA—MALAYSIA

Because this palm is often trunkless, its long leaves seem to erupt from the ground in a very spectacular fashion. The slender petioles may be more than 3.5 m long and are topped with a circular blade which may be nearly 2 m across. This is dark green and divided into about twenty diverging wedge-shaped segments, the basal ones of which are narrow, well separated and with three or four apical lobes. A much-branched inflorescence to more than 1 m tall is carried among the leaf-bases. Green, hairy flowers are followed by top-shaped, red fruit about 1 cm long. This startling palm is native to Burma and Malaysia, growing in shady places on the rainforest floor. It is grown on a limited scale but is very tropical in its requirements, although provenances originating from high altitudes in Malaysia (1000 m) may be more adaptable.

Licuala magna

(large)

NEW GUINEA

A beautiful palm which grows to about 8 m tall, having a narrow trunk about 9 cm across and a large crown of fronds. Slender petioles about 80 cm long are spiny at the base and the huge laminas

(nearly 1.5 m across) are semi-circular in outline with radiating segments. The large, much-branched inflorescence bears small, white flowers followed by globose, red fruit about 1 cm across. This species grows in inaccessible country along the Palmer River in the Western Province of Papua New Guinea. It is unknown in cultivation but from all accounts would be a worthy introduction. It would probably succeed best in tropical regions.

Licuala mattanensis

(from Mount Mattang, Borneo)

BORNEO

The fruit of this species are very unusual in the genus, being much longer than wide (about 20 mm x 4 mm), more or less cylindrical in shape, irregularly curved and yellowish-brown in colour. They radiate out conspicuously from the rachillae and are supported on short, fleshy stalks, sometimes in groups of two or three. Plants have a slender (2 cm thick) stem to about 1 m tall which bears an apical tuft of fronds. These have slender, unarmed petioles and the blades are divided into numerous (19–23) broadly linear segments which have four apical teeth. A real collector's item which is native to Borneo. Some highly desirable forms have attractively mottled leaflets.

Licuala micrantha

(with small flowers)

NEW GUINEA

This palm is from the Torricelli Ranges in Papua New Guinea where it grows in rainforest at about 1000 m altitude. Its fronds are divided into about thirteen, narrowly wedge-shaped, dark green segments and the petioles are armed with recurved spines. The sparsely branched inflorescences bear fleshy flowers about 3 mm long which are followed by red, obovoid fruit. This palm should grow well in a sheltered position in tropical and subtropical regions.

Licuala mirabilis

(wonderful, remarkable)

MALAYSIA

The central segments of the leaves of this palm are mostly united with the tips free, but the outer two

Licuala mattanensis. G. Cuffe

or three pairs of segments are completely free or nearly so and spread from the rest. All segments (there are 20–27 on each leaf) are narrow (to 7 cm wide) and have two to six short apical teeth. Inflorescences (to 45 cm long) are borne among the leaves and have three or four main branches. White, hairy flowers are followed by globose, pink to red fruit, about 12 mm x 8 mm. Although rarely seen in cultivation, this interesting Malaysian palm offers good prospects for tropical gardens.

Licuala modesta

(modest, unpretentious)

MALAYSIA

A small to medium-sized palm with a slender stem to 3 m tall and a crown of attractively divided leaves. The slender petioles (about 1 m long) have long spines towards the base and are unarmed near the apex. The blades are divided into numerous narrowly wedge-shaped segments (10–16) with the middle segments to 37 cm long and 11 cm wide at the apex. Pendulous, sparsely branched panicles, about 50 cm long, bear hairy, white flowers about 7 mm long which are followed by obovate, red fruit about 1.2 cm long. This species is native to Malaysia where it grows in highland rainforests at about 1400 m altitude. *L. wrayi* is a synonym.

Licuala moskowskiana

(after Dr Moskowski, original collector)

NEW GUINEA

This species is native to northern New Guinea where it grows in humid forests at low altitudes near streams. It is a robust palm growing to 19 m tall with a large crown of impressive fronds. These have a stout petiole, unarmed in the upper part and the blade is divided into about sixteen, wedge-shaped segments (to 90 cm x 18 mm), with the outer margin of each sharply toothed. The long, branched inflorescences bear rigid, leathery spathes and white flowers about 1 cm long. Details of the fruit are lacking.

Licuala moyseyi

(after L. Moysey, collector in Malaysia)

MALAYSIA

A solitary small palm with a short, slender stem and slender, unarmed petioles to 90 cm x 4 mm. Each has a bright green blade about 50 cm across which is divided into eight to ten segments, with the central four or five segments united into a broad fan. All segments are bluntly toothed on the outer margin. Each slender inflorescence, to 75 cm long, has two or three simple spikes towards the apex. The flowers are borne in clusters (up to three in a cluster) along the rachis and are followed by red, ellipsoid fruit. An attractive small-leafed species from Malaysia (rare and restricted to one area) which does not appear to be cultivated.

Licuala naumannii

(after Dr F. Naumann)

BOUGAINVILLE

This species grows in dense, lowland forests of Bougainville Island and may be prominent in coastal localities often near mangroves. It is a robust species growing 5–10 m tall, with a trunk about 2.5 cm thick and fibrous leaf sheaths. The blades are divided into about ten radiating segments (to 37 cm x 2.5 cm) which are broadly wedge-shaped, with shortly toothed outer margins. The central segment is larger than the rest (to 48 cm x 22 cm). All segments are glabrous on the upper surface and with brown scales on the underside. The inflorescences are much branched and

Licuala orbicularis. G. Cuffe

after flowering carry rounded, red fruit. This species would probably be very tropical in its requirements.

Licuala naumoniensis

(from Naumoni, northern New Guinea)

NEW GUINEA

A small palm with a slender stem and a graceful crown of fronds. Each has a slender petiole (to 70 cm x 5 mm) which is spiny at the base and a more or less circular blade divided into about twenty-three, narrowly linear segments (to 50 cm x 20 mm) with the outer margins sharply toothed. Branched inflorescences carry hairy, white flowers about 9 mm long followed by obovoid, red fruit. The species is native to northern New Guinea where it grows in lowland forests near streams.

Licuala olivifera

(olive-bearing)

BORNEO

The stem of this species is zigzagged or slightly twisted and grows to 80 cm tall. Slender petioles to 1.5 m tall are spiny towards the base and the blade is divided into about eleven, narrowly wedge-

shaped segments. The central segment is larger (80 cm x 17 cm) than the lateral segments (50 cm x 10 cm) and the outer margins of all are truncate and shortly toothed. Large, much-branched inflorescences carry white, hairy flowers about 4 mm long which are followed by olive-like, oblong to ellipsoid, red fruit, about 20 mm x 10 mm. A highly ornamental species from Borneo which grows naturally in swampy habitats.

Licuala orbicularis

(orbicular, rounded)

SARAWAK

A much sought-after dwarf palm which makes an excellent subject for container cultivation or for a warm, sheltered position in the garden. Plants have a subterranean or shortly emergent slender stem and entire, shiny green leaves. The petioles are very slender with spiny margins and the blades rounded to fan-shaped with attractive wavy and toothed margins. The inflorescence is sparsely branched and pendulous. Growth habit is compact and well-grown plants have an excellent appearance. The species is native to Sarawak where it grows in shady forests. The species is very similar to *L. cordata* but with the fronds tapered to a wedge-shaped base.

Licuala paludosa.

Licuala pahangensis

(from Pahang in Malaysia)

MALAYSIA

A solitary species with stems to 2 m tall and about
5 cm across. The slender petioles (to 1 m long)
have thorns in the basal third and the dark green
leaf blades are about 75 cm across. They are divided into twenty to twenty-six narrowly wedge-
shaped segments (to 5 cm wide), with the central
segment broader than the rest. The inflorescences
have two simple, thick spikes towards the apex,
with the white flowers (about 5 mm long) being
followed by globose, red fruit. An attractive palm,
from the lowland forests of Peninsular Malaysia,
which has excellent prospects for cultivation.

Licuala paludosa

(growing in swamps)

Swamp Fan Palm

SOUTH-EAST ASIA

This species is well named as it commonly grows in
peaty soils in swamp forests. It may be found in
lowland areas close to the coast but also extends
into the mountains. Specimens encountered in cultivation are mainly solitary but clumping forms
may occur. Plants develop a trunk to 4 m or more
tall and 20 cm across and have an open crown of
almost circular leaves about 1 m across. These are
borne on spiny petioles (black, curved spines) and
the blades are divided into six to ten, pleated,
wedge-shaped segments which are toothed on the
ends. Stout, branched inflorescences about 1 m
long bear white flowers about 6 mm long followed
by spherical, red fruit about 7 mm across. Although
grown in many countries, this palm has not
become popular and is uncommonly planted. It can
be grown in tropical and subtropical regions and
likes an open, sunny position with plenty of water.
The species is native to Indonesia (Borneo,
Sumatra), Malaysia, Thailand and Vietnam.

Licuala parviflora

(with small flowers)

NEW GUINEA

A palm from Papua New Guinea which occurs in
inaccessible areas of the north-east (at about 650 m
altitude) and which may not be in cultivation.

Plants grow to about 3 m tall with a short, thick
trunk, long slender petioles which are spiny at the
base and a crown of shiny green leaves. The leaf
blades divided into numerous spreading segments
(to 40 cm x 15 mm) with the middle segments
slightly wider (to 25 mm) than the rest. Sparsely
branched panicles at the end of a long peduncle
carry white flowers about 4 mm long, which are
followed by globose, red fruit about 9 mm across.

Licuala paucisecta

(with few segments)

NEW GUINEA

A small palm with a slender (1.5 cm thick) stem to
about 1 m tall and a crown of sparsely segmented
fronds. The leaf sheath and the petiole are densely
scaly. The petioles are very thin and have small
spines at the base and the semicircular lamina has
about five segments which have minute, brown
scales on the underside. The central segment is
longer and broader (to 32 cm x 4.3 cm) than the
outer ones (to 26 cm x 2.2 cm). Pendulous,
branched inflorescences, about 60 cm long, after
flowering carry red, globose fruit about 7 mm
across on stalks about 3 mm long. A collector's
item from lowland forests along the Fly River in
Papua New Guinea.

Licuala peltata

(with peltate leaves)

BURMA—INDIA—MALAYSIA

This very tropical palm forms impressive clumps
up to 5 m tall and spreading a similar width (sometimes plants have a solitary trunk). Its leaves are
orbicular, nearly 1.5 m across and are divided into
numerous (twenty to thirty), narrow, dark green,
widely spaced segments. These are broadest near
the apex, taper to the base, are deeply pleated and
cut squarely across at the ends where they are also
toothed, the whole arrangement imparting a splendid symmetrical impression. The blade is carried
on the end of a petiole over 2 m long, which is
triangular in cross-section, with very stout, sharp
marginal spines. The inflorescences arch well above
the leaves and carry greenish-white flowers about
12 mm long followed by bunches of orange, elliptical fruit each about 12 cm long. The species is
native to northern Malaysia, northern India and

Licuala peltata divided leaf form.

southern Burma, growing in hot, humid valleys
and ravines. The leaves are used locally to make
hats. Plants often flower when quite small. The
palm commonly grown as *L. elegans* is a form of
L. peltata with entire leaves.

A variant having magnificent, rounded
undivided, dark green leaves is moderately popular in cultivation. It occurs naturally in Thailand
and northern Malaysia and grows very well in a
shady position in the tropics. So striking is the
contrast between these variants (divided and undivided leaves) that it is difficult to believe they
are grouped together in the one species.

Licuala petiolulata

(with petiolules)

SARAWAK

Seed of this species has been listed in catalogues
for a few years, but plants are mainly to be found
in the collections of enthusiasts. It is an attractive,
small palm from humid forests of Sarawak. Plants
have a short trunk, slender petioles about 1.5 m
long and dark green blades divided into about
twenty narrow segments. The central segments,
being borne on prominent stalks (petiolules),
provide a useful means of identifying the species.
The inflorescences are long and much-branched.
Pink to reddish, globose fruit are about the size
of a cherry.

Licuala platydactyla

(with broad fingers or segments)

NEW GUINEA

A medium-sized palm growing to about 4 m tall
with sparsely divided leaves. These are broadly
wedge-shaped in outline (about 50 cm long) and
are each divided into three large segments which
are dark green on the upper surface and densely
brown-scaly beneath. The outer margins are
incurved and have numerous small teeth and the
central segment is deeply notched. White flowers,
about 4 mm long, are borne on a sparsely branched
inflorescence and are followed by spherical, red
fruit about 1 cm across. An interesting palm which
is native to the lowland areas of north-eastern New
Guinea, along the Fly River.

Licuala pulchella

(beautiful)

NEW GUINEA

A highly desirable small palm from northern parts
of Irian Jaya, growing in lowland forests adjacent
to mangroves. Plants have a short, slender stem
and thin (5 mm thick) spineless or sparsely spiny
petioles. The leaf blades are semi-circular in outline
(to 44 cm long) and divided into about nine,
narrowly linear segments (2.5–3 mm wide), which
are minutely scaly on the underside. Pendulous
inflorescences about 1.2 m long carry small, white,
stalked flowers which are followed by globose, red
fruit about 12 mm across. A choice palm which
awaits introduction to cultivation.

Licuala peltata undivided leaf form.

Licuala pumila

(dwarf, low)

INDONESIA

A highly desirable palm from the Indonesian slands of Sumatra and Java. Plants are stemless, or with very short, slender, erect stems with the petioles (about 1 m long) having recurved, brownish spines towards the base. The nearly orbicular leaf blades (about 45 cm across) are divided into ten to twenty, unequal, stiff, somewhat papery segments which radiate out like the spokes of a wheel. These are dark green and pleated on the upper face and bluish-green beneath. The outer margins of each segment have four to six shallow, blunt teeth. Inflorescences are rarely more than 40 cm long and bear whitish flowers followed by ellipsoid, pink to red, shiny fruit. An elegant, neat palm which is ideally suited to a shady tropical garden. *L. elegans* is a synonym. A form of *L. peltata* with entire leaves is sometimes wrongly cultivated as *L. elegans*.

Licuala pumila.

Licuala pygmaea

(dwarf, low)

BORNEO

The fronds of this palmlet erupt from the ground, as its stem is much reduced (about 2 cm across) and subterranean. The petioles are very slender (2.5 mm across), to 35 cm long, with a few recurved spines at the base. The blades are divided into three to five, stiff, papery segments, with the central segments wider (to 20 cm x 8 cm) than those on the outside (to 20 cm x 5 cm). The outer margins of all segments are divided into numerous blunt teeth. Erect, sparsely branched inflorescences after flowering carry small, ellipsoid fruit about 7 mm long. This little gem originates in Borneo where it grows in wet, humus-rich soil near streams in dense forests.

Licuala ramsayi

(after P. Ramsay, original collector)

Australian Fan Palm

AUSTRALIA

A spectacular palm which is native to north-eastern Queensland where it grows in swampy areas and along stream banks in the lowlands. Plants often grow in extensive colonies and the large, rounded fronds provide an impressive silhouette against the sky. They also creak and groan as they move in the slightest breeze. Plants have a solitary trunk to about 15 m tall and a crown of large leaves with the bright green blades about 2 m across. These appear to be circular but in fact consist of many closely spaced, narrowly wedge-shaped segments which radiate out from the apex of the petiole. Long, pendulous panicles bear attractive white flowers which are followed by clusters of brilliant orange-red fruit. Plants grow steadily in cultivation and are adaptable to various climates, from the tropics to warm–temperate regions. They require shelter, especially when young and should not be allowed to dry out. Young plants have been successfully used for indoor decoration.

Licuala reptans

(creeping)

BORNEO

A very unusual species which has a prostrate, creeping stem with a tuft of fronds at the apex. These have very long, slender, minutely spiny petioles and the orbicular blades are divided more or less into two halves, each with three or four radiating, narrowly wedge-shaped segments, the central two being united at the base. These segments are bright green on the upper surface and with numerous rusty-brown scales beneath.

253

Licuala rumphii.

Slender, whip-like inflorescences about as long as the leaves bear white flowers about 8 mm long followed by red, pear-shaped fruit. A collector's item from shady forests in Borneo.

Licuala robinsoniana

(after C.B. Robinson)

VIETNAM

A species from the humid forests of Vietnam which is most likely tropical in its requirements. Plants grow to about 5 m tall and have sparse, spiny petioles about 1 m long and the blades divided into numerous, more or less equal, linear segments (to 40 cm x 3 cm). The segment margins extend apically into small tails. Much-branched, paniculate inflorescences bear white flowers about 4 mm long which are followed by reddish, top-shaped fruit.

Licuala rumphii

(after G.E. Rumpf (Rumphius), 17th & 18th century German collector in Indonesia)

Celebes Fan Palm

MOLUCCAS—SULAWESI

This species is a popular plant for the tropics and is now widely planted in many countries. It is a suckering palm which forms a dense clump consisting of stems of different ages, a clump being leafy right to ground level. The stems are slender, grow to about 3 m tall and are covered in brown to black fibres. The petioles have spines towards the base and the semi-circular, dark green fronds have six to ten widely divergent segments to 50 cm long. These are broadly wedge-shaped, distinctly pleated and the outer margins are prominently toothed. Inflorescences as long as or longer than the leaves, carry small, white flowers followed by ellipsoid, red fruit about 1.5 cm long. Plants will grow in situations from shade to full sun in the tropics and subtropics. An easily grown and adaptable palm, excellent for planting in home gardens and parks. It is native to Sulawesi and the Moluccas.

Licuala sarawakensis

(from Sarawak)

SARAWAK

An attractive, small palm which is becoming popular in cultivation. It is extremely ornamental when well grown in a container and makes an excellent indoor plant. It also grows well in a shady location outdoors in the tropics and subtropics and is excellent for mingling with ferns and other shade-lovers. Plants grow about 1 m tall with a very short stem and long, slender petioles. The leaf blades are divided into six or seven, broadly wedge-shaped,

Licuala spinosa.

dark green segments, with the middle segment broadest. The inflorescences are pendulous and sparsely branched. The species is native to Sarawak where it may be locally common.

Licuala spathellifera

(bearing small spathes)

SARAWAK

A small palm growing about 2 m tall, with slender petioles and finely divided, dark green blades. Each has about twenty-five narrow segments presenting a crowded appearance. The inflorescence is erect, with numerous small, shortly hairy bracts. This is a handsome species from shady forests in Sarawak. It has great potential for cultivation in tropical and subtropical regions.

Licuala spicata

(in spikes)

BORNEO

This species has simple, unbranched spikes, with the flowers clustered in a short area at the apex. Its fruit are also obovate rather than rounded as are most members of the genus. Plants have a slender (2 cm thick) stem about 1 m tall which bears an apical tuft of fronds. These have slender, unarmed petioles to 50 cm long and the blades are divided into numerous (14–17) narrow segments which taper to the apex where there are three or four blunt teeth. This ornamental species is native to Borneo.

Licuala spinosa

(bearing spines)

Mangrove Fan Palm

SOUTH-EAST ASIA

In nature this species occurs in wet, sandy soils in coastal and near-coastal districts, often on the landward margin of mangrove communities. It is a vigorous palm which frequently grows in thickets. It has proved to be an excellent palm for cultivation and is commonly planted in public parks in many tropical countries. Plants will grow in a range of soils and will tolerate exposure to sun from an early age. They will withstand considerable coastal exposure without obvious detriment. They

Licuala spinosa in fruit. G. Cuffe

are relatively fast growing and respond strongly to fertilisers and watering. A well-grown clump is impressive, with crowded, slender (5–10 cm thick), fibrous stems and a dense canopy of dark green, orbicular leaves. Each blade may be close to 1 m across and is divided into about twenty narrow, radiating segments. These are narrowly wedge-shaped, pleated and with the central blades longest and broadest. Long inflorescences which arch well clear of the leaves bear white flowers about 1 cm long, followed by clusters of brilliant red, ovoid fruit each about 1 cm long. This palm is also excellent for home gardens. A widespread species, it occurs in Indonesia (Sumatra, Java, Borneo), the Philippines, Malaysia, Burma, Vietnam, Thailand, Andaman Islands and Nicobar Islands.

Licuala tanycola

(stretched out, in reference to the long peduncle)

NEW GUINEA

This species is native to Irian Jaya where it is reported as being common in rainforest on ridges and near rivers. It is a solitary palm to 2 m tall with a stem about 6 cm across and a large ligule (about 18 cm long) extending from the leaf sheath. The basal half of the petiole is well armed with recurved teeth and the blade is divided into numerous (13–20) segments. These are broadly wedge-shaped, dark green and shiny on the upper surface with numerous shiny red-brown scales beneath.

The central segments are larger (to 68 cm x 9 cm) than the outer segments (to 42 cm x 5 cm). The inflorescences (to 1.7 m long) have a very long peduncle and are branched only near the apex. Greenish-yellow flowers about 5 mm long are followed by red, globose to ellipsoid fruit (to 13 mm x 9 mm). An attractive palm for tropical and sub-tropical regions.

Licuala thoana

(after Dr Tho Yow Pong)

MALAYSIA

This species appears to be unique in the genus by having undivided fronds with a distinctly paddle-shaped blade (rarely divided into three segments). Plants have a subterranean stem and the leaves, including the spiny petiole may reach about 1 m long. The short inflorescence is sparsely branched and carries densely hairy flowers. Young fruit are pink. The species is native to the State of Johor, Malaysia and is locally common in some areas of lowland forest. It is not known to be in cultivation but would be a splendid addition to any collection.

Licuala tonkinensis

(from Tonkin)

VIETNAM

The fronds of this species are divided into six to eight unequal segments which have bluntly toothed outer margins. The segments are more or less wedge-shaped, with the central one broader than the rest. The basal part of the petiole is armed with conical spines. Erect inflorescences have somewhat rigid branches, with flowers about 4.5 mm long. The fruit are ovoid to oblong in shape. This species is native to western Vietnam where it grows in rocky areas.

Licuala valida

(strong, robust)

BORNEO

A Bornean species which grows in rainforest at low altitudes. Plants have a single trunk and large circular leaves divided to the base into about fifteen radiating segments. Dull, yellowish to brown fruit are carried on an inflorescence about 1 m long.

Licuala thoana. G. Cuffe

Essentially a palm for the collector, this species would need tropical conditions for its growth.

LINOSPADIX

(from the Greek *linos*, a net, *spadix*, a palm inflorescence)

A small genus of about eleven species of palm occurring in eastern Australia and New Guinea where they grow as understorey palms in rainforest. They are small, solitary or clumping, unarmed palms with very slender stems and entire or pinnate leaves. A crownshaft is present but poorly developed. The simple, unbranched inflorescence arises among the leaves and bears unisexual flowers of both sexes. Red fruit hang in long strings.

Cultivation: Excellent small palms for a shady garden or pot culture. Although relatively slow growing they are adaptable and long-lived. Single plants are capable of producing fertile seeds.

Propagation: Established suckers can be removed successfully but may be slow to re-establish. Fresh seed germinates readily three to six months after sowing. Seed has a short period of viability and should be sown quickly after collection. Each fruit contains a single seed.

Linospadix angustisecta

(with narrow segments)

NEW GUINEA

This species is distinguished by its regularly divided fronds (about 45 cm long) which have about twelve narrow, spreading segments (to 30 cm x 1 cm) each with a long-pointed apex. These are borne in a sparse crown atop very slender (about 5 mm diameter) stems. The inflorescence is a simple spike held stiffly erect. An attractive small palm from sheltered rainforests of central New Guinea.

Linospadix canina

(probably referring to the similarity between the fruit and a dog's incisor tooth)

NEW GUINEA

A small, graceful palm with stems to 1 m tall and about 5 mm thick and narrow, fan-shaped fronds which are deeply notched on the apex and taper to the base. Each frond is divided into two to four lobes 3–3.5 cm wide. Slender spikes much shorter than the fronds carry elongated, curved fruit (to 28 mm x 3.5 mm) which have prominent striae. An interesting species which is native to the Noord River of Irian Jaya where it grows in rainforests at about 800 m altitude.

Licuala thoana. G. Cuffe

Linospadix elegans

(elegant)

NEW GUINEA

An attractive small palm from Irian Jaya. Plants have wedge-shaped fronds about 20 cm x 12 cm, each divided into about eight narrow, pointed lobes. The basal lobes are about 1 cm wide, the rest about 4 mm across, with the terminal leaflets united. Slender, unbranched spikes about 45 cm long bear small, spindle-shaped fruit.

Linospadix longicruris

(with long auricles)

NEW GUINEA

The fronds of this species have a very short petiole and the sheathing base has two, prominent, triangular auricles. The fronds themselves can be entire with a deeply notched apex or are divided into two narrow lobes at the base, each with a long-pointed apex. The stem is recorded as being about 7 mm thick. Slender, striped fruit are borne on slender spikes about 50 cm long. A delightful small palm native to the Torricelli Mountains of Papua New Guinea, growing in rainforest at about 600 m altitude.

Linospadix microcarya

(with small nuts)

AUSTRALIA

A little-known species from the rainforests of north-eastern Queensland, Australia, where it is distributed from the lowlands to the highlands. It is a clumping species similar to *L. minor* but with much smaller fruit (about 8 mm compared with 16 mm in *L. minor*). It is virtually unknown in cultivation but has excellent potential.

Linospadix minor

(smaller, lesser)

AUSTRALIA—NEW GUINEA

A delightful, small clumping palm from the rainforests of north-eastern Queensland and which may extend to south-eastern New Guinea. Plants have very thin stems (about 1 cm thick) to 1.5 m tall, each with a loose crown of arching fronds about 1 m long. Irregularly shaped leaflets are toothed on

Linospadix monostachya in fruit.

the end. Pendulous strings of red fruit are an attractive feature. Plants grow well in shady, humid conditions in the tropics and subtropics.

Linospadix monostachya

(in a single spike)

Walking Stick Palm, Midgin-bil

AUSTRALIA

A small, solitary palm which is slow-growing but long-lived. Plants have a very slender stem and short pinnate fronds, with the leaflets irregularly distributed along the rachis and variable in width. Pendulous strings of small, red fruit are an attractive feature. Each fruit has a thin layer of edible flesh and the cabbage is also edible. A shade-loving palm well suited to subtropical and temperate regions. It is native to Australia where it is common in south-eastern Queensland and north-eastern New South Wales. The stems of this species were harvested in their thousands during the First World War and converted into walking sticks for returning wounded soldiers.

Linospadix palmeriana

(after Edward Palmer)

AUSTRALIA

An attractive small palm from highland rainforests of north-eastern Queensland, usually occurring above 800 m altitude. It is a clumping species with slender stems to about 1.5 m tall and a loose crown of leaves which have only two to four relatively large, dark green leaflets. Spikes of colourful red to orange fruit are a decorative feature. This attractive species, which is rarely grown, deserves to become better known in cultivation.

LIVISTONA

(after Patrick Murray, Baron of Livingston)

A genus of about twenty-eight species (including many undescribed) which occur in northern Africa, China, India, Ryukyu Islands, South-East Asia, Malaysia, the Philippines, New Guinea, Solomon Islands and Australia, where there is the greatest diversity of species. They grow in a range of habitats including swamp forest, rainforest, open savanna and rocky slopes and ridges, often in extensive colonies. They are solitary, usually tall, unarmed or spiny fan-leaved palms with an impressive crown. The inflorescence arises among the leaves and bears bisexual (rarely unisexual) flowers. The genus *Wissmannia* is now included in *Livistona*.

Cultivation: Impressive palms that are widely planted in parks and large gardens. One species or another will grow in most regions, from the tropics to temperate zones. They are excellent palms for avenue or driveway planting. Most species are hardy, sun-loving palms that grow easily, are adaptable to a wide range of conditions and are tolerant of neglect. Single plants are capable of producing fertile seeds.

Propagation: Seed germinates readily within one to four months of sowing. Each fruit contains a single seed.

Livistona alfredii

(after Alfred, Duke of Edinburgh)

Millstream Palm

AUSTRALIA

A very distinctive Australian palm with a restricted distribution, being found only around the headwaters of two rivers in Western Australia. Plants have a rounded crown of finely divided leaves which have stiff, spreading, blue-green segments. Clusters of flowers are yellowish. A real collector's palm suitable for subtropical regions. Plants are very slow growing and make an appealing subject in containers.

Livistona australis

(southern)

Australian Cabbage Palm, Fan Palm

AUSTRALIA

A very familiar palm, this Australian *Livistona* occurs naturally close to major centres of population and is now widely planted in many countries around the world. In its native state it grows in forests and swampy situations, sometimes as individuals or in small groups, often in extensive, dense colonies. Excellent for parks and large gardens—especially in group plantings—as well as for lining driveways and avenues. Grows very well in temperate regions and responds strongly to regular watering and the use of nitrogenous fertilisers. Frosts are generally tolerated with only a minor setback, even by young plants.

Livistona benthamii

(after George Bentham, 19th century English botanist)

AUSTRALIA

Plants of this species retain their older leaves for longer than in other Livistonas and hence have an extended head of foliage. This palm forms colonies in wet areas near the coast of northern Australia. Most of the trunk is covered with coarse leaf-bases and the leaves are an attractive shiny green. An impressive palm for tropical regions which lends itself well to planting in groups.

Livistona carinensis

(from the Oasis of Carin, Somalia)

ARABIAN PENINSULA—SOMALIA

A rare species which grows around oases in rocky areas of the Horn of Africa, north-eastern Somalia and adjacent areas of Arabia. A remarkably hardy, sun-loving palm which has been introduced into cultivation as a conservation measure. Natural populations have been greatly depleted by harvesting of the tough, woody trunk which is valued for building purposes. Plants seem to grow readily, if slowly,

Livistona australis is an excellent avenue palm.

and are adaptable to tropical and subtropical climates. Plants develop a slender trunk which may reach 30 m tall and the leaf blades are deeply cut into narrow segments which have thickened margins. The small, round fruit are brown when ripe. Previously known as *Wissmannia carinensis*.

Livistona chinensis

(from China)

Chinese Fan Palm

JAPAN—TAIWAN

This fan palm has prominently drooping tips which adorn the broad, glossy leaves of all but very young plants. The leaves often have a fresh, green lustre (sometimes yellowish) and this together with the weeping segment tips, provides a decorative combination. The fruit are an attractive bluish colour and hang in dense clusters. The Chinese Fan Palm is hardy and is a popular subject for cultivation in tropical and temperate regions. It makes an excellent tub plant and when young is very useful indoors, even tolerating dark positions. It tends to

Livistona chinensis.

be slow growing in temperate regions where plants rarely reach 4 m tall. The plants are tolerant of poor soil but respond to applications of fertiliser. They will grow in full sun or semi-shade and may flower while quite small. The species is native to Japan, Ryukyu Islands and Taiwan. A variety of this palm (var. *boninensis*) from the Bonin Islands has unusual, pear-shaped fruit which narrow to the base.

Livistona decipiens

(deceptive, misleading)

Weeping Cabbage Palm, Ribbon Fan Palm

AUSTRALIA

With its finely divided, weeping fronds, this Australian palm is easily one of the most attractive members of the genus. The fronds are often yellowish and the large clusters of bright yellow flowers are conspicuous. The species grows in colonies along stream banks and is also conspicuous in coastal districts. An excellent palm tolerant of a wide range of conditions, from the tropics to temperate regions and very useful near the coast.

Livistona drudei

(after Dr Oscar Drude)

AUSTRALIA

An attractive Australian palm with a relatively restricted distribution and becoming uncommon due to clearing of its habitat. Plants have a crown of shiny green fronds with shortly drooping tips and clusters of cream flowers. Well suited to tropical and subtropical regions. Looks attractive in a container and also when planted in groups of mixed ages.

Livistona eastonii

(after W.R. Easton, surveyor)

AUSTRALIA

Endemic to northern parts of Western Australia, this slender species grows in large colonies in open forest country. Plants develop a trunk to about 8 m tall and have a relatively small crown of bright green, shiny fronds which are shallowly folded along the midline. It is suitable for tropical areas with a seasonally dry climate but plants may be difficult to establish.

The unique dwarf *Livistona exigua*. J. Dransfield

Livistona exigua

(small, narrow, insignificant)

BRUNEI

Unusual in the genus, this palm grows as an understorey plant, with the largest specimens recorded having a trunk about 5 m tall. It is native to Brunei in Malaysian Borneo and grows on the top of ridges clothed in heath forest. Seedlings resemble a species of *Licuala* and mature leaves have very thin petioles and only about twelve widely separated leaflets in the blade. Seeds have been germinated at the Royal Botanic Gardens, Kew and it is hoped that the species will prove to be adaptable.

Livistona humilis

(of low growth)

Sand Palm

AUSTRALIA

An Australian palm which is very common in northern parts of the Northern Territory, growing in extensive colonies in sparse woodland. Plants have a very slender trunk to about 4 m tall and a small crown of stiff, dark green fronds. Yellow flowers and small black fruit are borne on long clusters which arch out clear of the leaves. This species has proved to be slow growing and is difficult to establish.

Livistona inermis

(unarmed)

Wispy Fan Palm

AUSTRALIA

A remarkably slender palm with a crop of hanging dead fronds and a very sparse crown of live fronds. These have very narrow, widely separated, grey-green segments which add to the wispy appearance of the plant. Endemic to the Northern Territory, Australia, this unusual palm grows in sandstone outcrops and open woodlands. Plants have proved to be very slow growing and are rather difficult to establish.

Livistona jenkinsiana

(after Father Jenkins who collected in India)

INDIA

A popular pot and glasshouse palm in Europe where it has been grown for a long period. It is native to northern India and has achieved some prominence through the unusual 'jhapee' hats which are made from its leaves and worn by the Assamese people. It is a typical *Livistona* with a thick, rounded crown atop a fairly slender trunk which may reach 10 m tall. Its large, fan-shaped leaves have numerous bilobed segments which are bluish-green on the underside. Fruit are carried in dense, hanging clusters and, when ripe, each are about 2.5 cm across and of a leaden blue colouration. In cultivation this palm succeeds in temperate and subtropical localities requiring soils and cultural conditions similar to other Livistonas.

Livistona humilis in flower.

Livistona mariae in its semi-arid habitat.

Livistona loriphylla

(with ribbon-shaped leaves)

Kimberley Fan Palm

AUSTRALIA

A slender fan palm from northern parts of Western Australia where it grows in rocky escarpments and woodland. Plants have a somewhat sparse crown of finely divided leaves and a hanging skirt of dead fronds. This palm is rare in cultivation and may be difficult to grow.

Livistona mariae

(after Maria, Duchess of Edinburgh)

Central Australian Cabbage Palm

AUSTRALIA

A truly relictual palm from Central Australia which is a leftover from a prior moist climatic regime. The prevailing climate of the area is semi-arid but the palm has survived as a few small colonies well established on permanent streams. These groves

are reproducing actively and are in a healthy state. The species has been widely planted in tropical, subtropical and temperate regions. Plants are generally very slow growing but have proved to be hardy and adaptable. Young plants commonly have attractive reddish leaves.

Livistona muelleri

(after Baron Sir Ferdinand von Mueller)

Dwarf Fan Palm

AUSTRALIA—NEW GUINEA

Widely distributed in north-eastern Queensland, the Torres Strait Islands and perhaps also southern New Guinea, this hardy palm grows in grassland and open woodland, sometimes in loose stands. Plants have a slender, usually blackened trunk to 10 m tall and a neat, rounded crown of stiff, dark green leaves (often greyish beneath). Yellow to brownish flowers are borne in large panicles are followed by clusters of bluish-black fruit. A slow-growing palm which needs tropical conditions.

Livistona rigida

(stiff, rigid)

AUSTRALIA

A large fan palm with a stout, brown trunk and a large crown of dull grey to blue-green leaves. The blades of the leaves may be about 1.5 m across and are divided for about half their length into narrow segments which may be stiff or droop slightly at the tips. The species, native to northern Australia, grows in colonies near water and is useful for group planting in the tropics. Young plants may develop a large, imposing crown and are surprisingly fast growing.

Livistona robinsoniana

(after C.B. Robinson, original collector)

PHILIPPINES

This collector's palm has a slender, green trunk which has contrasting broad, pale, annular rings at regular intervals along its length. The fan-shaped leaves are quite large and bright green with drooping tips and, at least when young, have a twist in the lamina. Unlike most *Livistona* species, the petioles of this species are smooth and unarmed. Native to the Philippines, it will grow in tropical and subtropical regions even tolerating some cold. For best appearance it needs a shady position, or at least protection from direct sun for part of the day. Mature fruit are yellowish-orange.

Livistona rotundifolia

(with round leaves)

Footstool Palm, Serdang

INDONESIA—MALAYSIA

The distinctive, round, glassy leaves of young plants earn this species its popular name. These leaves, when young, are only shallowly divided, with softly drooping tips. As a tub plant at this stage of growth the plant is excellent and can be successfully employed for indoor decoration. Leaves on mature plants, however, lose much of their appeal, especially when they have been tattered by the wind. Those plants grown in a deeply shady situation will retain the juvenile leaves much longer than plants grown in the open. The Footstool Palm is native to Malaysia and Indonesia where it grows in large impressive colonies. In its native state the cabbage is collected and eaten and the dark violet

Livistona rotundifolia in fruit.

Livistona rotundifolia in a picturesque setting.

fruit is eaten by children. It is an ideal palm for tropical climates and needs protection from direct sun when small. Rich soils promote strong, vigorous growth. In districts with a cold winter the plants struggle and are very slow growing.

A very striking variety is native to the Philippine island of Luzon. This is known as var. *luzonensis* and has a slender trunk with prominent white rings. If anything, this palm is more tropical in its requirements and tends to be slower growing than the typical variety. Fruiting plants are very decorative, bearing masses of brick-red fruit. Both variants can be propagated from seed which germinates within two months but seedling growth is very slow.

Livistona saribus

(from a Moluccan name)

Taraw Palm, Serdang

SOUTH-EAST ASIA

A widely spread South-East Asian palm which grows as an emergent plant in dense tropical jungles, its trunks reaching to more than 25 m tall. It is a typical fan palm, with deeply segmented green leaves and reddish to orange petioles which have large, very prominent thorns along the margins. One of its most distinctive and striking ornamental features are the large clusters of brilliant blue fruit which follow the flowers. Individually these are not

Livistona saribus.

much more than 1 cm long but in mass are very colourful. The palm is rather tropical in its requirements but will succeed in the subtropics and tolerate some cold in the winter. Young plants need protection from hot sun and are generally strong growers. *L. cochinchinensis* is a synonym.

Livistona speciosa

(beautiful, showy)

Mountain Serdang

BURMA—INDIA—MALAYSIA

A palm of the mountains of India, Burma and Malaysia, usually growing above 600 m altitude and forming extensive colonies. It has a slender, rough trunk to 18 m tall and a large, dense crown of leaves. These leaves are not flat but rather are irregularly folded, with stiff, spreading segments and are bright green above and bluish-green beneath. The inflorescence may be nearly 3 m long and is subtended by chocolate-brown spathes. The fruit, which are fairly narrow, grow to about 2.5 cm long and are jade green at maturity, ageing to black. Like other Livistonas, this species should prove to be reliable and hardy in cultivation.

Livistona tahanensis

(from Gunong Tahan)

MALAYSIA

A Malaysian palm which grows at high altitudes (1000–1600 m) in montane forest. This palm forms groves and may dominate the vegetation in some localities. It is a distinctive species which grows to about 8 m tall and has relatively small, stiff leaves which are dark green on top and greyish beneath. The blade is divided about half way into numerous, narrow segments which do not droop. The fruit are green with white spots. Seed of this palm has been distributed but its cultural requirements remain largely unknown. It may grow well in subtropical regions.

LODOICEA

(named for Louis XV of France, latinised to *Lodoicus*)

A monotypic genus of palm endemic to the Seychelles where it grows in forests on slopes and valleys. It is a palm from which many legends have sprung, mainly centred on the huge fruit and seeds which are of such a bizarre shape. It is a solitary, unarmed fan palm of large proportions. Massive pendulous inflorescences arise among the leaves and are unisexual. The huge fruit bears a strong resemblance to the human pelvis and has no

Livistona tahanensis in habitat, Malaysia. J. Dransfield

Above: *Lodoicea maldivica* young plants.
Right: *Lodoicea maldivica* in fruit. G. Cuffe

counterpart in any other group of plants.

Cultivation: A majestic palm worth trying for its novelty value. Plants are rare in cultivation and many of those planted do not survive. Some areas, such as the Peradeniya Gardens in Sri Lanka, achieve considerable success whereas other gardens report difficulties. A tropical climate is essential, as also is a deep soil, since the seed sends out a long radicle when germinating. Male and female plants are necessary to produce fruit and the fruit take five to seven years to mature on the plants.

Propagation: Seeds of *Lodoicea* must be sown in their permanent position since they produce a sinker over 4 m long. Germination may take from six to eighteen months.

Lodoicea maldivica

(from the Maldives, where nuts were frequently washed ashore)

Coco-de-Mer, Double Coconut

SEYCHELLES

Famed for its huge fruit which contains the largest seed produced by any plant in the world, this palm is endemic to two islands of the Seychelles (Praslin and Curieuse) where it grows in extensive colonies. It is a fan palm with huge, strongly costapalmate leaves which are dark green and glossy above and with stiffly spreading segments. The trunk is moderately stout (about 50 cm thick) and in nature may reach in excess of 25 m tall. Plants in cultivation, however, have proved to be exceedingly slow growing and a specimen of this height must be of great age indeed. Plants take from thirty to sixty years to flower. The base of the trunk fits into a round, wooden bowl and the roots grow through holes in this bowl into the surrounding soil. This bowl is extremely resistant to rotting and persists in the soil long after the plants have disappeared.

Its purpose is unclear but since the base of the trunk moves relative to it, it may ensure stability and cushion the effects of severe winds. The fruit are truly enormous and may weigh in excess of 20 kg. They are entirely taken up by the seeds which, when sliced crossways, resemble two coconuts joined in the middle, hence the common name of Double Coconut. Infertile and dead fruit float, while viable fruit sink and hence this palm is not distributed by the oceans currents as is the Coconut. In early times the fruit were regarded as mystical objects and fetched considerable sums of money. The seed was regarded as being a highly potent aphrodisiac and may still be regarded so

Lodoicea maldivica wild population. J. Dransfield

today. Today nuts are sold as curios to tourists, sometimes painted in strategic areas to more closely mimic the human anatomy.

LYTOCARYUM

(from *litus*, shore, *caryon*, a nut)

A small genus of three species of palm endemic to south-eastern Brazil where they grow as understorey plants in rainforest. They are solitary, unarmed palms with a slender trunk and arching pinnate fronds. The inflorescence arises among the leaves and bears unisexual flowers of both sexes. Palms formerly well known in the genus *Microcoelum* are included here. They are sometimes also sold under the incorrect names of *Cocos* and *Syagrus*.

Cultivation: Excellent ornamental palms suitable for a shady garden position and one species is also widely used for indoor decoration. They are generally adaptable and easy to cultivate. Single plants are capable of producing fertile seeds.

Propagation: Fresh seed germinates readily two to eight months after sowing. Each fruit contains a single seed.

Lytocaryum insigne.

Lytocaryum weddellianum.

Lytocaryum hoehnei

(after F.C. Hoehne)

BRAZIL

An attractive Brazilian palm which is grown on a limited scale. Plants have a slender trunk and a pleasant open crown of fronds similar to those of the other species in the genus, but the leaflets have brown hairs on the underside. They are also longer and broader (to 40 cm x 2 cm) than in the other species. This palm has very distinctive beaked fruit (to 3.3 cm x 2.3 cm) which split along two or three sutures at maturity, with the segments spreading to expose the seed. The species has also been known as *Syagrus hoehnei*.

Lytocaryum insigne

(remarkable)

BRAZIL

A small, neat palm that is adaptable in cultivation, growing from the warm, tropical regions to protected positions in the subtropics or as a pot plant in a heated glasshouse. The plants grow to about 2 m tall and have a rough trunk only 4.5 cm thick. The crown is composed of small, neat fronds that are silvery beneath. Once established, plants will tolerate considerable sun but while young they need protection. Potted specimens are useful for indoor decoration. This palm is usually seen only in the collections of enthusiasts but will probably become more widely grown. Seeds take six to eight months to germinate and benefit from the use of bottom heat. Seedlings of this species often appear in batches of seed of *L. weddellianum* and can be readily distinguished by their broad, undivided seedling leaves.

Lytocaryum weddellianum

(after H.A. Weddell, original collector)

BRAZIL

A small, graceful palm that is one of the best for indoor decoration. It will tolerate quite dark conditions and its slender, graceful, dark green, shiny leaves make it a very attractive addition to the decor. Plants can be grown in a large tub and appreciate a well-drained mix rich in organic matter. In the garden the species demands a semi-shady to shady position in a frost-free area. Plants are rarely seen over 2 m tall and the very slender trunk may be no more than 4 cm thick. It has the disadvantage of being extremely slow growing, especially in the first few years and advanced specimens are usually expensive.

METROXYLON

(from the Greek, *meta*, heart of a tree, *xylon*, wood)

Sago Palms

A small genus of five species of palm native to the Caroline Islands, Fiji, Samoa, New Hebrides, Solomon Islands, New Guinea and the Moluccas. They mostly form large colonies in lowland swamps. *Metroxylon sagu*, thought to originate in New Guinea, is very widely cultivated and is naturalised throughout the Pacific region and South-East Asia. They are large, solitary or clumping feather-leaved palms with the stems of four of the five species dying after fruiting. These monocarpic species produce a large, much-branched terminal inflorescence which carries male and bisexual flowers; the remaining species flowers within the crown. The fruit are covered in neat, vertical rows of overlapping scales. These palms are of major economic significance as a source of thatch and one species is a major source of sago. (For a recent treatment of the genus see J.B. Rauwerdink, 'An Essay on *Metroxylon*, the Sago Palm', *Principes* 30 (4): (1986) 165–180.)

Cultivation: Of limited ornamental appeal but widely cultivated as a source of sago and thatch. For these purposes sago palms are usually established in swampy situations where they thrive.

Plants, however, can be grown in well-drained soils providing they are watered and fertilised. They are sun-loving palms well suited to group planting. All species grow best in the tropics. Single plants are capable of producing fertile seed.

Propagation: The clumping species can be readily propagated by removal of suckers. Seeds germinate very slowly and sporadically, taking six to twelve months or even longer. Bottom heat speeds up the process and it is also reported that pre-soaking in water may be beneficial. Each fruit usually contains a single seed.

Metroxylon amicarum

(belonging to friends, a reference to the Friendly Islands where it was thought to originate)

Caroline Ivory Nut Palm, Rupung

CAROLINE ISLANDS—GUAM

Unlike the other species of *Metroxylon*, the trunk of this palm is long-lived and does not die after flowering. As well, the inflorescence arises among the leaves and is pendulous. The species occurs naturally on the Caroline Islands of Ponape and Truk and has become naturalised on Guam. It

Metroxylon amicarum in various stages of growth. A. Rinehart

Above: *Metroxylon salomonense*. Right: *Metroxylon salomonense* showing characteristic markings on leaf-base.

grows on forested slopes and valleys in the mountains. It is a solitary palm to 20 m tall with spiny leaves and large, handsome, apple-shaped fruit to 12 cm x 10 cm. These are chestnut brown when ripe and can be polished to make decorative ornaments. The hard kernel of the seeds has been used as a vegetable ivory for carving and to make buttons. An interesting palm for the tropics. It has previously been known as *Coelococcus carolinensis*, *C. amicarum* and *M. carolinense*.

Metroxylon sagu

(a native name)

Sago Palm

High-quality sago is obtained from the trunk of this palm which is felled just as it is about to flower. This sago is a staple part of the diet of millions of people in Asia and also generates revenue via its sale to Europe and other countries. Sago Palm, a large, impressive palm, is widely cultivated in various tropical countries of the world. A clumping species the thick trunk of which may grow to 20 m tall and bears an impressive crown of spreading, dark green pinnate fronds at its apex. At maturity a large, terminal panicle is thrust above the leaves and the bisexual flowers are followed by attractively patterned, round to apple-shaped, brown

fruits which may be up to 4.5 cm across. Once these fruit mature, the stem dies and is replaced by developing suckers. Sago Palms are very cold-sensitive and need the hot, humid conditions of the tropical lowlands. They are quite fast growing, taking fifteen to twenty years to flower. Plants like a sunny situation with plenty of water.

The Sago Palm is thought to occur naturally in New Guinea and the Moluccas but its exact origins are somewhat confused because it is now so widely cultivated. It is a variable species and four variants have been described as forms:
• forma *sagu* with the leaf sheath, petiole and rachis smooth and devoid of spines
• forma *tuberatum* with the leaf sheath covered in knoblike structures, the petiole and rachis smooth
• forma *micranthum* with the leaf sheath, petiole and rachis bearing short spines to 4 cm long
• forma *longispinum* with the leaf sheath, petiole and rachis bearing long spines to 20 cm long.

Metroxylon salomonense

(from the Solomon Islands)

MELANESIA

This *Metroxylon* does not grow in swamps but rather is a component of forests, growing in fertile, well-drained soils. It is native to eastern parts of

New Guinea, New Hebrides and the Solomon Islands and the leaves are harvested locally for thatch. It is a solitary species which dies after fruiting, the plants developing a stout trunk to about 20 m tall. Plants have a large crown of erect to arching leaves, each to about 10 m long and a hanging skirt of dead fronds. The terminal inflorescence, to 4 m long, bears both spreading and drooping branches. Cream flowers are followed by straw-coloured, scaly fruit about 9 cm x 6 cm. Young plants of this palm are handsome.

Metroxylon vitiense

(from Fiji)

FIJI

Native to the island of Viti Levu, Fiji, this palm grows in *Pandanus* swamps and the leaves are used as thatch but the trunks are apparently not harvested for sago. It is a solitary species with a large trunk to 15 m tall and persistent dead fronds. Live fronds, to 5 m long, are held erect in an arching crown. The trunk is terminated by an inflorescence

Metroxylon vitiense. C. Goudey

about 4 m long which has spreading and pendulous branches. The top-shaped fruit, to 7 cm x 5 cm, are yellowish-brown when ripe. This monocarpic palm appears to be rarely grown outside of Fiji.

Metroxylon warburgii

(after Otto Warburg, 19th & 20th century German botanist)

NEW HEBRIDES—WESTERN SAMOA

A solitary palm which dies after fruiting. Plants develop a trunk to about 7 m tall which tapers from near the middle to the apex. Dark green leaves, to about 3 m long, are held in an erect crown. The terminal inflorescence is about 3 m long with all its branches spreading or erect. The flowers are followed by brown, pear-shaped fruit to 12 cm x 9 cm. The species is native to the New Hebrides and Western Samoa and is often planted as a source of thatch.

NANNORRHOPS

(from the Greek *nannos*, dwarf, *rhops*, a shrub)

A monotypic genus of palms native to Afghanistan, Pakistan, Iran and the Arabian Peninsula. It grows in semi-arid regions where the plants can tap ground water. The species is a clumping, unarmed fan-leaved palm with subterranean or emergent trunks. The inflorescence terminates a stem and bears bisexual flowers.
Cultivation: A hardy, sun-loving palm which is uncommon to rare in cultivation. Plants are slow growing and must have excellent drainage. Single plants are capable of producing fertile seeds.
Propagation: Clumps can be successfully divided, although divisions may be slow to establish. Seed germination is reported to be slow and erratic. Each fruit contains a single seed.

Nannorrhops ritchieana

(after David Ritchie, original collector)
Mazari Palm

MIDDLE-EAST—PAKISTAN

Mostly the trunk of this palm is subterranean and it branches freely so that the plants form dense clumps. Occasionally, however, the trunk is erect, and while it still branches the plants may exceed

Nannorrhops ritchieana.

5 m in height. Native to desert areas, this very hardy palm is abundant in scattered colonies in infertile, stony soils often on treeless plains. It is of considerable importance to the local inhabitants who eat the very young leaves as a vegetable and the flesh of the fruit; and use other parts for building, thatching, weaving and as fuel for fires. The Mazari Palm extends into the mountains of Afghanistan and plants may be covered with snow for long periods during the winter. An interesting feature of its growth is that each crown is monocarpic, producing a single terminal inflorescence to 2 m tall. As the fruit of an inflorescence matures, the stem dies back down the trunk to the next branch, which replaces it. Each trunk is thick and densely covered with old leaf-bases which carry orange wool. The leaves are fan-shaped, stiff to rigid and an attractive grey-green colouration. In cultivation this species has proved to be very slow growing but hardy and extremely cold-tolerant. It will grow in a sunny position in temperate and subtropical regions. *N. naudeniana* is regarded as being a dwarf-growing variant of *N. ritchieana*.

NENGA

(from a native name)

A small genus of five species of palm occurring in Vietnam, Burma, Malaysia and Indonesia where they grow in rainforests. They are solitary or clustered unarmed palms with pinnate leaves and a prominent crownshaft. A few species have stilt roots. The inflorescence arises below the leaves (among the leaves in one species) and bears uni-

sexual flowers of both sexes. (For a recent study on the genus see E. S. Fernando, 'A Revision of the genus *Nenga*', *Principes* 27 (2) (1983): 55–70.)

Cultivation: Admirable palms of manageable size which are excellent for tropical gardens. Plants need shade, at least when young, and loamy, well-drained soils. Single plants are capable of producing fertile seeds.

Propagation: Clumping species can be propagated by division. Fresh seed germinates readily two to four months from sowing. Each fruit contains a single seed.

Nenga banaensis

(from Mount Bana, Vietnam)

VIETNAM

A poorly known species which has apparently not been collected in fruit. It is a solitary palm with a slender trunk to about 3 m tall and a long crownshaft topped by a crown of finely divided pinnate fronds. It grows in humid forests at Mount Bana near Da Nang, Vietnam. Although not cultivated, this would be an interesting species for a shady tropical garden.

Nenga pumila var. *pumila*.

Nenga pumila var. *pachystachya* in fruit. J. Dransfield

densely covered with large (about 4.5 cm x 2 cm), deep red to purplish-black fruit, each with a distinct long beak. The species is native to the State of Johore, Malaysia, where it grows in dense forests up to 500 m altitude.

Nenga gajah

(from the local name Pinang Gajah)

SUMATRA

This species is native to Sumatra where it grows on forested slopes and in valleys, often near streams. It is a solitary palm with a stout trunk to 2 m long suported by a buttress of prominent stilt roots. The yellowish-green crownshaft is not prominent and about eight leaves spread widely in a graceful crown. The inflorescence arises among the leaves (unusual for the genus) and is erect and sparsely branched. Fruit are spindle-shaped, beaked, large (to 8 cm x 2.5 cm) and ripen to dark purplish-brown. These are borne in a characteristic club-like cluster.

Nenga grandiflora

(with large flowers)

MALAYSIA

An attractive, vigorous palm which has a slender, solitary green trunk (with brown rings) and a prominent, inflated crownshaft (sometimes purplish). The pinnate fronds spread widely in a graceful crown (seven to eight per crown), with the leaflets being relatively widely spaced. Pendulous, sparsely branched inflorescences after flowering are

Nenga macrocarpa

(with large fruit)

MALAYSIA

An attractive Malaysian palm which could be readily confused with a species of *Pinanga* but is distinguished by its long pendulous spikes. It occurs in mountainous areas to 1400 m altitude and would be an excellent palm for the subtropics. Plants have a solitary, stout trunk to 3 m tall, a large, inflated, purple-green crownshaft and a crown of spreading leaves with stiff, dark green leaflets. Purple-black, shiny fruit hang in strings.

Nenga pumila var. pumila

(dwarf, of low stature)

INDONESIA

A specimen of this small palm from West Java resembles a cluster of dwarfed bangalow palms crowded together. A graceful, arching crown of dark green pinnate leaves (1–1.5 m long) tops a prominent, bright purple-green crownshaft on each stem. Clusters of oval fruit each about 2 cm x 1 cm are orange-brown when ripe and are a decorative feature. This is a graceful and highly ornamental palm which, because of its manageable size, is ideally suited to garden cultivation. In nature it grows on a range of soil types, including those derived from limestone. Altitudinally it ranges from about 150 m to about 1300 m and genetically offers much scope for selection of plants suitable for a range of climates. Young plants require protection from direct, hot sun.

Nenga pumila var. pachystachya

(with thick spikes)

SOUTH-EAST ASIA

This is the most widely distributed *Nenga*, being found in south Thailand, Malaysia, Singapore and Indonesia. Plants are similar in general appearance to *N. pumila* var. *pumila* but have larger (to 2.4 cm

x 1.5 cm) fruit which ripen bright red. An attractive small palm for the tropics.

NEODYPSIS

(from the Greek *neos*, new, *Dypsis* another genus
of palm)

A genus of fourteen species of palm all endemic to
Madagascar. They grow in open forest and rainforest at low to high altitudes. They are small to tall,
solitary or clumping, unarmed palms with pinnate
leaves, sometimes arranged in three ranks and with
or without a crownshaft. The inflorescence arises
among the leaves or below the leaves (crownshaft
species) and bears unisexual flowers of both sexes.
Cultivation: The genus contains some excellent
horticultural subjects which are gaining popularity.
Those species originating in high altitudes have
good prospects for subtropical zones. Species tested
to date have proved to be very adaptable to cultivation and are highly ornamental. Single plants are
capable of producing fertile seed.
Propagation: Fresh seed germinates readily, often
within a month or two of sowing. Each fruit contains a single seed.

Neodypsis decaryi.

Neodypsis decaryi infructescence.

Neodypsis baronii

(after M. Baron, original collector)

MADAGASCAR

Plants of this species are reminiscent of the Golden
Cane Palm but are much greener. It is a clumping
palm from Madagascar which grows in moist
forests at 900–1800 m altitude. Grey, smooth
trunks grow 2–6 m tall and 6–10 cm across and the
graceful, curved leaves are about 2 m long. Each
leaf has about eighty, narrow, spreading segments.
The fruit are ovoid, to 18 mm x 14 mm, and greenish-brown when ripe. This is a very decorative
palm which is becoming popular in cultivation.
Plants have proved to be easy to grow in tropical
and subtropical regions.

Neodypsis decaryi

(after R. Decary, original collector)

Triangle Palm, Three-sided Palm

MADAGASCAR

A spectacular and distinctive palm which can be
readily recognised by the prominent three-sided
trunk, with the pinnate leaves arranged in three
rows. As well, the leaves are held stiffly erect,
drooping only at the tips and the lowermost leaflets
are developed into filamentous extensions (reins)
which hang to the ground. The chalky-white bloom
on the trunk and the grey leaflets add to the character of this most unusual palm. It has become very
popular and is now common in cultivation,

Above: *Neodypsis leptocheilos*. Above Right: *Neodypsis leptocheilos* crownshaft. G. Cuffe

although quite restriced in its natural state. Plants succeed best in tropical and subtropical regions, although they have also proved to be somewhat cold tolerant. They require a sunny or partially sheltered position. Garden-grown plants have proved to be adaptable and quite fast growing.

Neodypsis lastelliana.

Neodypsis lastelliana

(after M. de Lastelle)

Col Rouge

MADAGASCAR

A tall, solitary-stemmed, emergent palm which occurs in the moist forests of Madagascar, ranging from sea-level to about 1000 m altitude. Plants grow to about 20 m tall with a moderately strout trunk (about 40 cm across) and a reddish to reddish-brown crownshaft which is covered with thick hairs and has a felty appearance. The fronds are stiff, with widely spreading, dark green segments to about 80 cm x 3.5 cm. The rachises of young plants may be pinkish. The obovoid, greenish-brown fruit are about 20 mm x 12 mm when mature. This palm has become popular since its introduction only a few years ago.

Neodypsis leptocheilos

(with a narrow lip)

Redneck Palm

MADAGASCAR

A recently described species from Madagascar which has become a very popular item in cultivation. It is a highly ornamental palm which has been confused with *Neodypsis lastelliana* but is recognised by its bright red crownshaft which develops significant colour from an early age. Plants are very fast growing and will adapt to a wide range of soils

273

and climates (tropical to subtropical). They respond strongly to regular watering and the use of fertilisers and will tolerate sun when quite young.

Neodypsis loucoubensis

(from Lokobe Forest, the original area of collection)

MADAGASCAR

This species is native to the island of Nosy Be off the north-west coast of Madagascar. Plants have a distinctly ringed, slender trunk to 10 m tall and a prominent, silvery-green crownshaft below a somewhat untidy crown of dark green fronds. These have narrow pendulous segments to 90 cm long. Small inflorescences carry cream coloured flowers and the ovoid fruit are about 1 cm long. Experience with this species in cultivation seems to be limited.

NEOVEITCHIA

(from the Greek *neos,* new, *Veitchia,* another genus of palm)

A monotypic genus of palm endemic to Fiji, where it has been reduced to rarity and is now a protected species. It grows in moist forests and the crowns of the mature palms emerge above the forest canopy. It is a robust, solitary, feather-leaved palm with a crownshaft. The inflorescence arises below the crownshaft and bears unisexual flowers of both sexes.
Cultivation: An ornamental palm suitable for tropical districts. Plants dislike coastal exposure and must be regularly watered during dry periods. Young plants may need some protection but older plants are tolerant of sun. Single plants are capable of producing fertile seeds.
Propagation: Seeds may be slow and erratic to germinate, not appearing for six months or more. Each fruit contains a single seed.

Neoveitchia storckii

(after Jacob Storck, botanical collector in Fiji)

FIJI

An endangered palm which is now conserved in a reserve on the island of Viti Levu, Fiji. The species was formerly widely distributed on the banks of the Rewa River, but has now been reduced by human exploitation to an area of about 2 ha which contains about 200 mature plants. It is a

handsome palm reminiscent of some species of *Veitchia,* but with a heavy crown of fronds which have relatively broad, pointed leaflets and a loose crownshaft. Numerous inflorescences with hanging spikes, carry white flowers which are followed by reddish-yellow fruit. The trunks of this palm were used to form the supporting poles of dwellings so characteristic of the local area and the young fruit may have been eaten. This species has been introduced into cultivation and is suitable for the tropics.

NEPHROSPERMA

(from the Greek *nephros,* kidney, *sperma,* a seed)

A monotypic genus of palm endemic to the Seychelles where it is distributed on various islands. The species grows in regrowth and on rocky slopes in sparse forest, often on cliffs or ravines up to about 400 m altitude. It is a solitary, unarmed or slightly spiny feather-leaved palm lacking a crownshaft. The inflorescence arises among the leaves and bears unisexual flowers of both sexes.
Cultivation: An attractive palm which is well established in cultivation. Plants will tolerate sun from an early age and need excellent drainage. They are very sensitive to the cold and are best suited to the tropics. Single plants are capable of producing fertile seed.
Propagation: Fresh seed germinates within one to three months of sowing. Each fruit contains a single seed.

Nephrosperma vanhoutteanum

(after Louis Benoit Van Houtte, 19th century Belgian nurseryman)

Latanier Millepattes

SEYCHELLES

Another of the remarkable palms native to the Seychelles. It has become a popular palm in cultivation, being used as a conservatory plant in Europe and grown as a garden plant in many tropical countries. The species develops a slender grey trunk to 12 m tall and has a graceful crown of arching or drooping, dark green fronds. Young plants have prominent, black, bristly spines on the petioles but these are much less obvious on larger plants. Long inflorescences hang clear of the crown and after flowering carry small, round- to kidney-shaped fruit which ripen from green through yellow to red. An adaptable palm easily grown in the tropics.

NORMANBYA

(after the Marquis of Normanby)

A monotypic genus of palm endemic to north-eastern Australia where it grows in rainforests at low elevations. It is a distinctive, solitary, unarmed feather-leaved palm with a prominent crownshaft and plumose leaflets. The inflorescence arises among the leaves and bears unisexual flowers of both sexes.

Cultivation: A handsome palm deserving of wider cultivation. It grows well in tropical and subtropical districts in warm, humid conditions and needs to be kept regularly moist. Single plants are capable of producing fertile seed.

Propagation: Seed germinates readily with most seedlings appearing two to three months after sowing, but others straggling up for twelve months or so. Each fruit contains a single seed.

Neoveitchia storckii.

Normanbya normanbyi

(after the Marquis of Normanby)

AUSTRALIA

This distinctive species can be recognised by the slender, light grey trunk, pale grey crownshaft and an attractive crown of arching fronds with the leaflets irregularly arranged to give a whorled appearance. The leaflets of young plants are broader than in mature plants and the distinctive, silvery-white undersurface is very noticeable. Clusters of whitish flowers are followed by dull pink to purplish fruit each 3–3.5 cm long. This species is native to wet, coastal rainforests of north-eastern Queensland, Australia. It is an excellent palm for group planting and is relatively fast growing in the tropics. Young plants need a sheltered position for the first three to five years. Best growth is achieved in well-structured, loamy soil and plants respond to mulches, fertilisers and regular watering.

NYPA

(from the Malay name, *nipah*)

A monotypic genus of palm which is widely distributed in Asia and the western Pacific, extending from India and Sri Lanka to the Ryukyu Islands and northern Australia. It is a true mangrove which grows in coastal estuaries; in some countries dominating the mangrove vegetation and forming extensive stands. It is a feather-leaved palm with a forking, prostrate or subterranean trunk. A unique inflorescence arises among the leaves and bears unisexual flowers of both sexes, the males crowded in catkin-like structures, the females aggregated in a dense, rounded head. The seeds, which germinate in the head before shedding, float and are distributed by ocean currents.

Cultivation: Cultivated on a limited scale as an ornamental. Best suited to estuarine conditions such as salt marshes or tidal mudflats. In the tropics it can also be grown around freshwater swamps and often takes readily to these situations. One colony in Bogor has thrived for at least 100 years and, more recently, clumps have been succesfully established at the Palmetum, Townsville, Australia. Plants are also growing in the waterlily pool in a tropical glasshouse at Kew Gardens, UK. Single plants are capable of producing fertile seeds.

Propagation: Seeds germinate in the head before

shedding. Seedlings can be readily established in suitable conditions.

Nypa fruticans
(shrubby or bush-like)

Nipah, Mangrove Palm

ASIA—WESTERN PACIFIC

In the Philippines and some other countries, this palm forms extensive stands to the exclusion of all other vegetation. It is important locally for a range of products, including sugar, vinegar and alcohol from the flower stalks, leaf stems for building, leaflets for thatch (leaflet squares or shingles called 'atap') and the young seed is edible. Although ornamental, this palm is rarely cultivated except in botanic gardens where it is usually included as a novelty item and for unique botanical features. It is exclusively tropical in its requirements.

Mangrove palm *Nypa fruticans* in habitat with coconut palms. G. Cuffe

OENOCARPUS
(from the Greek *oinos*, wine, *carpos*, fruit)

A genus of nine species occurring in Central America and South America, extending south to Bolivia and east to French Guiana. They are essentially riverine palms growing in rainforest, often in extensive stands. They are solitary or clumping, unarmed feather-leaved palms which lack a crownshaft. The inflorescence arises among the leaves and bears unisexual flowers of both sexes.

Recognition of *Oenocarpus* species is aided by features of the inflorescence. Developing inflorescences are characteristically flattened with a sharp, two-edged prophyll. Mature inflorescences, with their prominently drooping segments, are reminiscent of a horse's tail. Palms of this genus have important oils in the pericarp of their fruit and interest has been shown in growing these plants commercially for this product. Species of the genus *Jessenia* are now included in *Oenocarpus* (see *O. bataua*). (For a recent study of the genus see M.J. Balick, 'Systematics and Economic Botany of the

Oenocarpus—Jessenia (Palmae) complex', *Advances in Economic Botany* 3 (1986): 1–140.)

Cultivation: Interesting ornamental palms suitable for cultivation in tropical regions and growing best in well-structured loams. They would be excellent as ornamental specimen plants in parks and may also have prospects for avenue planting. Single plants are capable of producing fertile seeds.

Propagation: Fresh seeds germinate readily two to four months after sowing. Each fruit contains single seed.

Oenocarpus bacaba

(a local name)

Bacaba Palm

SOUTH AMERICA

A widely distributed palm occurring in Colombia, Venezuela, Guyana, Surinam, French Guiana, Peru and Brazil. It often grows in extensive colonies, sometimes in areas prone to flooding, and ranges from the lowlands to about 1000 m altitude. The fruit are used to prepare nourishing drinks and the people of various countries frequently cut mature palms down solely to collect ripe fruit. Plants have a solitary, slender trunk and a rounded crown of arching fronds, with the leaflets more or less pendulous. The inflorescence is similar to that of *O. distichus* and the fruit are light purple when ripe. This attractive palm, which is best suited to tropical regions, deserves to become more widely cultivated.

Oenocarpus bataua ssp. bataua

(a local name)

Bataua Palm, Seje Palm

CENTRAL AND SOUTH AMERICA

This handsome palm, which occurs naturally in Panama and in many areas of northern South America, is an excellent ornamental for tropical regions. Plants have proved to be fast growing and respond strongly to cultural practices such as mulching, regular watering and the use of fertilisers. It is a solitary species with a columnar trunk and a crown of large fronds (to about 10 m long), in which the glossy, dark green leaflets assume a drooping stature. Inflorescences, each somewhat reminiscent of a horse's tail, bear separate, creamy white, fragrant, male and female flowers and later clusters of heavy,

Developing infructescence of *Nypa fruticans*.

purple-black fruit. In its native countries this palm is used by local people for a variety of purposes: as thatch, fibres and darts and arrows for blowpipes and bows; the palm heart is edible; and a nutritious drink is prepared from the ripe fruit. In its native state this species may form extensive stands in low-lying areas subject to periodic inundation, or grow as scattered individuals within rainforest. *Jessenia bataua* is a synonym. Two variants are known from the Amazon region, having either white or reddish-pink fruit. One subspecies has been described.

Oenocarpus bataua ssp. oligocarpus

(with few fruit)

This subspecies occurs in Trinidad, northern Venezuela, Guyana and Suriname. It can be recognised by its leaflets which have a distinctly whitish to glaucous or grey undersurface, as well as distinctive inflorescence and floral details.

Oenocarpus distichus

(in two ranks or rows)

BRAZIL

A little-known palm from the Amazon region of Brazil and possibly extending into Venezuela. Of very distinctive appearance, this striking palm can be readily recognised by the leaves being arranged alternately in two ranks (*distichous*). Thus from side-on the crown of the palm appears narrow whereas when viewed from the front a full crown of arching fronds is apparent. Plants have a solitary, slender trunk to 20 m tall and each leaf has a prominent, swollen, olive-green sheathing base and numerous, narrow leaflets arranged in irregular groups of up to seven. Inflorescences, which arise from below the lowest leaf, have strongly pendulous segments which bear creamy white, unisexual flowers of both sexes. Two variants can be distinguished based on the colour of the fruit pulp, one having white pulp and the other reddish. The pulp is used to manufacture a drink and it is interesting to note that the drink made from fruit with white pulp is considered superior to the other.

Oenocarpus mapora ssp. mapora

(a local name)

CENTRAL AND SOUTH AMERICA

A clumping palm which develops 2–12 slender trunks which may grow to about 20 m tall, although many plants are much shorter than this. The trunks are smooth and unarmed and superficially resemble bamboo. Each trunk bears 6–8 graceful, arching fronds which have crowded, spreading to drooping, dark green leaflets. Ellipsoid, mauve to purple fruit are borne in heavy clusters following the creamy white flowers. A widespread palm which occurs in Central America and South America where it is distributed from the lowlands to about 1000 m altitude. Trunks are harvested for their edible cabbage and interest has been shown in growing this species commercially since harvested trunks are naturally relaced by suckers. This attractive palm deserves to be more widely planted as an ornamental but at present is mainly found in botanical collections. Seed is often sold as *O. panamanus* which is a synonym for this species. One subspecies has been described.

Oenocarpus mapora ssp. dryanderae

(after E.I. Dryander, original collector)

This subspecies from Colombia differs in having 1–5 trunks in a clump, dense silvery hairs on the leaflets and larger fruit (about 3 cm x 2 cm).

Oenocarpus minor ssp. minor

(lesser, smaller)

BRAZIL

The typical subspecies of this Brazilian palm develops only a single trunk. It grows in the Amazon region where it is found as scattered individuals in shady forests. Gracefully arching fronds bear evenly arranged, spreading, dark green leaflets, presenting an attractive appearance and making this a desirable species for cultivation. The fruit ripen to a light purple colour. Although this palm deserves to become widely planted in tropical regions, it is at present mainly an item for the collector. One subspecies has been described.

Oenocarpus minor ssp. intermedius

(intermediate, in the middle)

This subspecies is also known from the Amazon region and differs from the typical subspecies by its clumping habit.

ONCOSPERMA

(from the Greek *oncos*, humped, *sperma* seed)

A genus of five species of palm occurring in Sri Lanka, South-East Asia, Malaysia, the Philippines and Indonesia. They frequently form communities on hilly slopes and one species grows on the landward fringe of mangroves. They are tall, densely clumping, very spiny feather-leaved palms with a well-defined crownshaft. The inflorescence arises below the crownshaft and bears unisexual flowers of both sexes. The cabbage of most species is highly esteemed by local people and the trunks provide hard, desirable wood.

Cultivation: Although most organs are armed with long, black, brittle spines, these palms form impressive clumps and are worthy subjects for cultivation. They are essentially palms for the tropics and should be given room to develop to their potential. They make excellent subjects for lawns but because of their spines they are not suitable for planting near paths. Single plants are capable of producing fertile seed.

Propagation: Suckers can be removed successfully and used for propagation. Fresh seed germinates readily one to three months from sowing. Each fruit contains a single seed.

Oncosperma tigillarium.

Oncosperma fasciculatum

(arranged in clusters)

SRI LANKA

Endemic to Sri Lanka, this palm is basically similar to the more widely cultivated *O. tigillarium* but grows in narrower clumps with fewer stems and rarely reaches more than 15 m tall. Each stem grows to about 15 cm thick and is liberally armed with long, black, brittle spines. The leaves, which are about 2.3 m long, have spiny petioles and fairly broad leaflets (to 5 cm) which droop at the tips. These are characteristically arranged in irregular clusters (and held at different angles) along the rachis. Inflor- escences are not spiny. Large clusters of black fruit, which are only about 0.5 cm across, are a feature of this palm. Like the other species in this genus, it is essentially for parks and large gardens in the tropics. The species occurs naturally up to 1600 m altitude and provenances from high altitude may succeed in subtropical zones.

Oncosperma gracilipes

(with a slender foot or stalk)

PHILIPPINES

This species grows in thickets on inland hills on the Philippine island of Luzon. Plants have slender stems (about 4 cm thick) to 10 m tall and the rusty brown leaf sheaths are well armed with shiny black,

brittle spines. The leaves have equally spaced, papery leaflets (to 70 cm x 3 cm) which are dark green on the upper surface and paler beneath. Clusters of bright red fruit (each globose and about 12 mm across) are highly noticeable. An interesting lawn specimen for the tropics.

Oncosperma horridum

(horrible, prickly)

SOUTH-EAST ASIA

Like the other species in the genus, this palm forms impressive, but much sparser clumps, often with only four to six stems (sometimes solitary). Native to Malaysia, the Philippines and Indonesia (Sumatra and Borneo), it occurs naturally on hillsides at elevations of 500–1000m. Each stem is liberally armed with black spines and bears a relatively sparse crown of leathery fronds, with the leaflets spreading widely and drooping only at the tips. The fruit, which are 2–2.5 cm across, are spherical and purple-black when ripe. Outdoor cultivation of this palm is restricted to the tropics as it is very sensitive to cold and likes warm, humid conditions.

Oncosperma platyphyllum

(with broad leaves or leaflets

PHILIPPINES

This species has stems to about 8 m tall and the rusty brown leaf sheaths have numerous shiny, black, brittle spines. A few spines also occur on the petiole and on the underside of the rachis. The leaves have equally spaced, sword-shaped leaflets (to 60 cm x 4.5 cm), which are pleated on the upper surface. An interesting clumping palm which grows on inland hills on the island of Negros in the Philippines.

Oncosperma tigillarium

(with small stems)

Nibung Palm

SOUTH-EAST ASIA

An eye-catching palm which grows in tall, slender, very dense clumps which may reach more than 25 m high. Each trunk is only 10–15 cm across (and covered with black spines) but up to fifty may be together in a clump and are so crowded that they almost touch at the base. The upper part of the clump is a mass of the dark green pinnate fronds which crown each trunk. These have characteristic, leathery, drooping leaflets and the petioles (and inflorescence) are covered with long, black spines. The palm is very widespread in South-East Asia and grows in coastal swamps and on the fringe of mangroves. It is a handsome species but because of its size and rather formidable nature is suitable only for large municipal parks and gardens. It thrives in the hot, humid conditions of the tropical lowlands. Seeds lose their viability quickly after collection. The species was previously known as *O. filamentosum*.

ORANIA

(after William, Prince of Orange, later King of Holland)

A genus of about seventeen species of palm occurring in South-East Asia, Malaysia, the Philippines, Indonesia and New Guinea with a disjunct species found in Madagascar. They grow in humid forests, particularly rainforests and a couple of small species are understorey palms. They are solitary, unarmed feather-leaved palms which lack a crownshaft. The inflorescence arises among the leaves and bears unisexual flowers of both sexes. Some species are very poisonous and the cabbage and seeds of all species should be treated with caution. (For a treatment of New Guinean species see F.B. Essig, 'The genus *Orania* in New Guinea', *Lyonia* 1 (5) (1980): 211–233.)

Cultivation: Interesting palms which are rarely seen in cultivation. Most species appear to be best suited to the tropics. Some species are lofty palms suitable for parks and large gardens, whereas the lower-growing types are more suited to general garden culture. Single plants are capable of producing fertile seed.

Propagation: Fresh seeds germinate three to four months after sowing, although some species may germinate sporadically. Bottom heat may be beneficial. Each fruit contains one to three seeds.

Orania archboldiana

(after Richard Archbold, entrepeneur who financed botanical expeditions)

NEW GUINEA

This species is readily distinguished by the narrow, drooping leaflets being clustered in irregular groups along the rachis and oriented in different planes to present a plumose appearance. Plants have a slender stem only 10 cm thick and an elongated crown of graceful fronds. The species is endemic to Papua New Guinea where it occurs in the Fly River Basin of the south-west growing in rainforest.

Orania disticha

(in two ranks)

NEW GUINEA

Although relatively unknown in cultivation this Papua New Guinean palm is to be found close to centres of population such as Port Moresby, where it grows on forest margins and stream banks. It is a tall species (to 15 m) with a moderately slender trunk (10–15 cm thick) and a large, graceful crown. The fronds, which are dark green, arch out in a pleasant fashion and have prominently drooping leaflets. The basal leaflets of each frond are the

Orania sylvicola in habitat Malaysia. G. Cuffe

largest and overlap to give the centre of the crown a crowded appearance.

Orania glauca

(bluish-green)

NEW GUINEA

A Papua New Guinean palm which is characterised by its erect, stiff leaves which seem to thrust through the surrounding vegetation. The crown consists of relatively few leaves but each is large and quite broad with attractively drooping leaflets. The trunk is fairly stout and prominently ringed. When in flower the inflorescence is most impressive, being nearly 4 m long, with the stems glaucous from a waxy covering (from which the species gets its name). *O. glauca* is native to the West Sepik District of Papua New Guinea, growing in lowland forest. It is an impressive palm that is probably very tropical in its cultural requirements.

Orania lauterbachiana

(after C.A.G. Lauterbach, German plant collector)

NEW GUINEA

This species and *O. macropetala*, both native to New Guinea, are very similar in general appearance but are separated on botanical features (*O. macropetala* has whitish hairs on the inflorescence, *O. lauterbachiana* has red). Both are widespread throughout lowland areas growing in rainforests. They have a slender, grey trunk which may grow to 20 m tall and a large crown of obliquely erect, dark green fronds which have prominent drooping leaflets. The round fruit are up to 6 cm across and orange when ripe. Like most palms from the tropical lowlands, they are very tropical in their cultural needs.

Orania palindan

(a native name)

PHILIPPINES

Native to the Philippines, this palm is common on Luzon, growing in forested valleys and ascending from the lowlands to mountains at medium altitudes. The plants grow to about 5 m tall with a slender, prominently ringed, very hard trunk and a graceful crown of pinnate leaves, the leaflets of

which are drooping and with shredded tips. The spherical fruit, 5–6 cm across, are yellowish when ripe. This is a very decorative palm suitable for tropical and warm, subtropical regions. Young plants are slow growing and require protection from direct sun for the first few years. The fruits and cabbage of this species are poisonous.

Orania sylvicola

(growing in forests)

Ibul Palm

INDONESIA—MALAYSIA

A lofty palm which can grow to 20 m tall, with a rounded crown somewhat reminiscent of the Coconut, but much finer. Young leaves are yellowish but mature green and the leaflets are greyish beneath. The fruit and cabbage are poisonous, with the seeds being especially potent. This palm is native to Malaysia and Indonesia (Sumatra) where it forms colonies in rainforest at low altitudes.

ORANIOPSIS

(resembling an *Orania*)

A monotypic genus endemic in north-eastern Australia where it grows in rainforests at low to high altitudes. It is a solitary, unarmed feather-leaved palm which lacks a crownshaft. The inflorescence arises among the leaves and bears unisexual flowers.

Cultivation: A handsome palm which is extremely slow and difficult to grow to perfection. Plants need rich, loamy soils and appreciate humid conditions. Male and female plants are necessary for fertile seed production.

Propagation: Seeds are extremely slow and erratic to germinate, with the first seedlings appearing after about six months and others straggling through two to four years after sowing. Each fruit contains a single seed.

Oraniopsis appendiculata

(with an appendage)

AUSTRALIA

In general appearance this palm resembles a dwarf Coconut. Plants have an attractive, erect to arching crown of greyish fronds, often with a few dead fronds hanging as a skirt. The inflorescences are rather short and the whitish flowers have prominent triangular appendages on the petals. The globular fruit, 3–3.5 cm across, are yellow when ripe. This species is very slow growing and in the wild plants seem to remain quiescent for decades, producing new fronds but not developing a trunk. The species is native to north-eastern Queensland, Australia where it grows in dense rainforests in coastal ranges and adjacent tablelands.

ORBIGNYA

(after A.D. d'Orbigny, 19th century French naturalist)

A genus of twenty species of palm which occur in Mexico, Central America and South America. They grow in a range of habitats, from open woodland and savanna to rainforest. They are large to massive, unarmed solitary palms with subterranean to emergent and erect woody trunks. The large pinnate leaves have the leaflets regularly arranged or crowded along the rachis. The inflorescence arises among the leaves and bears either unisexual flowers of one sex or of both sexes. These palms are used for a wide range of purposes, including the production of valuable oils extracted from the fruit. (For a spectus of the genus see S.F. Glassman, 'Preliminary Taxonomic Studies in the palm genus *Orbignya*', *Phytologia* 36 (2) (1977): 89–115.)
Cultivation: Large, majestic palms of impressive proportions. They are generally suitable for tropical regions although some species have proved to be adaptable and withstand some cold. Single plants are capable of producing fertile seed.
Propagation: Fresh seeds germinate readily, taking two to four months from sowing. Seed of some species may lose its viability quickly. Each fruit contains from one to about seven seeds.

Orbignya cohune
(a native name)
Cohune Palm
CENTRAL AMERICA

This massive palm is frequently included in the genus *Attalea*. Native to Mexico, Honduras, Belize, El Salvador and Guatemala, it is characterised by an enormous crown of dark green pinnate fronds which are held stiffly erect. These crown a trunk which may be more than 15 m tall and the result is a very impressive palm which,

because of its size, is best suited to parks and public gardens. In some tropical countries the species may also planted in rows to line streets and driveways. In young, trunkless plants the erect fronds seem to erupt out of the ground. The large, egg-shaped fruit yield cohune oil, valued for its lubricating properties. This palm is best suited to the hot, humid tropics and will withstand direct sun when quite small. Plants may be slow growing until they form a trunk.

PARAJUBAEA

(similar to or near the genus *Jubaea*)

A small genus of two species which occurs naturally in Ecuador, Bolivia and Colombia. They are tall, solitary, unarmed feather-leaved palms which

Parajubaea cocoides.

lack a crownshaft. The inflorescences arise among the leaves and bear unisexual flowers of both sexes. The fruit of both species has a layer of sweet edible flesh. (For a recent treatment see R. Monica Moraes and A. Henderson, 'The genus *Parajubaea* (Palmae)', *Brittonia* 42 (2) (1990): 92–9.)

Cultivation: Both species are highly ornamental palms which deserve to become well known in cultivation. One species is fairly well known, the other seems to be rarely grown, if at all. They are well suited to temperate areas which have a cool, somewhat dry climate. Single plants are capable of producing fertile seeds.

Propagation: Seeds are usually slow and difficult to germinate. Some growers advocate leaving them in a warm, dry place for one or two months prior to sowing. Germination may be improved by pre-soaking in water for two days or carefully cracking or slicing the woody outer coat. It is recommended that the seeds be sown shallowly in an open mix and covered with a layer of sphagnum moss. Germination is usually erratic, with some seedlings appearing a few years after sowing. The seed produces a long sinker before leaves appear. Each fruit contains one to three seeds.

Parajubaea cocoides

(like a *Cocos*)

SOUTH AMERICA

A high-altitude palm which has been reported to occur at elevations of about 3000 m in the Andes Mountains of Ecuador and Colombia. The palm is cultivated in these areas for the fruit, which are relished by the local people who eat the nut-like seed kernel. The species is, however, unknown in the wild and there is speculation that it is of cultivated origin. It is a hardy, relatively fast growing, very handsome palm which is eminently suitable for temperate regions. Plants have a slender, dark grey trunk to about 20 m tall and a rounded crown of graceful leaves which are bright green to dark green above and silvery beneath. This palm is becoming a popular item in coastal California and is also grown on a limited scale in Sydney, Australia. Plants require well-drained soil and are tolerant of exposure to sun when quite small. Young plants may become stunted if underpotted or held in a container for too long.

Parajubaea torallyi

(after Torally, 19th century medical doctor in Chuquisaca, Bolivia)

Palma Chico

BOLIVIA

A palm of montane regions in Bolivia where it grows in the ravines of sandstone mountains at altitudes of 2400–2700 m. The palm grows in dense, extensive stands of many thousands of plants. The prevailing climate is dry with no rain for eight to ten months of the year. To survive, the palms must tap underground water and their survival is assisted by the very humid atmosphere in the ravines. The species is described as a handsome palm averaging about 14 m tall, with a crown of graceful, arching fronds each 4.5–5 m long. Large, much-branched inflorescences arise among the leaves. The greyish-green, smooth fruit, about 5 cm long, are broadly ovoid in shape and have a yellow kernel which is sweet and eaten avidly by the local people. This interesting palm has been introduced into cultivation but it remains a very scarce item. Plants should grow very well in temperate regions tolerating cold periods and dryness.

PELAGODOXA

(from the Greek *pelagos*, sea, *doxa*, glory, the glory of the sea)

A genus of one or perhaps two species. A well-known species, *P. henryana* is endemic to the Marquesas Islands north-east of Tahiti, with a second possible species on the Vanuatu island of Vanua Lava, in the Banks Group. Both species grow in dense rainforest. *P. henryana* is a very rare species which grows in narrow ravines in humid valleys at low altitudes. It is a solitary, unarmed palm with large, simple, bifid leaves and unusual tuberculate fruit. The inflorescence arises among the leaves and bears unisexual flowers of both sexes.

Cultivation: A highly prized palm which is much sought-after by collectors. It is still very rarely grown and requires warm to hot, humid conditions. Single plants are capable of producing fertile seed.

Propagation: *P. henryana* has proved to be difficult to germinate. A successful method used in Tahiti is to leave the fruit on the ground until the corky outer layer rots away. The hard seed is recovered and the outer endocarp cracked with a hammer to

Pelagodoxa henryana young plant.

Phoenicophorium borsigianum. G. Cuffe

obtain the kernel. The seed is placed on the surface of the mix (not buried) with the embryo in contact with the mix but not covered. Each fruit contains a single seed.

Pelagodoxa henryana

(after Augustine Henry, botanical collector in Asia)

MARQUESAS ISLANDS

A spectacular but very rare palm. Its most handsome feature is its large, entire leaves which may be 2 m long and 1 m wide. These are bright green above and whitish beneath, but unfortunately are readily split by the wind and only those plants in sheltered positions retain entire leaves. The trunk grows to 7 m tall and is fairly slender. The more or less round fruit are large (up to 10 cm across) and are covered with rough, tubercular, corky projections. A very rarely grown palm seen only in occasional collections, but much sought-after. Gardens in Tahiti probably have the most plants in cultivation and also have an active propagation program underway to increase numbers. This palm

is very cold-sensitive and the large leaves are subject to wind damage.

PHOENICOPHORIUM

(from the Greek *phorios*, stolen, thieved and *phoenix*, a palm, alluding to the theft of a plant of this palm from Kew Gardens)

A monotypic genus of palm endemic to the Seychelles, where it occurs on the islands of Silhouette and Mahé. It grows as an understorey plant in mountainous areas in dense rainforest, sometimes on steep slopes. A very distinctive, solitary, spiny palm with large entire leaves which are pinnately veined. The inflorescence arises among the leaves and bears unisexual flowers of both sexes. The leaves are used locally for thatch. The genus was previously known as *Stevensonia*.
Cultivation: A highly prized collector's item. At one time this species was a popular glasshouse plant in Europe. Now it is cultivated widely in the tropics but is still not common. It is very cold-sensitive and suitable only for the tropics. Single

Phoenicophorium borsigianum young plant showing attractive undivided leaves.

plants are capable of producing fertile seed.
Propagation: Seed may be slow and erratic to germinate. Each fruit contains a single seed.

Phoenicophorium borsigianum

(after Latanier Feuille Borsig, German horticulturist)

SEYCHELLES

A magnificent palm valued for its arching, simple leaves which may be more than 2 m long and nearly 1 m wide. Young plants are particularly distinctive, although spiny and make excellent tub specimens. The margins of the leaves are deeply indented, giving a toothed appearance. The petioles and trunk are very spiny when young but become smooth with age. In the forests to which it is native, this palm may reach more than 15 m tall, but cultivated plants are much less. In cultivation this species is quite slow growing and demands warm, shady, moist conditions.

PHOENIX

(from the Greek name for the date palm)
Date Palms

A distinctive, familiar group of about seventeen species of palm which are commonly known as date palms. They are found in Africa, Crete, various countries of the Middle East, Asia, South-East Asia, the Philippines and Indonesia. They are solitary or clumping (sometimes with creeping stems) feather-leaved palms which lack a crownshaft. Some species are large palms with massive woody trunks, others are small. The lowermost leaflets on each leaf are reduced to long, stiff, very pungent spines and the leaflets themselves are induplicate. The leaves also have a terminal leaflet. The inflorescence arises among the leaves and bears unisexual (rarely bisexual) flowers. The fruit is a berry. Many species are widely planted as ornamentals and the Date Palm is of major commercial significance.

Cultivation: Very hardy, adaptable, sun-loving palms which can be grown in a wide range of climates and situations. They are very popular subjects for cultivation, despite the spiny nature of their lower leaflets and are familiar in the parks, gardens and avenues of many countries. Male and female trees are essential for fertile seed production. Species hybridise freely and this should be borne in mind when collecting seed from cultivated plants.

Propagation: Clumping species can be propagated readily by the removal of suckers, although those lacking roots can be slow and difficult. Seed germinate readily, usually taking one to three months, but some species may take up to six months. Each fruit usually has a single seed, but sometimes up to three.

Phoenix acaulis

(without a stem)

BURMA—INDIA

The trunk of this palm is very short, swollen (sometimes grotesquely) and densely covered with persistent petiole bases. At its apex it bears a crown of arching pinnate fronds, 1–2 m long. The leaflets, which are dark green, are 30–45 cm long and arranged in small groups. The inflorescence (30–60 cm long) often trails over the ground and may be

Phoenix canariensis in fruit.

partially buried. After flowering it carries clusters of bright red, oval fruit which have a sweet, edible flesh. The palm is widespread in parts of India and Burma, usually growing on poor, stony ground. An attractive, very hardy palm which will grow well in subtropical and tropical regions.

Phoenix canariensis

(from the Canary Islands)

Canary Island Date Palm

CANARY ISLANDS

The Canary Islands are the home of this familiar palm which is now widely cultivated throughout the world. It is a familiar sight and has proved to be especially hardy in inland districts, tolerating the heat and dry conditions without setback. It is especially adaptable since it can be grown in temperate, subtropical and tropical regions and in coastal as well as inland areas. It is very frost-

hardy and will thrive in quite poor soils although it does not succeed where the drainage is poor. The stout, woody trunk grows up to 20 m tall and is crowned by long, pinnate, light green fronds which are spiny at the base. Large clusters of golden fruits are especially colourful in inland districts.

Phoenix dactylifera

(date bearing or finger bearing)

Date Palm

The Date Palm is famous for its succulent, edible fruit but these are only produced on plants growing in areas with a hot, dry climate. Despite this drawback, the palm is widely planted through-out temperate and tropical regions and is a very hardy, ornamental plant. It is an excellent coastal plant, tolerating full exposure and heavy buffeting by salt-laden winds. It also succeeds very well in hot inland districts especially where ground water is available. In cold–temperate regions the plants grow happily but rarely produce flowers. Young Date Palms produce suckers both on the trunk and around the roots, but in parks and gardens these suckers are rarely allowed to persist. Mature plants of flowering age produce suckers much less frequently than young plants. The crown is a graceful spreading one and the leaves are usually grey-green. Trees are either male or female and if fruit are required at least one of each sex is necessary. Date Palms can be propagated from seed or suckers.

Phoenix farinifera

(flour bearing)

Flour Palm

INDIA—SRI LANKA

In nature this palm grows close to the sea and frequently forms impenetrable, spiny thickets. It is native to parts of India and the northern area of Sri Lanka. The trunk grows to about 1 m tall and has a floury pith which may be harvested and eaten after baking or other suitable preparation. The fronds are short, deep green, with the lower pinnae very sharp and spiny. The fruit, which are about 1.5 cm long, have a sweet, mealy flesh. Because of its spiny nature, this palm has limited horticultural appeal, however, it can be successfully grown in a sunny position in tropical or subtropical regions.

Plantation of *Phoenix dactylifera*.

Phoenix humilis

(of low growth)

INDIA

A variable palm which is native to India where it
is widely distributed in a range of habitats from
coastal and lowland plains to the mountains. Plants
from lowland areas where they grow in sandy soils
are usually trunkless, whereas on rocky hills plants
develop a moderately stout trunk to about 2 m tall.
Arching leaves 1–2 m long, have long (to 50 cm),
narrow, grey to green leaflets arranged in groups of
up to four along the rachis. The fruit, which are
about 1.5 cm long, ripen from orange to black.
They have a sweet, edible flesh and are carried in
long clusters. One interesting variant (var. *peduncu-
lata*) from about 2000 m altitude in the Western
Ghats Mountain Range has very long inflorescences
which exceed the leaves and the plants are multi-
trunked.

Phoenix loureirii

(after João Loureiro, 18th century Portuguese naturalist)

INDIA—SOUTH-EAST ASIA

This species, although named in 1841, has been
much confused with *P. roebelenii*, especially in horticulture. Both species are very similar in general
appearance and in many important features.

Both grow in Indochina but *P. loureirii* is more
widespread, extending also to India and Hong
Kong. Plants of *P. loureirii* can be distinguished
from the very commonly grown *P. roebelenii* by the
leaflets which are carried in several planes (those
of *P. roebelenii* are flat and in one plane). In nature
plants may grow to 4 m tall but this is unusual in
cultivation. Fruits are oblong, about 12 mm long
and dark purple when ripe. This palm has also
been known as *P. humilis* and *P. hanceana*.

Phoenix paludosa

(growing in swamps)

Mangrove Date Palm

INDIA—SOUTH-EAST ASIA

As the common name suggests, this palm which is
native to India, Indochina and Malaysia, commonly
grows in low-lying, wet areas adjacent to mangrove
communities and along river deltas and estuaries.
It also grows inland and frequently forms dense,
almost impenetrable, thorny, thickets. It is a suckering species, very reminiscent of *P. reclinata* and
like that species makes a very decorative lawn
specimen for areas such as parks and large gardens.
The trunks often do not grow more than 2 m tall
(but may sometimes double that height) and the
leaves are spreading, with somewhat flaccid leaflets
which are green above and grey beneath. Leaf
sheaths are covered with coarse fibre and the

Phoenix pusilla.

petioles have long, yellowish thorns. The fruits, which are about 1 cm long, may be orange or black when ripe. The species is best suited to tropical and subtropical areas although in southern parts it has proved to be rather cold-sensitive.

Phoenix pusilla

(small, puny)

Ceylon Date Palm, Inchu Palm

SRI LANKA

Although a native of hot, moist, lowland tropical areas, this palm has proved to be surprisingly cold-tolerant and will grow quite successfully in temperate regions. It is a fairly slender but coarse palm, the trunk of which is rough and covered with densely packed leaf-bases. In cultivation, plants do not grow much more than 3 m tall. The fronds are green or more commonly glaucous and spread in an arching crown. They are distinctive because the leaflets radiate in many planes. The fruit are interesting, ripening to a red or violet colour and with a sweet pulp. This palm is very easily grown and

adapts to most well-drained soils. A sunny position is essential. This species has also been well known as *P. zeylanica*.

Phoenix reclinata

(turned downward)

Senegal Date Palm

AFRICA

This handsome palm forms a large clump consisting of many slender trunks curving away from each other and crowned with bright green, attractively curved pinnate fronds. Numerous basal suckers form a green rosette around the bottom of the trunks. The palm is native to tropical Africa along the margins of streams and soaks, often close to the coast and is a most impressive species for cultivation. It grows readily in tropical or temperate regions and seems quite tolerant of cold and frosts. Excellent for park planting, it looks particularly attractive when sited near water. Young plants respond markedly to regular watering, mulches and side dressings with fertiliser rich in nitrogen. A sunny position is most suitable. The brown fruit are edible but rather astringent.

Large clump of *Phoenix reclinata.*

Group planting of *Phoenix roebelenii*.

Phoenix robusta

(vigorous, robust)

Slender Date Palm

INDIA

The slender, tessellated trunk of this palm is a useful feature for its identification. The pattern is created by the persistent bases of fallen leaves and these are arranged in a loose spiral. The relatively short leaves (to 2 m long) and the shiny, thinly textured leaflets arranged in groups along the rachis are also useful identifying features. The fruit, to about 2 cm long, are brown when ripe. A native of India, this palm grows in isolated hilly districts and may be locally common. Plants seem to be relatively rare in cultivation.

Phoenix roebelenii

(after Carl Roebelen the original collector who discovered the species in Laos)

Dwarf Date Palm

LAOS

The origin of this palm has been uncertain for many years but has recently been established as being from Laos. It has become extremely popular in cultivation and is commonly grown, particularly in coastal districts of the subtropics, although it will also grow happily in temperate regions. Plants will grow in a sunny or shady position and are quite hardy and drought-tolerant once established.

When young it makes an excellent tub specimen and is also useful for indoor decoration, although its spiny leaflets are somewhat of a drawback. The trunk is slender and rarely grows more than 2 m tall, while the dark green fronds are generally curved or arch gracefully. Regular watering and side dressings with nitrogenous fertilisers promotes strong, lush growth. The fruit is edible but has only a thin layer of flesh.

Phoenix roebelenii inflorescence.

Phoenix rupicola

(growing on cliffs)

Cliff Date Palm

INDIA

This small- to medium-sized palm is native to India where it grows among rocks on cliffs and gorges. It is similar in many respects to *P. roebelenii* but is much more robust and with longer fronds which arch in a graceful crown. It also bears purple-red fruits that are very decorative in mass. This palm is not as commonly grown as other species of *Phoenix*, although occasional specimens are seen in tropical and subtropical regions. The trunk grows to 8 m tall and this, combined with the neat crown, makes this palm excellent for gardens. It will grow in a sunny or shady postion and is ideal for mingling with ferns in a shade house.

Phoenix sylvestris

(growing in forests)

Silver Date Palm, Sugar Date Palm, Khajuri

INDIA

A native of India where it is common in scattered stands, this palm is similar in general appearance to the Canary Island Date Palm, *P. canariensis*. It is

Phoenix sylvestris.

Phoenix rupicola.

generally much faster growing, however, and has a crown of grey-green or glaucous fronds. In its native state the flower stalks are tapped for their sap, which is boiled down to make date sugar. The yellow fruit are reputedly edible but are very acid and if at all green are also astringent. This hardy palm will grow in tropical or temperate regions and in inland and coastal districts. It requires a sunny position in well-drained soil.

Phoenix theophrasti

(after Theophrastus, Ancient Greek philosopher and botanist)

Cretan Date Palm

CRETE—GREECE—TURKEY

There has been some speculation that this palm has evolved from the Date Palm, *P. dactylifera,* or perhaps it itself is the wild predecessor of the Date—in fact the two species are botanically distinct. Early botanists have classified the two together but modern botanists have recognised the Cretan Date Palm as a distinct species because of differences in its flowers, the erect inflorescence and its small fruit which lack any edible flesh. A native of Crete, Greece and Turkey this species, restricted in number, forms small colonies on rocky

Pigafettia filaris.

regions but success has also been gained in the warm subtropics. Male and female trees are essential for fertile seed production.

Propagation: Fresh seed germinates rapidly, often within a month of sowing. Not all of the seed germinates together, with sporadic seedlings appearing months after the main batch has germinated. Seed loses its viability quickly if allowed to dry out. Each fruit contains a single seed.

Pigafetta filaris

(like a thread or string)

Wanga Palm

SULAWESI—MOLUCCAS—NEW GUINEA

One of the most graceful and beautiful of all palms, this species is very fast growing and with a slender, straight trunk which is greenish with prominent, grey rings. In nature plants are reported to reach a towering 50 m tall. The pinnate fronds are arching in a large spreading crown, with the dark green leaflets held obliquely erect. Golden spines cover each leaf base, petiole and rachis. The round, yellowish fruit are about 1 cm across and are patterned with overlapping brown scales. Studies by John Dransfield have shown that the species is a pioneer palm colonising disturbed

slopes in semi-arid areas. The climate is harsh with unreliable rainfall and the plants are very drought-resistant. This hardy palm has similar cultural requirements to those of *P. dactylifera*.

PIGAFETTA

(after Antonio Pigafetta, historian)

A monotypic genus of palm which occurs naturally in New Guinea, the Moluccas and Sulawesi. It is a common colonising palm of disturbed sites, with the plants being fast growing and long lived. The species is a very tall, spiny, solitary feather-leaved palm which lacks a crownshaft. The inflorescence arises among the leaves and bears unisexual flowers. The fruit is very small and covered with rows of reflexed scales. (For details of this species in the wild see J. Dransfield, 'A Note on the habitat of *Pigafetta filaris* in North Celebes', *Principes* 20 (20) 1976: 48.)

Cultivation: A stunning palm which is becoming very popular in cultivation. It grows best in tropical

Pigafetta filaris showing ornamentation of spines.

earth, embankments, roadsides and so on, but being absent from dense forests. Seedlings in fact die if they are too shaded. *Pigafetta* is rapidly gaining popularity in cultivation and is excellent if given ample room to develop. Young plants require an open, sunny position and are very sensitive to shading. The species was previously known as *P. elata*.

PINANGA

(from the Malayan *pinang*, a palm)

Pinang Palms

A large and varied genus of palm consisting of about 120 species occurring in southern China, northern India, South-East Asia, the Philippines, Indonesia and New Guinea. Significant areas of diversity are found in Malaysia, the Philippines and Borneo (about 45 species). These palms are mostly understorey plants which grow in rainforests and other humid forests and shady habitats. Some species are widespread whereas others are extremely restricted and may only grow in highly specialised habitats. They range from very small palmlets (some are acaulescent) to tall and robust plants and may have a solitary stem or develop a clumping habit (sometimes both growth types occur in the one species). The stems are usually distinctly ringed and the crownshaft (if present) is often prominent and colourful. Leaves range from being undivided (and commonly with an apical notch) to pinnate, with the leaflets arranged either regularly or irregularly on the rachis. Often the leaves, especially those of seedlings or sucker growths, exhibit interesting variegations or patches of colour other than green, these being displayed in an irregular pattern (termed mottling—see below). The inflorescence mostly arises below the crownshaft and bears unisexual flowers of both sexes. The rachillae may change colour between flowering and fruiting. The fruit are often highly coloured and may contrast with either a fleshy calyx or the swollen, colourful branches of the infructescence. Frequently the fruit ripen through a range of colours before reaching maturity. Some species produce very few fruit in the wild and there is a danger that regeneration of such species will be adversely affected by excessive seed collection. (For recent studies into members of the genus see J. Dransfield, 'Systematic Notes on *Pinanga* (Palmae) in Borneo', *Kew Bulletin* 34 (1980): 769–88; E.S.

Pinanga acaulis. J. Dransfield

Fernando, 'The Mottled-leaved Species of *Pinanga* in the Philippines', *Principes* 32 (1988): 165–74; and J. Dransfield, 'Notes on *Pinanga* (Palmae) in Sarawak', *Kew Bulletin* 46(4) 1991; 691–8.)

Cultivation: Without doubt this is one of the most exciting genera of palms for horticulture and it has attracted an ardent band of enthusiastic disciples. These palms offer a tremendous range of growth habit, stem size, stem colour and patterning, leaf shape and colouration and colourful, sometimes vivid, fruit displays. The phenomenon of mottled leaves in some species of *Pinanga* has attracted a great deal of attention. This mottling shows up as light areas on a dark green background and sometimes the mottling is associated with brown or reddish tinges. Frequently the mottling is spectacular on the developing new leaves which may have silvery, whitish or reddish tonings. Mature leaves may retain some mottling or become green. The mottled effect is also much more noticeable in juvenile plants and on sucker growths rather than on mature specimens.

Most specimens of *Pinanga* are of modest dimensions and are well suited to garden culture. Quite a number are small palmlets which make excellent pot subjects. The genus is well suited to the warm, humid conditions of the tropics but species originating in highland areas have performed very well in the subtropics. As a general rule these palms need shade, good drainage, abundant moisture and mulching. Species originating in the Bornean heath forests (kerangas) may have special cultural requirements as a number of them have proved to be difficult to grow. Some growers warn against excessive use of fertilisers for *Pinanga* and suggest that regular light doses are best. Heavy fertilising may reduce the colouration of mottle-leaved species. Certainly the colour and contrast of such leaves shows up best in plants which are grown relatively hard and with low nutrient levels. Single plants of all *Pinanga* species are capable of producing fertile seed, although some species are shy of flowering.

Propagation: Clumping species can be propagated by removal of suckers but these are often slow to establish. Seedlings of the larger species establish vigorously but some of the smaller-growing types may be slow. Fresh seeds germinate readily one to three months from sowing. Each fruit contains a single seed.

Pinanga acaulis

(without a stem or trunk)

MALAYSIA

A true dwarf palmlet with stems apparently absent or reaching about 10 cm tall. Plants have a rosette of entire, wedge-shaped leaves which are deeply notched at the apex. Sometimes these are divided into four or six leaflets. The leaves are dark green above and light bluish-green beneath. A simple, undivided spike bears white male flowers and pink female flowers. Black fruit contrast with the swollen red infructescence. A fantastic collector's palm from Malaysia.

Pinanga angustisecta

(with narrow segments)

BORNEO

This species can be recognised by its very narrow leaf segments which are held stiffy erect and are drawn out into long points. Plants have a short slender trunk and a crown of fronds each about 1.5 m long. The terminal leaflets are united and fishtail-like. Inflorescences have six to eight spreading branches and the ovoid fruit are about 17 mm x 8 mm, with a distinct point. An interesting species from the forests of Borneo.

Pinanga aristata showing mottled leaves. J. Dransfield

Pinanga aristata

(awned)

BORNEO

A colourful species which has been introduced into cultivation via seed collected in the Meratus mountains in Borneo at about 600 m altitude. It is widely distributed on the island growing on better soils between 200 m and 1000 m altitude. Newly emerged leaves have an attractive purplish colouration. The mature leaves have four to seven relatively broad, spreading, curved leaflets each with several veins. These leaflets are strongly mottled, with dark green on the upper surface and uniformly pale green beneath. The infructescence is densely covered in white, woolly hairs and the small fruit ripen through bright yellow to scarlet. The crownshaft is pale green and the slender stem (to 2 m long) is anchored to the ground by long stilt roots. Although of a clumping habit, plants of this species usually consist of a single stem and a few quiescent basal suckers. Previously known as *Pseudopinanga aristata*.

Pinanga auriculata

(ear-like)

SARAWAK

A native of Sarawak where it grows as an understorey palm in shady forests. Plants have a prominent crownshaft atop a slender trunk to about 2 m tall (20 cm thick) and a crown of pinnate fronds each about 1 m long. The fronds are divided into about twelve narrow, curved, pointed leaflets to

Pinanga bicolana.

35 cm x 8 cm. The downcurved inflorescence has about four branches and after flowering bears obovate fruit about 17 mm x 9 mm.

Pinanga bicolana
(from Bicol National Park, Luzon, the Philippines)

PHILIPPINES

Described in 1988, this species is related to *P. maculata* and *P. copelandii*. As with these taxa it is a solitary species but smaller growing, with a stem to 3 m tall and 4 cm across and a light green crownshaft covered with brown scales. The leaves are about 1 m long and have about twenty leaflets which are dark green and lightly mottled on the upper surface and greyish-white beneath. The leaflet tips are deeply incised. The pendulous infructescence carries two rows of fruit which ripen through red to purplish-black. This species is endemic to the island of Luzon and grows in forest at low altitudes. This species is uncommonly cultivated and is an excellent palm for the the tropics.

Pinanga brevipes
(with short stalks)

BORNEO

A solitary species which grows less than half a metre tall and has simple, pleated leaves deeply notched at the apex. This notch is flanked on each side by a short margin which is irregularly toothed.

Because this palm is so short, the inflorescence is often buried in the forest litter. An attractive small palm suitable for containers or a sheltered position in a tropical garden.

Pinanga caesia
(blue-grey, lavender blue)

SULAWESI

This slender palm has a solitary trunk to about 3 m tall, a reddish to red-brown crownshaft and a crown of graceful pinnate leaves which have an unusual bluish to violet coloured rachis. The infructescence has red rachillae and red fruit. The species is native to the island of Sulawesi and is found at low to moderate altitudes. It grows well in shaded tropical gardens.

Pinanga capitata var. capitata
(in heads, head-shaped)

BORNEO

A clumping palm which is very common in the montane rainforest of Borneo. Plants have bamboo-like stems to 2 m tall and 2.5 cm across and pinnate leaves with fourteen to twenty-four leaflets. The stems, leaf sheaths and petioles have a dense covering of scales. The leaflets are attached to the rachis at an acute angle and the terminal pair are

Pinanga caesia.

Pinanga capitata var. *capitata*. J. Dransfield

joined for part of their length. The inflorescence is much branched, with reddish rachillae supporting purple-black fruit. A robust species worth trying in subtropical regions. One variety has been described.

Pinanga capitata var. divaricata

(widely spreading apart)

BORNEO

Differs from the typical variety by the leaflets, those in the basal half of the leaf being irregularly arranged and reflexed whereas those in the upper part are regularly arranged and spreading at right angles from the rachis, with the apical pair not joined at all. This Bornean palm grows in rainforest at about 1600 m altitude.

Pinanga chaiana

(after Paul Chai, forest botanist in Borneo)

BORNEO

A distinctive undergrowth palm from Borneo where it grows on ridges at low altitudes. It can be recognised by its dull green crownshaft densely covered

with russet-brown hairs and its large entire leaves which have a whitish rachis. The leaves are deeply notched at the apex, more than 1.5 m long, pale dull green in colour and form a large rosette atop the slender (2.5 cm across) solitary stem which can grow in excess of 4 m tall. Young plants of this species would look particularly attractive in a pot.

Pinanga copelandii

(after E.B. Copeland, 20th century botanist specialising in ferns)

PHILIPPINES

This species is commonly confused with *P. maculata* in cultivation. It is larger growing than *P. maculata*, with a solitary stem to 10 m tall and 10 cm across and a long (to 1 m) dull green crownshaft. The leaves grow to 3 m long and the leaflets of both species are similar in size and shape, but those of *P. copelandii* are a uniform dark green on the upper surface or only slightly mottled. The leaflet tips are deeply incised. The fruit, which are borne in two rows (three rows in *P. maculata*) are widest above the middle (below or at the middle in *P. maculata*) and ripen through red to purplish-black. This robust palm is native to the islands of Luzon, Bohd, Negros, Leyte, Mindanao and Basilan in the Philippines, growing in forests from 100–250 m altitude (see also entry for *P. maculata*).

Pinanga caesia showing characteristic crownshaft and petioles.

Left: *Pinanga coronata*. Above: *Pinanga coronata* inflorescence.
Above Right: *Pinanga coronata* developing infructescences.

Pinanga coronata

(with a crown)

SULAWESI

An attractive cluster palm with slender trunks
each of which is topped by a prominent crownshaft
and a graceful crown of fronds. Each leaf is bright
green and pinnate, with moderately broad segments
which have prominent ribbed veins. Clusters of
fairly small fruit are red when ripe. Native to
Sulawesi, this palm grows very readily in tropical
and subtropical regions. Plants grow best under the
canopy of shady trees and like an abundance of
water, especially during dry times. In the wild this
species intergrades with *P. kuhlii* (see that entry
for further details).

Pinanga crassipes

(with a thick stalk)

BORNEO

A dwarf species from Borneo which has a very
short, relatively thick trunk (6–7 cm across). The
pinnate fronds have a thick, leathery sheathing
base and slightly curved leaflets, to 40 cm x 2.5 cm,
dark green above, slightly glaucous beneath and
with long-pointed tips. Branched infructescences

carry ellipsoid to oblong, fruit about 16 mm x
9 mm. Flowers and fruit may be present together,
making an impressive colourful display.

Pinanga cucullata

(hooded, hood-shaped)

SARAWAK

An attractive palmlet which forms colonies by
short, spreading stolons. The erect, slender stems
(to 60 cm x 8 mm) have the surface covered with a
mixture of brown scales and white hairs and aerial
branches sometimes arise from the internodes.
Each stem has an elongated crown of about eight
leaves which are deeply divided into two narrow
lobes (to 25 cm x 3 cm). These have a fleshy,
almost succulent texture and are dark green and
shiny on the upper surface and grey-hairy beneath.
Spreading or pendulous, simple or bifid inflorescences (white at flowering, red in fruit) are about
8 cm long. The ellipsoid fruit, to 13 mm x 5 mm,
are black when mature. *P. cucullata* is a highly
attractive palm which grows naturally in heath
forest (kerangas) in the State of Sarawak, Borneo.
It may be locally common but is shy of flowering.
This splendid ornamental has much to offer
horticulture.

Pinanga curranii

(after F.B. Curran, original collector)

PHILIPPINES

Another large species from the Philippines where it grows in lowland rainforests on the islands of Palawan, Dumaran and Busuanga. Plants commonly have a clumping habit but are occasionally solitary and with a long, dark green crownshaft covered with brown scales. Pinnate leaves, to 2.5 m long, have about fifty leaflets (to 60 cm x 4 cm) scattered unequally along the rachis. These are dark green above, greyish beneath and densely covered with woolly scales. The broad terminal leaflets are united and fishtail-like. The branched inflorescence has deflexed rachillae and bears bright red fruit, about 1.5 cm long, in two rows. This species is one for the tropics.

Pinanga densiflora

(with densely clustered flowers)

SUMATRA

One of the most colourful of all palms, this beautiful species is native to Sumatra where it grows in rainforest at about 500 m altitude. Developing leaves are of a pinkish colouration with chocolate brown mottles and blotches and a crimson petiole. As they mature, the main colour of the leaf

Pinanga crassipes inflorescence and colourful fruit. G. Cuffe

Pinanga densiflora. G. Cuffe

becomes bluish-green patterned with brown-centred dark green mottles and blotches. These colourful patterns are enough to whet the appetite of any serious palm collector but when the intense colour of the seedlings and young sucker growths is taken into account, the species becomes an even more desirable horticultural subject. The palm has a clumping habit and grows to about 2 m tall. Plants have been introduced into cultivation in Bogor and it is hoped that the species will eventually become widely grown in tropical countries.

Pinanga dicksonii

(after James Dickson, 18th & 19th century British botanist and nurseryman)

INDIA

An Indian palm which is common in cool, humid, mountainous gullies. Plants may be solitary or clumping, with smooth, slender, green stems to 5 m tall and less than 5 cm thick. The pinnate leaves are about 1 m long and carry numerous, broad, sessile leaflets which may be up to 60 cm long. The fruit are about 2 cm long and reddish when ripe. This *Pinanga* has proved to be adaptable in cultivation and more cold-tolerant than many of its relatives. It will succeed in both tropical and sub-tropical areas. A shady position seems preferable.

297

Pinanga disticha published as *Pinanga maculata* (*Botanical Magazine*, Vol. 1 of the 4th series, plate 8011)

Pinanga disticha

(in two rows)

MALAYSIA—SUMATRA

Native to Malaysia and Sumatra, this dwarf palm is commonly encountered in dry rainforest, growing in extensive, spreading patches. Its shiny, brown stems are thin and sinuous and grow to about 1 m tall. Plants sucker freely and the clumps spread quickly under good conditions. The leaves are simple with a deep apical notch or may be divided into a few small, narrow leaflets. These are very dark green and are attractively marbled with paler blotches and spots. Small, red, elliptical fruit are carried in two rows on slender, simple spikes about 10 cm long. This little gem of a palm would appear to have tremendous potential for garden culture. It should succeed in subtropical as well as tropical areas in a shady position. A variant having intensely mottled leaves has been collected in North Sumatra. Another with whitish, purple-blotched leaves is known from Malaysia.

Pinanga dumetosa

(bushy, shrub-like)

BORNEO

A clumping species from lowland areas of Borneo which may grow in thickets. Plants develop slender stems to about 1 m tall and the surface of the stems, crownshaft and leaf sheaths is covered with reddish-brown scales. The pinnate leaves, to 1.5 m long, have twelve to sixteen lanceolate leaflets to 40 cm long. The inflorescence is branched and, at fruiting, the rachillae become orange-red and the fruit ripen from crimson to purplish-black. This species appears to be rarely cultivated but has good prospects for tropical regions.

Pinanga elmeri

(after A.D.E. Elmer, 20th century American botanist specialising in the Philippine flora)

PHILIPPINES

This elegant species is very similar to *Pinanga philippinensis* but can be immediately distinguished by the rusty-brown scales on the leaf sheaths. It also is a small to medium-sized palm with slender, ringed stems that reach 2–4 m tall and 2–5 cm thick. Plants commonly form clumps but solitary

Pinanga dumetosa infructescence. J. Dransfield

specimens are known. Its fronds have numerous, narrow, pointed, dark green pinnae and spread in a graceful crown. Native to lowland rainforests of the Philippines, it is an attractive palm for the tropics and subtropics, apparently succeeding best in a sheltered position.

Pinanga geonomiformis

(like a *Geonoma*)

PHILIPPINES

A small palm from the Philippines which is uncommon to rare in its natural habitat, but is now becoming fairly widely grown in the tropics. It forms a neat, suckering clump of thin stems to about 1.5 m tall and the small, bright green fronds are entire or bear one or two pairs of leaflets which contrast nicely with the unusual purple, bamboo-like stems. The palm makes an excellent pot plant and in the garden likes a moist, shady position. It grows well in the tropics and can also be grown as a glasshouse plant further south. Seed germinates readily but may take up to five months to appear.

Pinanga gracilis

(slender)

INDIA

A graceful palm from India where in parts it is reported to be common in moist forests. It is a clumping species with stems only 2 cm thick and 3–5 m tall. The pinnate leaves are about 1 m long, dark green and with deeply cleft leaflets which are about 30 cm long. Plants in fruit are very decorative, having clusters of scarlet to orange fruit each about 1 cm long. This is a slender palm with many ornamental features and thus deserving of wide cultivation in tropical and warmer subtropical regions. It needs a protected position in moist soil.

Pinanga heterophylla

(with more than one type of leaf)

PHILIPPINES

This species from the island of Negros in the Philippines is cultivated on a limited scale. Plants have a slender stem to 3 m tall and 1.5 cm across, topped by a crown of spreading pinnate leaves. The crownshaft is covered with rusty-brown scales. The curved, widely spaced leaflets are dark green on the upper surface and paler beneath. Fruits are borne in three rows along the branches of the infructescence.

Pinanga hexasticha

(in six rows)

BURMA

A poorly known species which is native to Burma where it is reported to grow in marshy situations. It is a clumping palm with stems to 8 m tall and 3 cm across. The pinnate leaves have numerous, crowded leaflets, to 45 cm x 5 cm, with two or three ribs prominent. The inflorescence is a simple, undivided, stout, fleshy spike.

Pinanga hookeriana

(after J.D. Hooker)

INDIA

A species from the Khasia Hills, northern India, growing between 600 and 1200 m altitude. It is a solitary palm with a slender stem about 1 m tall and an arching crown of pinnate leaves each about

Pinanga javana in fruit. J. Dransfield

1.5 m long. The leaflets are crowded, to 30 cm x 2 cm and curve towards the apex of the frond. The inflorescences have four or five hanging branches and after flowering carry blackish, ellipsoid fruit, about 12 mm x 6 mm, narrowed at the apex. Apparently rare in cultivation but should be suited to tropical and subtropical regions.

Pinanga inaequalis

(unequal)

SULAWESI

An attractive small species with a solitary, slender stem and a prominent, white crownshaft. The leaves are less than 1 m long, have a very short petiole and six or eight, erect to spreading leaflets (to 25 cm long) with drawn-out tips. The petiole and rachis are covered with loose, woolly, brown scales. Inflorescences have three branches and after flowering carry blackish, ellipsoid, blunt fruit about 13 mm long. This species is native to Sulawesi, Indonesia.

Pinanga insignis

(remarkable)

PALAU ISLANDS—PHILIPPINES

Pinangas are mostly medium-sized, slender palms but this species is an exception, developing a large, woody trunk to about 10 m tall and 25 cm thick. It has a prominent crownshaft and its fronds are also relatively large (3 m. long) and have numerous, sword-shaped pinnae (scaly beneath) to 40 cm x 4 cm, which are distributed evenly along the rachis. These are stiffly spreading to almost rigid and have deeply notched, acuminate apexes. The ovoid fruit are about 2.5 cm long and red when ripe. This species is occasionally encountered in collections and is suitable mainly for tropical regions. Seed takes three to four months to germinate. It occurs naturally in the Philippines and on the Palau Islands, growing in wet forests at low altitudes.

Pinanga isabelensis

(from the Isabela Province, Luzon Island, the Philippines)

PHILIPPINES

An attractive species which is cultivated on a limited scale by enthusiasts. Plants have a solitary, ringed stem about 2 cm across and soft hairs on new fronds and leaf sheaths. The leaves have about twelve widely spaced, curved segments. The infructescence has flattened branches which bear two rows of fruit each about 2 cm long. The species grows in lowland districts often close to the sea on the island of Luzon in the Philippines.

Pinanga javana

(from Java)

JAVA

A palm which has become moderately popular in cultivation, growing well in tropical and subtropical regions. It is a solitary, very tall species with a brilliant green crownshaft and graceful crown of dark green pinnate fronds. Clusters of fruit are colourful and highly ornamental. Well-established young specimens look very attractive in a large pot and can be used for decoration indoors in a well-lit position. As the specific name suggests, the species is native to Java where it grows in rainforest at moderate to high elevations.

Pinanga kuhlii

(after Heinrich Kühl, German botanist who died in Java)

Ivory Cane Palm

INDONESIA

A handsome cluster palm which forms a neat clump of slender, yellowish trunks to about 8 m tall. Each bright green, prominently ringed stem has a short, swollen crownshaft and the leaves are scattered along the trunk rather than in a dense crown. The pinnate leaves are just over 1 m long and bear six to eight pairs of variably shaped, rather broad, falcate, prominently ribbed leaflets. It is native to Indonesia (Sumatra and Java) and grows in shady rainforest. It has proved to be rather cold-hardy, although plants need some protection, especially when young. It grows well in the tropics and subtropics and can perhaps be induced to grow in warm, protected positions further south. Given plenty of water and fertiliser the plants can be quite fast growing and prefer a shady position rather than full sun.

This is a very attractive and desirable palm which in cultivation has been confused with *P. coronata*. Indeed in the wild this species intergrades with *P. coronata* and John Dransfield regards them as synonymous, with *P. coronata* being the accepted name. The two extremes (*P. kuhlii* as one extreme and *P. coronata* as the other) do look very different and both are in cultivation. The situation becomes more complex because intermediate forms between the extremes may also be in cultivation and, as well, undescribed species of *Pinanga* of similar general appearance are also grown (from Sumatra, for example).

Pinanga latisecta

(with broad segments)

SUMATRA

A palmlet with a very short stem, 3–4 cm thick and dark green pinnate fronds which have brown scales on the sheathing base. The fronds have curved, lanceolate segments to 50 cm long and an apical flabellum about 10 cm wide. Erect inflorescences with two to four branches carry ovoid, blackish fruit, to 17 mm x 9 mm, each with a distinct, apical point. The species is native to Sumatra where it grows in rainforest.

Pinanga maculata.

Pinanga ligulata

(ribbon-like, strap-like)

BORNEO

An elegant species from the forests of Borneo. Plants have very finely divided fronds with numerous (about forty) narrow crowded segments to 30 cm x 1 cm. These are straight, have a long-pointed apex and are green above and paler beneath. The inflorescence has a few short, thick, deflexed branches. The ellipsoid, blackish fruit are about 2 cm x 8 mm.

Pinanga limosa

(growing in boggy places)

MALAYSIA

Although close to *P. paradoxa*, plants of this species can be readily distinguished by the pale greyish stems, paler leaves and waxy pink, yellow or white fruits. It is a solitary species with a slender trunk which is about 1 cm thick and yet can reach more than 4 m tall. The leaves, which are about 40 cm long, can be entire with a deeply forked tip or are

more usually divided into broad, curved, pale green leaflets. The fruit are about 1 cm long and are carried in two opposite rows on a simple spike 10–12 cm long. This species is a common, conspicuous palm of Malaysian rainforests growing from the lowlands to about 1000 m altitude.

Pinanga maculata

(spotted)

Tiger Palm

PHILIPPINES

Plants of this species are very rare in cultivation, with most of those grown as *P. maculata* actually being *P. copelandii*. The identity of *P. maculata* has been the subject of confusion with one authority incorrectly synonymising it with *P. copelandii*. More recently the taxon *P. barnesii* has been shown to be identical to *P. maculata*. Much of the confusion arose from an illustration which accompanied the original description of *P. maculata*. This beautiful painting depicts a young palm that was cultivated in a nursery in Belgium. The plant, which had heavily spotted foliage, had been originally collected in the Philippines in the nineteenth century by the French explorer Marius Porte. It is now known that *P. maculata* is widely distributed in the Philippines, being found on the islands of Luzon, Polillo, Mindoro, Mindanao, Panay and others. It grows in humid forests from low levels to high altitudes (300–1500 m).

    P. maculata is a solitary species with a stem to 5 m tall, 5 cm across and a purplish brown to orange crownshaft. Leaves to 2 m long have about thirty leaflets unequally spaced along the rachis. These are dark green above with large, irregular blotches of lighter green and greyish-white beneath. The leaflet tips are deeply incised. The infructescence has pendulous branches and bears dense clusters of fruit, each about 2.5 cm long, which ripen through red to purplish black (see also entry for *P. copelandii*).

Pinanga malaiana

(from Malaya)

SOUTH-EAST ASIA

A slender, clumping palm (occasionally solitary) with ringed, bamboo-like trunks which may reach more than 5 m tall and 3.5 cm across. Its crown consists of a few short pinnate leaves with equally spaced, broad leaflets. In young plants the leaflets are distinctly notched at the apex but this feature disappears in older plants. The crownshaft is prominent, being flushed with bronze or pink tonings. The infructescence is pendulous and particularly showy, having pinkish-red rachillae and purplish-red fruit set in a black calyx. This species is an excellent garden palm for the tropics. It likes a shady or protected position and appreciates rich soil, plenty of organic mulches and water during dry periods. The species is native to Malaysia where it is common in lowland rainforests and apparently also extends to Sumatra and Borneo.

Pinanga minor
(smaller, lesser)

SULAWESI

This species is known from the vicinity of Gorontalo in northern Sulawesi where it grows in lowland rainforest. Plants have a solitary, slender trunk, to 3 m x 3 cm and a spreading crown of pinnate fronds each about 1.5 m long. The leaves have about twenty pairs of narrow-lanceolate leaflets to 40 cm x 3 cm. The branched inflorescence is curved downwards and after flowering bears ellipsoid, blackish fruit each about 15 mm x 8 mm.

Pinanga mirabilis form with divided leaves. G. Cuffe

Pinanga mooreana infructescence. J. Dransfield

Pinanga mirabilis
(wonderful)

BORNEO

With its crown of large, arching fronds somewhat reminiscent of a shuttlecock, this species is an impressive palm much in demand by collectors. The fronds, which may be either entire or pinnately divided, are deeply notched at the apex, have coarsely toothed apical margins and taper narrowly into the dark petiole. Plants usually consist of a single erect stem with a few quiescent basal suckers, but sometimes lack any suckers. The species is native to lowland rainforests in Borneo. Plants of this species from the State of Sarawak have large, undivided leaves which arch in an impressive rosette.

Pinanga mooreana
(after Professor H.E. Moore Jr, noted 20th century American palm botanist)

SARAWAK

The ripening fruit of this palm change from greenish to yellowish then orange before maturing purplish-black and at the same time the rachillae of the infructescence change from yellowish-green to yellow then orange. Plants form dense clumps, with the stems growing to 8 m or more tall. The crownshaft and leaf sheaths are a dull purplish colour and densely covered with chocolate-brown scales. The pinnate leaves, to 3 m long, have about fifty, dark green, leathery, stiff leaflets with the apical pair united. An interesting palm for the tropics, originating in the State of Sarawak, Borneo, growing in lowland forests.

Pinanga negrosensis

(from the island of Negros, the Philippines)

PHILIPPINES

A large clumping member of the genus with pale green to whitish stems, to about 8 m tall and 7 cm across and a pale green to glaucous crownshaft. Pinnate leaves to 2 m long form a dense crown with the rigid leaflets held erect. These are dull green with two prominent yellowish veins on the upper surface. The infructescence bears many pendulous branches with the ellipsoid fruit (about 12 mm long) ripening through yellowish to red then velvety black. Essentially a collector's palm, this distinctive species grows on the island of Negros in the Philippines, in damp ravines at about 1000 m altitude.

Pinanga pachyphylla

(with thick leaves)

SARAWAK

A rare species which grows in the heath forests (kerangas) in the State of Sarawak, Borneo, at low to moderate altitudes. Plants may have a solitary stem or develop into clumps with stems to 3 m tall and 2 cm thick. A well-developed, succulent, green crownshaft supports a crown of about five pinnate leaves. These are about 1.2 m long and have up to twenty, bright green, succulent, divergent, narrow leaflets. The rachillae of the branched, pendulous inflorescences are green in flower and orange in fruit. An interesting collector's palm which is not known to be in cultivation.

Pinanga paradoxa

(contrary to the usual form)

MALAYSIA

A small species from the mountains of Malaysia where it grows in moist forests at altitudes between 300 m and 1400 m. Plants have a thin (6 mm), dark, shiny, brown stem 1–2 m tall and a crown of small pinnate fronds each about 50 cm long. These are divided irregularly into narrow, curved leaflets which are dark green above and grey-green beneath. Simple pendulous spikes carry red, curved fruit arranged in two ranks. An attractive small palm for tropical and perhaps subtropical regions.

Pinanga patula

(slightly spreading)

SOUTH-EAST ASIA

This clumping palm has become moderately popular in cultivation, being most suitable in tropical regions. Plants will withstand exposure to some sun but for best appearance should be protected from extremes. An attractive palm which forms crowded clumps to about 3 m tall with arching pinnate leaves, each having a few, broad, widely spaced, dull green leaflets. The rachillae of the infructescence and the ripe fruit are both bright red. The species occurs naturally in Malaysia, Sumatra and Borneo at low to moderate altitudes.

Pinanga pectinata

(divided like a comb)

MALAYSIA

A clumping species from lowland forests of Malaysia which often grows in low-lying areas subject to periodic inundation. Plants have ringed stems to 4 m tall and pinnate leaves about 1 m long, which are regularly divided into pointed leaflets to 30 cm x 3 cm. The sparsely branched infructescence is held erect and carries dark red to blackish fruit arranged in two ranks. An interesting palm for the tropics.

Pinanga perakensis

(from the State of Perak)

MALAYSIA

The prominent zigzag appearance of the rachillae and the erect habit of the infructescence are useful features for the identification of this Malaysian palm. It is a clumping species with crowded, slender, somewhat sinuous stems to about 3 m tall and arching fronds with numerous narrow leaflets. The ellipsoid fruit (about 1 cm long) are borne alternately in two ranks along the rachillae. They mature through orange and red colorations to ripen black. In some areas of Malaysia this species may form extensive thickets. It grows at moderate elevations and favours dry ridges in rainforest. *P. densifolia* is a synonym.

Pinanga philippinensis

(from the Philippines)

PHILIPPINES

The trunks of this Philippine palm do not get any more than 3 cm thick and have prominently stepped nodes. The crownshaft is very conspicuous, being pale-coloured and somewhat swollen and the short fronds are held erect in a manner reminiscent of Betel Nut Palm. The leaf sheaths are covered with a thick, grey wool and the leaves have numerous, broad, pointed, dark green segments. Its fruit, which are about 1.5 cm long, are carried in two opposite rows on the rachillae. This is an ornamental, small- to medium-sized palm suitable for tropical conditions.

Pinanga pilosa

(hairy)

BORNEO

A small palm from Borneo which grows in highland rainforest in montane situations, such as on Mount Kinabalu. A clumping species, the plants having very slender (less than 1 cm thick), almost willowy stems and spreading to arching pinnate leaves. The leaflets, which are relatively broad and multiveined, are widely spaced on the rachis and curve towards the apex of the leaf. An attractive species which may succeed in subtropical regions.

Pinanga polymorpha

(with many forms)

MALAYSIA

As the specific name suggests, this palm is somewhat variable and can be found in many different forms. Its most variable feature is its leaves, which may be entire or divided into narrow or broad leaflets. The leaves of this species are generally short and crowded into a dense, somewhat untidy crown. They are dark green above with conspicuous, attractive yellow mottling and greyish-green beneath. Its stems, which are generally less than 2 cm thick, grow to about 3 m tall and lean and twist through the surrounding vegetation. Conspicuous aerial roots are produced at the base of each stem. The fruit are black. This palm is native to the highlands of Malaysia and is very common, growing in extensive thickets in moist

Pinanga pilosa infructescence. J. Dransfield

Pinanga polymorpha. G. Cuffe

areas. Plants in fruit are very decorative as the fruit are dark bluish green contrasting with the red rachillae. The fruit are edible. A useful ornamental palm, this species grows best in a protected, shady position in the tropics.

Pinanga rivularis. J. Dransfield

Pinanga pulchella

(pretty, beautiful)

SUMATRA

The stems of this palm are very slender (about 1.5 cm across) and grow to about 1.5 m tall, with the larger ones often falling over. The widely spaced leaflets are marbled and mottled with yellowish-green and dark green, the colours being much intensified in sucker growths and newly emerging leaves. Fruit are black and attached to crimson rachillae. This species, a native of northern Sumatra, has been introduced into cultivation.

Pinanga punicea

(scarlet, carmine)

NEW GUINEA

A little-known species from eastern Papua New Guinea where it grows in lowland rainforest. It is a single-stemmed palm, the trunks of which are fairly slender (4–8 cm thick) and may grow to 6 m tall. The crown consists of a few finely divided fronds atop a bright green crownshaft. Ripe fruit are bright red. Cultural requirements of this species are largely unknown but are probably similar to those of other tropical species.

Pinanga ridleyana

(after H.N. Ridley, 20th century Malaysian botanist)

SARAWAK

A truly dwarf palm with slender stems to about 30 cm tall and a rosette of spreading, narrow, shiny, undivided leaves. Each leaf has a blade which is more or less oblong in shape, tapered to the base

with a notched apex and a very short petiole. A delightful palm from Sarawak which would make an excellent potted subject.

Pinanga rigida

(stiff, rigid)

PHILIPPINES

A native to the island of Negros in the Philippines, this species grows in mossy woods at about 2000 m altitude. It is similar to *P. woodiana* but with a single vein in the leaflets. Plants have a solitary, prominently ringed, brown trunk to 5 m tall and a crown of dark green, stiff leaves. The rigid leaflets have a papery texture and a stiff, pointed apex. The inflorescence is much branched with recurved branches and the fruit are dark red and velvety when mature.

Pinanga riparia

(growing beside streams)

MALAYSIA

An interesting species which grows in wet situations beside streams, sometimes with the base of the plant submerged. It spreads by long stolons which may appear some distance from the parent plant. Stems grow to about 5 m tall and the leaves are divided into about twelve curved leaflets which are broadest in the middle and pointed at each end. The fruit are carried in two opposite rows on a branched infructescence. The species is native to some lowland areas of Malaysia.

Pinanga rivularis

(growing by rivers)

BORNEO

A rheophytic species which grows in dense clumps along stream banks in Borneo. It is similar to *P. tenella* in that it has slender stems 1.5 m tall or less and leaves with narrow, grassy leaflets. Its stems, however, commonly produce aerial branches from the nodes and the leaflets have a shorter, less tapered apex. In addition, the inflorescence is unbranched and the female flowers have a short, tubular calyx with three blunt lobes. Its fruit are bright red to purple-black, rounded with an apical depression and have unusual, hooked fibres which may aid in the dispersal of the species by water. An interesting small palm that deserves to become well known in cultivation.

Pinanga rupestris

(growing among rocks)

SARAWAK

A remarkable palm which grows in the crevices of sandstone rock faces and cliffs. It is known only from Bako National Park in Sarawak and, to date, fruit have not been collected. It is a diminutive, clumping palmlet, with slender, usually pendulous stems to 60 cm long (usually much less) and only 6 mm thick. Plants sucker freely and aerial growths are often produced from the internodes. The narrow leaves (to 50 cm x 6 cm) are entire but deeply notched at the apex, with the tips of the lobes being shallowly toothed. This restricted, highly specialised palm is not known to be in cultivation.

Pinanga salicifolia

(with willow-like leaves)

BORNEO

This palm has interesting, dimorphic leaves. Those on seedlings and sucker growths have finely divided, fern-like leaves with numerous small, one-veined leaflets, whereas mature stems have leaves with three to six, broad, S-shaped leaflets. Between these extremes is a wide range of size and leaf division. The rachis, petiole and leaf-base are also distinctive, being densely covered in a mixture of narrow, brown hairs and grey, inflated hairs. The inflorescence is branched and the fruit are long, narrow and curved. This species is widespread and common in Borneo where it grows in a wide range of habitats, including steep ridge tops and on mounds in swamps. An attractive palm for cultivation. *P. canina* is a synonym.

Pinanga scortechinii

(after Father Scortechini, 19th century collector in Malaysia)

MALAYSIA

A solitary or clumping palm to 4 m tall with a prominent, somewhat swollen, orange crownshaft and large pinnate fronds which spread in a graceful

Pinanga simplicifrons in habitat. G. Cuffe

crown. The leaflets are glaucous-green beneath. The erect, branched infructescence has attractive red rachillae and waxy, ivory-white fruit flushed with pink. The species is widespread in moist forests of Malaysia, from the lowlands to about 1000 m altitude.

Pinanga simplicifrons

(with undivided leaves)

MALAYSIA—SUMATRA

As the specific name suggests, the leaves of this palm are simple and undivided (although they are occasionally pinnate). They are oblong in shape, with a deeply forked apex and are scattered along the stems. The stems are very slender and sinuous and the palm forms spreading clumps to about 1 m tall. It is native to Malaysia and Sumatra and is locally common in lowland forest, frequently along the floodplains of streams where it thrives in the alluvial soils. It is an attractive small palm for a pot or a shady tropical garden. Small, scarlet, horn-shaped fruit are borne sporadically throughout the year. This species can be propagated from seed, suckers or by aerial layers.

Pinanga speciosa
(beautiful)

PHILIPPINES

A slender solitary palm from the island of Mindanao in the Philippines where it grows in shady rainforests. The plants have a relatively sturdy, ringed trunk of unusual blackish-green hues, which grows to about 10 m tall. The crownshaft is of similar dark colouration below a crown of dark green pinnate leaves each about 2.5 m long. Branched infructescences carry small, blackish fruit each about 4 mm long in two opposite rows. A sturdy species suitable for cultivation in tropical regions. Plants will tolerate exposure to some sun if mulched and kept moist.

Pinanga tenella var. *tenella*. J. Dransfield

Pinanga tenacinervis infructescence and stilt roots. J. Dransfield

Pinanga stricta
(stiff, upright)

BORNEO

This is one of a group of small *Pinanga* which hold their leaflets stiffly erect along the rachis. *P. stricta* is readily distinguished from the others by the stiffly erect (almost rigid) fronds (50–75 cm long) which have few (eight to ten) segments each to 25 cm x 2.5 cm. Inflorescences have two to four flexuose branches and bear blackish fruit about 14 mm x 6 mm. An interesting species from the forests of Borneo.

Pinanga tenacinervis

(with tough nerves)

SARAWAK

A very unusual species from Sarawak where it grows at low altitudes on shallow, alluvial soils overlying limestone. Plants have very short stems (no more than 10 cm tall), supported by conspicuous stilt roots. They apparently produce subterranean suckers which emerge some distance from the parent and so form loose colonies. Leaves about 1 m long bear about twelve, dark green, glossy leaflets. The pendulous inflorescence, which may be entire or forked, carries shiny, red fruit at or near ground level. This novelty palm is not known to be in cultivation and would probably be most suitable in the tropics.

Pinanga tenella var. *tenuissima*

(very slender)

BORNEO

This intriguing little palm is common along stream banks throughout Borneo. It is a true rheophyte, growing within the flood zones of the rivers in silt, on pebbly banks and in rock crevices. Plants form dense clumps of sixty or more stems to 2 m tall, with the older stems tending to trail in the direction of the water flow. Plants branch only from the base. The pinnate leaves are dark green, with very narrow (grass-like), leathery leaflets, which provide minimum resistance to water flow. Each leaflet tapers to a long acuminate point. The inflorescence is branched and the female flowers have three unattached, but closely overlapping, sepals. The fruit is spindle-shaped and pointed towards the tip. This species would make an excellent palm for cultivation in shady areas among ferns and other plants. One variety has been named.

Pinanga tenella var. *teunissima*

(very slender)

BORNEO

This variety can be distinguished by having leaflets with two folds rather than a single fold, as in var. *tenella*. It grows along streams in Borneo, in zones subject to inundation by floodwaters. *P. calamifrons* is a synonym.

Pinanga veitchii showing colourful mottled leaves.
J. Dransfield

Pinanga tomentella

(finely covered with felt-like hairs)

SARAWAK

A clumping palm with slender, erect stems 50–80 cm tall and distinctive entire, paddle-shaped leaves in a graceful, arching crown. These are dark shiny green on the upper surface with a silvery central stripe, silvery beneath; and the apex has a narrow cleft flanked by a series of sharp teeth. It is native to the State of Sarawak in Borneo.

Pinanga variegata

(of more than one colour)

SARAWAK

A palmlet with a single, slender trunk to 1 m x 1 cm, and a crown of small, obovate fronds deeply notched at the apex. Some fronds are divided into about four curved segments, most are entire. They are green with patches of lighter colour. Flowers are borne on a simple, undivided spike and are followed by ellipsoid, black fruit about 17 mm x 7 mm arranged in two rows. The species is native to Sarawak, Borneo.

Pinanga veitchii

(after John Veitch, 19th century English nurseryman)

BORNEO

The mottled leaves of this small palm are well camouflaged among the litter of the Bornean forest floor where it occurs naturally. A small clumping species with slender stems to about 50 cm tall and entire leaves which are deeply notched at the apex. The dark veins and mottling is prominent against the pale background of the leaf lamina. An excellent small subject for container cultivation.

Pinanga woodiana

(after Major General Leonard Wood)

PHILIPPINES

A solitary species with slender (4 cm), ringed stems to 4 m tall and a finely hairy crownshaft. The pinnate leaves, to 2 m long, have about twenty-five, well-spaced leaflets to 55 cm x 4.5 cm. The leaflets are stiff and papery, dull green and with a stiff, rigid point. The inflorescence has rigid branches and after flowering bears ovoid fruit about 12 mm long which ripen through red to dark purple. This attractive species occurs in the Philippines on the island of Mindoro, growing at about 1000 m altitude.

Pinanga yassinii

(after Haji Mohammad Yassin bin Ampuan Salleh, Director of Forestry, Brunei)

BORNEO

A noteworthy species from Brunei and Sarawak which grows in heath forests (kerangas) in shallow hollows and areas just below the crests of ridges. Plants form dense clumps to about 1 m across and have many sucker growths, but few erect stems develop. The leaves of suckers have distinct petioles and frequently they exhibit a faint mottling, whereas those on mature stems are sessile and lack any markings. The leaves themselves (to 50 cm x 8 mm) are remarkable in the genus, being very stiff and deeply pleated with a grey undersurface. They are entire (rarely divided into a few leaflets) with an apical notch and the two lobes are sharply toothed. This intriguing palm appears rarely to fruit and as with other species from this specialised habitat may be difficult to grow.

POLYANDROCOCOS

(from the Greek *poly*, many, *andros*, male (stamen) and *Cocos*, another genus of palm, literally a many-stamened *Cocos*)

A small genus of two species of palm endemic to Brazil. They grow in open habitats and along forest margins in dry sandy soil. They are tall, solitary, armed or unarmed feather-leaved palms which lack a crownshaft. The large, unbranched, pendulous inflorescences arise among the leaves and bear unisexual flowers of both sexes.

Cultivation: Interesting palms which are often grown in botanical collections but are worthy of wider general use. Seedlings and young plants are particularly attractive but older plants are less appealing. Although usually seen in tropical regions, these palms have proved to be adaptable to the subtropics and perhaps will survive in warm temperate regions. Drainage must be excellent. Single plants are capable of producing fertile seed.

Propagation: Seeds are usually slow and erratic to germinate. Pre-soaking in warm water and the use of bottom heat may be beneficial practices. Each fruit contains a single seed.

Polyandrococos caudescens

(in the form of a tail)

Buri Palm

BRAZIL

Interesting features of this palm include a large, graceful crown of dark green, finely divided leaves with the leaflets clustered, showy yellow flowers borne on long, thick, sausage-like spikes and clusters of greenish-yellow to orange, edible fruit. It is a hardy, adaptable but slow-growing palm which will tolerate sun from a very early age. In its natural habitat of eastern Brazil, plants commonly grow in dry, sandy soils close to the coast and the species could have potential for coastal planting. Plants are also apparently tolerant of periods of dryness once established. Previously known as *Diplothemium caudescens*.

Polyandrococos pectinata

(divided like a comb)

BRAZIL

Although described in 1901, this elegant palm was generally overlooked until a well-established specimen that appeared to fit the original description was found growing in the palm house at Kew Gardens. It has a slender trunk topped with an attractive, spreading crown of dark green fronds (whitish to silvery beneath). These have widely spreading leaflets regularly arranged along the rachis. The inflorescence and fruit are similar to those of *P. caudescens*. The species is native to Brazil and grows at higher altitudes than *P. caudescens*. Whether this poorly known species is really distinct from *P. caudescens* needs further research. Plants have been successfully grown in Sydney, Australia.

PRITCHARDIA

(after W.T. Pritchard, 19th century British official in Polynesia)

Loulu Palms

A genus of thirty-seven species of palm, with all but four being endemic to the Hawaiian Islands. The others occur in Fiji, Tonga, the Cook Islands, Solomon Islands and the Danger Islands. Many of the species from the Hawaiian Islands are reduced to great rarity and some, such as *P. montis-kea* and *P. macrocarpa*, are apparently extinct in the wild. The species from Polynesia grow commonly on coral atolls whereas the Hawaiian species are found on the windward side of volcanic islands in dense, wet forests and often precipitous valleys, ranging from near sea-level to moderately high altitudes. These palms are very popular with the native Hawaiians who call them *lo'ulo* and use the leaves for thatch, hats, baskets and fans and prize the immature fruits, called *hawane,* as a delicacy.

Species of *Pritchardia* are solitary, unarmed fan palms which are renowned for their heavy, rounded crown of large, stiff, pleated leaves. The inflorescence, which arises among the leaves and carries bisexual flowers, is unique with its long peduncle having the floral branches clustered at the apex.

Cultivation: Handsome palms of exotic appearance which epitomise the essence of the tropics and which are unrivalled for adding that tropical flavour to a garden or palm collection. Some species are widely planted, many are hardly known, but all have ornamental features and are worthy of planting. Unfortunately these palms hybridise very freely so seed collection from mixed plantings is not recommended. Plants need excellent drainage and a sheltered location which is exposed to sun for part of the day but protected from extremes of sun and wind. Most species grow best in the tropics but some Hawaiian species from higher altitudes succeed in the subtropics. Single plants are capable of producing fertile seed.

Propagation: Fresh seed germinates readily, usually within two to five months of sowing. Each fruit contains a single seed.

Pritchardia affinis

(similar to another, related)

HAWAIIAN ISLANDS

This species is one of five to occur on the island of Hawaii which is the youngest island of the Hawaiian group. It grows as individuals or in scattered colonies on the west coast from sea-level to about 600 m altitude. Whether these plants are truly wild or survivors from ancient cultivation is conjectural. The species is an attractive palm which is popular in cultivation, being most suitable for tropical regions. The leaves are bright green and when examined closely, are found to be dotted with small yellowish scales. The flowers are yellowish and the globose fruit 20–25 mm across.

Pritchardia arecina

(like an *Areca*)

HAWAIIAN ISLANDS

A large, distinctive species which grows to about 12 m tall and with a trunk to 50 cm across. Young leaves are yellowish and the mature leaves retain a layer of golden, fringed scales on the undersurface. The large, rounded to ovoid fruit, about 4.5 cm x 4 cm, are dark brown and shiny when ripe. The species inhabits the wet, forested slopes of the eastern end of the Hawaiian island of Maui, between 600 m and 1200 m altitude.

Pritchardia beccariana

(after Odoardo Beccari)

HAWAIIAN ISLANDS

This beautiful, very symmetrical palm is native to the island of Hawaii where it occurs in rainforest on the windward slopes up to 1200 m altitude. Plants have a slender stem to 18 m tall and large, handsome leaves which are bright green but somewhat scaly on the underside, the scales being distinctly fringed. The tips of the leaflets are not rigid but droop slightly. A useful diagnostic feature is the large, shiny black fruit which is variable in shape (ovoid to globose), from 2.5–3.5 cm long. Plants of this species grow well in the tropics and in warmer areas of the subtropics in a sunny position in well-drained soil.

Pritchardia gaudichaudii.

Pritchardia eriostachya

(with woolly spikes)

HAWAIIAN ISLANDS

A native of the island of Hawaii, this species can be immediately recognised by the dense covering of woolly, salmon-coloured hairs on the inflorescence, especially the bracts. The leaves are also distinctive, with the lower surface of young leaves being covered with soft, woolly scales. The fruit, which is ovoid to obovoid, is 3.5–4 cm long when mature. The species grows in dense rainforest at about 1000 m altitude.

Pritchardia gaudichaudii

(after Charles Gaudichaud-Beaupre, 19th century French botanist)

HAWAIIAN ISLANDS

Native to the Hawaiian island of Molokai where it grows in large colonies on vertical cliffs near the sea, this palm is a very attractive subject for cultivation. The plants have large, heavily pleated fan leaves which are more than 1 m across. Young leaves and the upper part of the crown are covered with attractive, white wool and as the leaves mature, the wool is shed and the leaves are bright green. The spherical, shiny fruit are about 4 cm

across. This fan palm succeeds admirably in subtropical regions. A sunny aspect is essential and the plants prefer soils with free drainage.

Pritchardia hardyi

(after W.V. Hardy who discovered the species)

HAWAIIAN ISLANDS

A tall-growing species (to 20 m or more) with an impressive crown of large, leathery leaves which have conspicuous drooping segments. The lower surface of the leaves is covered with yellow, fringed scales and the ellipsoid, black fruit are about 2 cm long. The species is native to the Hawaiian island of Kauai where it grows on ridges in rainforest at about 600 m altitude.

Pritchardia hillebrandii

(after W.F. Hillebrand)

HAWAIIAN ISLANDS

A handsome Hawaiian palm which will withstand some coastal exposure although not severe conditions. It succeeds best in tropical districts but can also be induced to grow in warmer subtropical areas. Interestingly, this species was described from material cultivated by Hawaiian Islanders on the

Pritchardia hillebrandii.

dry leeward coast of Molokai and to this date has not been discovered in the wild. The plants grow 5–7 m tall and their leaves have bluish-green petioles which are densely woolly beneath. The blade itself lacks scales and is bluish-green from a thin coating of waxy material, especially noticeable on the underside. The leaf segments are rather stiff and do not droop. The globose fruit, about 2 cm across, is shiny bluish-black at maturity.

Pritchardia kaalae

(from Mount Kaala, Oahu, Hawaii)

HAWAIIN ISLANDS

Native to the island of Oahu, this fan palm grows in rocky areas in the drier forests of the western side. It can be distinguished by its long inflorescences (more than 2 m long) which hang well clear of the crown, and its small, black fruit which are usually about 2.5 cm across. The plants grow to about 10 m tall and have a sturdy crown of large leaves which are deep green on both surfaces and with drooping segment tips. Tropical and warm-subtropical climates are suitable for its cultivation. It is a slow-growing species but plants are decorative from a young age.

Pritchardia lanaiensis

(from the Hawaiian island of Lanai)

HAWAIIAN ISLANDS

The green leaf blade of this species is almost paper-thin but with a tough, leathery texture and the lower surface is closely spotted with rusty-brown scales. The leaf segments are slender and may droop at the tips. The dull brown fruit are spherical, about 2 cm across and with the remains of the style on the top. The species grows on precipitous cliffs and gorges on the Hawaiian island of Lanai. It has been introduced into cultivation and is grown on a limited scale.

Pritchardia lanigera

(bearing wool)

HAWAIIAN ISLANDS

A native of the island of Hawaii, this species is recognised by its very large leaves (about 1 m across) which are shallowly divided into rigid

segments. In nature it grows in patches of rainforest which clothe the margins of steep valleys and it often survives on precipitous sites. Although the species is cultivated on a limited scale, it has not become commonly grown, possibly because of a lack of propagating material. It is a very handsome palm, especially those plants in early trunk development.

Pritchardia maideniana

(after J.H. Maiden, NSW government botanist, 1896–1924)

HAWAIIAN ISLANDS

For many years only two specimens of this palm were known, these being found in the Royal Botanic Gardens, Sydney, Australia. It is believed that the original seeds were collected somewhere in the Hawaiian Islands by J.H. Maiden but the palm has not been relocated in the wild, probably

Pritchardia maideniana.

because of the destruction of its habitat. The species was described in 1913 by Beccari from plants growing in Sydney. Of the two palms in Sydney one has fruited, but this plant was subsequently damaged in a storm. The existence of this species is very tenuous indeed but seed has been distributed from the Sydney plants. Despite its apparent tropical origin, the plant seems well suited to the Sydney climate and soils. The species is characterised by leathery, glaucous, waxy, deeply segmented leaves, short, rigid inflorescences and small globular fruits. Two fruiting plants in the Jardin Botanique in Tahiti have a close affinity with this species.

Pritchardia martii
(after Carl F.P. von Martius, father of palm botany)

HAWAIIAN ISLANDS

A dwarf fan palm which rarely grows more than 3 m tall, although it has a stout trunk and fairly large, stiff leaves which spread in a rounded crown. The leaf undersurface is densely clothed with felty scales. Native to the Hawaiian island of Oahu, it is uncommonly encountered in cultivation and would seem to be a palm for the collector. Subtropical and warm–temperate regions offer suitable climatic conditions for its cultivation. It is a slow-growing

Pritchardia martii.

species and seedlings will tolerate sun from an early age. Small plants make a very decorative container specimen.

Pritchardia minor
(lesser, smaller)

HAWAIIAN ISLANDS

A relatively small species which grows to about 10 m tall. Young leaves emerge yellowish and this colour may be retained on the undersurface of mature leaves. The leaves are leathery, with short, rigid segments which may be almost pungent. The shiny black fruit, to 2 cm x 13 mm, are spindle-shaped to ovoid or ellipsoid. This species grows in wet to swampy forest on the Hawaiian island of Kauai at about 1400 m altitude. *P. eriophora* is very similar to *P. minor* and there is doubt about whether the two are distinct.

Pritchardia napaliensis
(from the No Pali Coast, Kauai Island)

HAWAIIAN ISLANDS

This species is somewhat similar to *P. remota* but with smaller flowers and smaller, black fruit which have a thiner pericarp. The species is native to Kauai Island in the Hawaiian group where it grows in valleys close to the coast. Plants grow to about 6 m tall and have a yellowish-brown trunk about 20 cm across. The leaf blades, which are about 85 cm long, are green and glabrous on the upper surface and with numerous fringed scales on the underside. The fruit are obovoid, to 23 mm x 18 mm and black when ripe. It is uncertain whether this species is cultivated.

Pritchardia pacifica
(from the Pacific)

Fiji Fan Palm

TONGA

Although commonly called the Fiji Fan Palm, this species probably originated in Tonga and was introduced into Fiji very early on. It is a very handsome fan palm which is associated with many islands of the Pacific and is now widely cultivated in tropical and, to a lesser extent, subtropical regions. It is an excellent palm for coastal districts and lends itself

Pritchardia pacifica.

well to group planting. The leaves on young plants are especially impressive, the blades being up to 1.8 m long and nearly as wide, deeply pleated and with a brown, hairy surface when young. Clusters of small, round black fruit are carried within the crown. The plants may grow to about 9 m tall but are generally rather slow growing even in the tropics. They need very well-drained soil and some protection when small. In Fiji, the leaves of this palm were once made into fans which were for the exclusive use of the chiefs.

Pritchardia pericularum

(of danger, that is from the Danger Archipelago)

FRENCH POLYNESIA

Apparently this species is very similar to *P. vuylstekeana* but with smaller, spherical fruit. It is native to the Danger Islands in French Polynesia and was cultivated as a hothouse plant in Europe during the nineteenth century.

Pritchardia remota

(remote, isolated from the others)

HAWAIIAN ISLANDS

As suggested by the epithet, this species is rather islolated from other members of the genus, being restricted to Nihoa Island (commonly called Bird Island) some 320 km north-west of Oahu. This is a relatively low island and is drier than many others of the Hawaiian group. The palms dominate the vegetation in two valleys, growing in deep soil at the foot of basalt cliffs where there is water seepage. Plants grow to about 6 m tall and have a crown of about forty bright green, waxy leaves. The inflorescence is shorter than the leaves and after flowering bears clusters of greenish-brown to dark purplish-black fruit (to 20 mm x 19 mm). Plants adapt readily to cultivation and are reported to be more tolerant of dryness than other members of the genus.

Pritchardia rockiana

(after J.F. Rock, former Professor of Botany, Hawaii)

HAWAIIAN ISLANDS

One of the smaller species in the genus, *P. rockiana* grows to about 5 m tall and has a grey trunk about 30 cm wide. Its leaves are more than 1 m across and the lower surface is covered with yellow to golden, fringed scales. The large fruit are reverse pear-shaped and grow to about 5 cm x 3 cm. This species is apparently rare in the wild and although it has been introduced into cultivation, it is mainly the province of collectors. It is native to the Hawaiian island of Oahu, growing on forested ridges and slopes at about 700 m altitude.

Pritchardia schattaueri

(after George Schattauer, original collector)

HAWAIIAN ISLANDS

This species was discovered in 1960 during clearing operations in the District of South Kona, Hawaii and was described in 1985. It is a very rare species known only by twelve plants, some of which are in decline and not regenerating due to alienation of the habitat and seed predation by feral animals. Plants have been introduced into cultivation and it is hoped that conservation measures can be initiated to ensure the survival of the species in the wild. This palm is very tall, with a trunk to 30 m or more long and a rounded crown of glossy green leaves which have drooping segment tips. The fruit, which are up to 5 cm across, are brown to black with brown spots.

Pritchardia thurstonii

(after J.B. Thurston, one-time governor of Fiji)

FIJI

An excellent palm for coastal planting which thrives in tropical conditions but which can be induced to grow in a warm, sheltered position in the subtropics. The species grows in colonies on the Lau Group, Fiji, in shallow coralline limestone, where the climate is hot and dry and buffeting salt-laden winds are frequent. Plants commonly grow to about 10 m tall and have a large, impressive crown of light green, stiff, pleated leaves. The long inflorescences hang well clear of the leaves. The small, rounded fruit are dark red when ripe. Plants must have excellent drainage and need a sunny position. They are well suited to calcareous soils but will grow in other light soils, however do not succeed in clays.

Pritchardia vuylstekeana

(after Charles Vuylsteke, 19th century Belgian nurseryman)

FRENCH POLYNESIA

This little-known species is native to the Danger Islands and Tuamotu Archipelago in French Polynesia. Despite the near inaccessibility of this locality, it is recorded that the species was cultivated as a hothouse plant in Europe during the nineteenth century. Whether it is still in cultivation today is uncertain. The species has a slender trunk and dark green, semi-circular leaves, divided into numerous stiff segments. It is recorded that the oblong fruit are about 2.5 cm long.

Pritchardia waialealeana

(from Mount Waialeale, island of Kauai, Hawaii)

HAWAIIAN ISLANDS

A recently described (1988) species which grows on Mount Waialeale, reputedly the wettest place on earth. It is a robust species which grows in

colonies. Plants develop a large, stout trunk, to 50 cm across and 20 m tall and have a large, rounded crown of about forty leaves. The blades, which are about 1 m across, are distinctly pale green to glaucous. The small fruit (about 12 mm long) are somewhat pear-shaped. Seed of this species has been distributed but its performance in cultivation at this stage is unreported.

PSEUDOPHOENIX

(from the Greek *pseudo*, false and *phoenix*, a palm)

Cherry Palms

A small genus of four species of palm occurring in the Caribbean region and distributed from Florida to Mexico and Belize. They grow in open habitats on well-drained alkaline or saline soils near the coast and inland on dry hills, always in areas of low and erratic rainfall. They are solitary, unarmed feather-leaved palms with an often swollen trunk and a short, prominent, swollen crownshaft. The crowns are sparse, consisting of

Pseudophoenix lediniana.

few leaves and the inflorescences arise among the leaves. The flowers are either bisexual or the inflorescences comprise a mixture of bisexual and unisexual flowers.

Cultivation: Hardy, adaptable palms which make attractive ornamentals. They are suited to coastal conditions of the tropics and subtropics and will also grow inland. Excellent drainage is essential as also is a sunny situation. Plants may be slow growing but are hardy to dryness and neglect. They are, however, sensitive to poor drainage and root-rotting fungi. Single plants are capable of producing fertile seeds.

Propagation: Seeds retain their ability to germinate for up to two years after maturity. This is probably an adaptation to the low, erratic rainfall of their natural habitat. The seeds also float when dry and may be dispersed by ocean currents. Each fruit contains one to three seeds. The fruit contain oxalate raphides and may cause a burning sensation or irritation if handled excessively.

Pseudophoenix sargentii ssp. *sargentii*.

Pseudophoenix ekmanii

(after E.L. Ekman, original collector)

DOMINICAN REPUBLIC

A poorly known species which occurs on the Barahona Peninsula of the Dominican Republic. It grows in arid regions and a drink can be squeezed from the pith of the trunk and consumed fresh or fermented into wine. Plants are reported to have a stoutly bulging trunk (to 5 m tall) which contracts to each end. Leaves are about 1.5 m long and the leaflets are whitish beneath. The fruit are reported to produce pains in the throat after being eaten, probably caused by oxalate raphides. This species has ornamental qualities but is apparently not cultivated.

Pseudophoenix lediniana

(after Dr R.B. Ledin, 20th century horticulturist and palmologist)

HAITI

A large species somewhat similar to *P. vinifera* but readily distinguished by the broad, much-branched inflorescence which arches out well clear of the trunk. The trunk, which is mostly columnar or slightly swollen near the middle, may grow to 25 m tall and supports an open crown of fronds which

have glossy, dark green leaflets. The species is endemic on the south-western peninsula of Haiti where it grows on limestone cliffs among cacti.

Pseudophoenix sargentii ssp. sargentii

(after Charles S. Sargent, original collector)

Cherry Palm

FLORIDA KEYS—CARIBBEAN—CENTRAL AMERICA

The most widespread member of the genus, this species is distributed from the Florida Keys to Mexico and Belize. It is essentially a coastal species and a subspecies is also common on Cuba and various Caribbean islands, particularly the Bahamas (see below). All variants grow on limestone and sandy soils in situations protected from excessive wind; and plants will tolerate salt water inundation of their roots. Plants have a relatively slender, tapered trunk and a sparse crown of arching fronds which have stiff leaflets which are dark green above and grey or silvery beneath. Dense clusters of bright red fruit are held erect and are an attractive feature. This species is a hardy, yet ornamental palm which is able to grow in extreme conditions and is a decided acquisition wherever it is planted. It grows well in subtropical regions. One subspecies consisting of two varieties has been named.

Pseudophoenix vinifera young plants.

Pseudophoenix sargentii ssp. saonae

(from Saona Island, near Hispaniola)

Differs from ssp. *sargentii* by its pendulous infructescence, heavier trunk and larger fruit. This subspecies, the commonest variant cultivated, is abundant on various Caribbean islands, including Cuba. The typical variety of this subspecies has leaflets with a grey-green lower surface, whereas, var. *navassana* has leaflets white or silvery beneath. It occurs on the island of Navassa off the coast of Haiti.

Pseudophoenix vinifera

(yielding wine)

Buccaneer Palm, Wine Palm

HISPANIOLA

A striking palm which has a stout, bulging trunk to 25 m tall, abruptly narrowing below the crown. The arching, feathery leaves have dark green leaflets and the inflorescence hangs straight down close to the trunk (compare with *P. lediniana*). Huge clusters of large, bright red fruit are a most decorative feature. At maturity these fruit are covered with a thin layer of wax which is readily rubbed off. In cultivation this palm has proved to be very hardy and ideal for drier subtropical and tropical regions. As the specific name suggests, this palm has been used for wine-making on the island of Hispaniola where it is native. It grows on dry, limestone hills in areas where rainfall is low and erratic.

PTYCHOCOCCUS

(from the Greek *ptychos*, wrinkled, folded and *coccos*, grain or seed)

A small genus of seven species of palm, six endemic to New Guinea and the other found in the Solomon Islands. They grow as understorey plants in rainforest and along the margins of streams. They are solitary, unarmed feather-leaved palms with a long crownshaft. The inflorescences arise below the crownshaft and bear unisexual flowers of both sexes. The various species are poorly understood botanically.

Cultivation: Attractive palms which appear to have excellent horticultural prospects but remain little known. Plants respond to mulches and regular watering and require unimpeded drainage. Young plants need protection from excess sun and wind. A single plant is capable of producing fertile seed.

Propagation: Fresh seed takes about three months to germinate. Each fruit contains a single seed.

Ptychococcus lepidotus

(dotted with small scurfy scales)

NEW GUINEA

A small- to medium-sized solitary palm native to New Guinea where it is found in rainforest at moderate altitude. The plants grow to about 5 m tall with a slender, grey trunk and a spreading crown of dark green fronds which have blunt leaflets. The crownshaft is quite prominent and greyish from a coating of soft hairs. The large fruit (3.5–5 cm long) are bright red when ripe and follow green flowers. A succession of inflorescences in various stages from buds to ripe fruit may be present on a plant simultaneously. In New Guinea, the local people use the trunk for making spears, bows and tipping their arrows. Although poorly known in cultivation, this palm has potential for planting in tropical and subtropical areas.

PTYCHOSPERMA

(from the Greek *ptychos*, wrinkled, folded and *sperma*, seed)

A genus of twenty-eight species of palm found principally in New Guinea and the surrounding islands and archipelagos, extending to the Solomon Islands, Caroline Islands and northern Australia. They grow in a range of habitats, including highland rainforests and swamp forests of the lowlands. They are solitary or clumping, unarmed feather-leaved palms with slender, often grey stems and a prominent crownshaft. The leaflets, which have an uneven, jagged apex, may be arranged evenly along the rachis or unevenly clustered into groups. The inflorescence arises below the crownshaft and bears unisexual flowers of both sexes. (For a study of the genus see F.B. Essig, 'A Revision of the genus *Ptychosperma* Labill.', (Arecaceae), *Allertonia* (1978): 415–78.

Cultivation: The genus includes many handsome palms of moderate dimensions which have become well entrenched in cultivation. It would seem that most species in this genus have ornamental potential and are worthy of trial. As a group they are tropical in their requirements, demand good drainage but with an abundance of water and will grow in sunny or semi-shady positions. Mulches are beneficial and plants respond strongly to fertilisers. Hybridisation is common in this genus, especially among cultivated plants and seed should be collected with care. Some erroneous names have been attached to hybrids. Solitary plants of this genus are capable of producing fertile seed.

Propagation: Fresh seed germinates within three to five months of sowing. Each fruit contains a single seed.

Ptychosperma ambiguum

(doubtful, uncertain)

NEW GUINEA

Although restricted to West Irian, this small solitary palm has been introduced to cultivation and hopefully will become more freely available in the future. In its native state it grows on limestone hills and is locally common. In cultivation it grows readily enough but is very tropical in its requirements. Because of its compact growth habit, plants make an excellent subject for containers or tubs. The stem will eventually grow to 5 m tall but is less than 2.5 cm thick. The species can be readily recognised by its narrow leaflets arranged in clusters along the rachis. The leaves grow to about 1.3 m long. Infructescences can be very colourful, with the black fruit contrasting with the bright red rachillae and cupules.

Ptychosperma angustifolium

(with narrow leaves)

Although this name is widely applied to plants in cultivation, the true identity of the species cannot be determined because of inadequate descriptions of true specimens. As such it is a confused entity with no botanical standing.

Ptychosperma bleeseri

(after Florenz A.K. Bleeser, botanical collector)

NORTHERN AUSTRALIA

A rare palm which is endangered in its natural state, suffering badly from the damaging effects of feral animals and wildfires. It is endemic to the Northern Territory, Australia, where it is known from about seven sites in patches of monsoon rainforest. In all, fewer than 350 adult plants are known in the wild state. Plants have been introduced into cultivation and grow readily in tropical regions and are also successful, but slower, in the subtropics. The species is a sparse clumping palm with very slender stems which, as they age, may lean or become procumbent. The crown is sparse, typically consisting of three to five leaves, each about 1.5 m long. Fruit are bright red, with a short apical point.

Ptychosperma caryotoides

(resembling a *Caryota*)

NEW GUINEA

The mountains of New Guinea are the home of this palm which may be locally common in rainforest. It is a dwarf to small-growing species (1–9 m tall) with a single, slender stem and dense, white, woolly scales on the crownshaft. Arching leaves have narrow to broad, wedge-shaped pinnae with the apex irregularly lacerated and notched. The flowers are green to yellowish and are followed by brilliant-red fruit which ripen through yellow and orange. It is unfortunate that this handsome palm is rare in cultivation, as it would be an excellent acquisition for tropical and subtropical gardens. This species is rather variable, with plants from some areas—for example, Kokoda—producing mature seed on plants about 1 m tall. Others, especially from the coastal lowlands are much more robust.

Ptychosperma cuneatum

(wedge-shaped)

NEW GUINEA

This species was described in 1935 from a plant cultivated in Bogor Botanic Gardens. The original seeds were collected at Lake Sentani in central-north New Guinea. It is a solitary species to 5 m tall and with a very slender stem, 2–3 cm across. The leaves each contain about thirty narrowly wedge-shaped leaflets, with those towards the base of the frond being more crowded than the rest. Brownish-purple flowers are followed by black fruit. An attractive palm which is essentially a collector's item.

Ptychosperma elegans.

Ptychosperma elegans

(elegant, graceful)

Solitaire Palm

AUSTRALIA

This palm is well-named for indeed it is elegant and graceful. A tall solitary species, it is native to eastern Queensland, Australia, where it is prominent in coastal districts and it is also known from the Northern Territory, Australia. Branched inflorescences bear white, fragrant flowers which are followed by bright red, globose to ellipsoid fruit. A popular palm for tropical and subtropical regions, growing especially well on the coast. Plants are fast growing and highly responsive to fertilisers and regular watering. The species has been used for indoor decoration but demands high light intensity.

Ptychosperma gracile

(graceful)

BISMARCK ARCHIPELAGO—NEW GUINEA

A solitary palm which has a slender stem to 15 m tall and a crownshaft densely covered with white, woolly scales. The fronds, to 2.5 m long, are arranged in a graceful crown and have the narrowly wedge-shaped leaflets evenly arranged along the rachis. Green or yellowish flowers are followed by ellipsoid fruit which mature through yellow and orange to red. This attractive palm is native to eastern parts of New Guinea, New Ireland and New Britain, growing in rainforests from the coast to the foothills. It is strictly tropical in its requirements.

Ptychosperma hentyi

(after E.E. Henty, 20th century botanist in New Guinea)

NEW BRITAIN

A distinctive species which can be recognised by its graceful downcurved or weeping fronds which have broadly wedge-shaped leaflets with unevenly jagged tips. The longest leaflets are found towards the middle of the frond and reduce in size towards each end, with the apical pair being quite small (2–5 cm long). Plants have a solitary stem to about 8 m tall and a prominent crownshaft. The leaflets are evenly distributed along the rachis. The globose fruit, about 13 mm across, are bright red when ripe. This highly ornamental species which is

Ptychosperma lauterbachii.

grown on a limited scale in some tropical regions, deserves to become more widely cultivated. It is native to New Britain where it grows along streams in the lowlands on the eastern side of the island.

Ptychosperma hosinoi

(after Hosino Shutaro, former director of the Ponape Biological Laboratory)

Kattai Palm

CAROLINE ISLANDS

A little-known species from the island of Ponape in the Caroline Islands where it grows on volcanic slopes. It is cultivated on a limited scale in Hawaii and in Florida and has proved to be of easy culture. Plants appear to grow best in a sheltered location. The species is an attractive solitary palm to about 15 m tall. It has been confused with P. ledermannianum but has smaller, less-branched inflorescences and smaller (28–32 mm long) fruit.

Ptychosperma hospitum

(hospitable—usually to pests)

This palm was originally described from a plant cultivated at Bogor Botanic Gardens. It is now considered to be merely a delicate form of P. macarthurii, with narrower leaflets and is not specifically distinct. Plants are still grown and sold under this name.

Ptychosperma lauterbachii

(after C.A.G. Lauterbach, German plant collector)

NEW GUINEA

A robust species which forms clumps of stems to 12 m tall, each stem to 9 cm across and with a densely woolly white crownshaft. About fifty leaflets, regularly arranged along the rachis, impart a crowded appearance to each leaf. Green flowers are followed by attractive orange, ovoid fruit. This palm is essentially one of coastal forests and often also occurs along estuaries and adjacent to mangrove communities. It occurs naturally in New Guinea and is very tropical in its requirements.

Ptychosperma ledermannianum

(after C.L. Ledermann, 19th & 20th century Swiss botanist)

CAROLINE ISLANDS

This palm is one of three species of *Ptychosperma* to be found in the Caroline Islands. It is a tall (to 15 m), solitary species with a moderately stout

Ptychosperma ledermannianum. A. Rinehart

stem (12 cm across) and a woolly, white crown-shaft. The fronds have long (to 90 cm), tapered leaflets, each with an obliquely ragged apex. Pale-coloured flowers are followed by red spindle-shaped fruit about 4 cm long. Native to the islands of Ponape and Kusaie growing on slopes in rainforest, it is apparently not in cultivation.

Ptychosperma lineare

(linear)

NEW GUINEA

Plants of this palm usually form clumps but occasionally they remain solitary and are then very slender. In nature the trunks have been recorded as being up to 15 m tall and with a diameter of only 4 cm. Each trunk bears a crown of spreading fronds about 2.5 m long. These have the narrow leaflets evenly arranged throughout, with the basal ones being short and somewhat crowded. Young leaf sheaths are woolly white. Purplish-black fruit, each 1–1.5 cm long, are borne in large clusters. This palm is native to New Guinea where it grows in lowland swamp forests and is basically very similar in appearance to *P. macarthurii*. It can be grown in tropical and subtropical regions in similar conditions to that species.

Ptychosperma macarthurii

(after Sir W. MacArthur of New South Wales, Australia)

MacArthur Palm

AUSTRALIA—NEW GUINEA

This is one of the most widely grown and popular members of the genus, being extensively planted as an ornamental in many tropical countries. It is valued for its dense clumping habit and attractive, arching fronds which are dark green in shady positions and yellowish-green when exposed to the sun. A versatile, undemanding palm which adapts well to many situations, but is generally unsuccessful for indoor decoration unless given bright light. Plants are fast growing and respond strongly to the application of nitrogenous fertilisers. This species hybridises freely with other members of the genus and a confusing array of hybrid progeny are to be found in cultivation. The species is native to north-eastern Queensland, islands of the Torres Strait and south-central New Guinea growing in areas subject to periodic inundation.

Ptychosperma microcarpum

(with small fruit)

NEW GUINEA

A New Guinean clumping palm of generally similar appearance to *P. macarthurii* but readily distinguished by its pinnae being strongly clustered along the rachis. It is a frequent species of central New Guinea (common around Port Moresby) where it grows in swamps and lowland rainforest, often along streams. A very ornamental palm which is very cold-sensitive and best suited to the hot humid tropics. Young plants make attractive tub specimens.

Ptychosperma macarthurii.

Ptychosperma mooreanum

(after Harold E. Moore Jr, noted American palm botanist)

NEW GUINEA

This solitary species has a very slender stem (2–4 cm across) which may grow to 8 m tall. The crownshaft is sparsely covered with white, woolly scales and the leaves have about thirty narrowly wedge-shaped leaflets, with the basal ones being crowded together. The inflorescences carry dark red flowers which are followed by fruit which mature through dark red to purplish black. An interesting understorey palm from lowland rainforests of New Guinea. It is suited to cultivation in tropical regions.

Ptychosperma nicolai

(probably after Tsar Nicholas of Russia)

Essig (1978) has concluded that this palm is of hybrid origin and was originally described from juvenile material in cultivation. As such, it is a confused entity with no botanical standing. Plants are still grown and sold under this name.

Ptychosperma propinquum

(related to another)

ARU ISLANDS—MOLUCCAS

This attractive clumping palm is native to the Aru Islands of Indonesia, possibly the Moluccas and may extend to other islands near West Irian. It is very similar in appearance to *P. macarthurii* but the stems of the inflorescence are densely covered with dark hairs, its fruit are larger (17–20 mm long) and the pinnae are irregularly clustered along the rachis. This species demands more tropical conditions than *P. macarthurii* but otherwise its requirements are similar. It is reported to grow well close to the coast.

Ptychosperma salomonense

(from the Solomon Islands)

BOUGAINVILLE—SOLOMON ISLANDS

This palm, which is common in lowland and highland rainforests of Bougainville and the Solomon Islands, is very tropical in its requirements and likes hot, humid climatic conditions. It is a handsome, but variable, solitary species which strongly resembles *P. elegans* but has broader, relatively shorter, arching pinnae and arching fronds. The trunk may reach 12 m tall and 8 cm thick. Young leaf sheaths are covered with white, woolly scales but these are quickly shed as the leaves mature. Leaves have been measured at more than 3.5 m long and are carried in a graceful crown. The basal leaflets are crowded and this gives the crown a denser appearance than that of *P. elegans*. When ripe, the fruit are bright red and very decorative. A number of variants exist within this species, probably the best known of which is the palm cultivated as *Strongylocaryum latius* (treated as a synonym of *P. salomonense*). This is a small-growing variant, with spreading leaflets and dark orange fruits on purple rachillae.

Ptychosperma microcarpum.

Ptychosperma waitianum.

Ptychosperma sanderianum

(after Henry Sander, 19th century English nurseryman)

NEW GUINEA

This palm is apparently only known from cultivation and is believed to have originated in the Milne Bay area of Papua New Guinea. It is a clumping palm similar in most respects to *P. macarthurii* but with shorter leaves (1.75 m long) and much narrower leaflets (about 1.5 cm across). The red fruit are about 2 cm long. This species is comparatively rare in cultivation but grows readily in tropical regions.

Ptychosperma schefferi

(after R.H.C.C. Scheffer, 19th century Dutch botanist)

NEW GUINEA

Many plants of this species that are in cultivation may be of hybrid origin and are not typical of the wild plants which are to be found in wet areas on the north coast of Papua New Guinea and West

Irian. Plants may be solitary or clumping, with stems to 7 m tall and up to 6 cm across. Young leaf sheaths are covered with white scales which impart a woolly appearance. The leaves are 1.5–1.8 m long with about twenty-five pairs of regularly arranged pinnae which are dark green. Mature fruit (1.5–1.8 cm long) are dark purple and contrast with the yellow-orange rachillae. In cultivation, this palm requires hot, humid conditions and is most suitable for the tropics.

Ptychosperma streimannii

(after Heinar Streimann, 20th century collector in New Guinea)

NEW GUINEA

This species is similar in many respects to *P. ambiguum* but the narrow leaflets are arranged regularly along the rachis. It is a small solitary palm with a slender stem to 4 m tall and a woolly, white crownshaft. Red flowers are followed by purplish-black fruit. A very attractive tropical palm which grows naturally in rainforests near Port Moresby, Papua New Guinea.

Ptychosperma waitianum

(after Lucita H. Wait, palm enthusiast)

NEW GUINEA

A small solitary palm from the Milne Bay district of Papua New Guinea which promises to become widely grown for its neat, compact habit. Large plants may grow to 5 m tall but those in cultivation are usually much less. Leaves are less than 1 m long and spread in a graceful crown. They have bright green, broad, wedge-shaped leaflets arranged evenly along the rachis but with those at each end being clustered. The flowers are very distinctive, being deep red and densely scaly. They are followed by fleshy, black fruit nearly 2 cm long. Plants grow well in Florida, USA and like a shady position at least when young, good drainage and plenty of water.

RAPHIA

(from a Latinisation of a Malagasy name—*raffia, rofia, ruffia*—for one species)

A genus of about twenty-eight species of massive palms with all but one being confined to Africa and Madagascar; the other occurring in tropical

America. Most species grow in freshwater swamps and take in oxygen via specialised roots which grow upwards against gravity (pneumatophores). They are solitary or clumping, spiny palms with either a subterranean trunk or an erect trunk. They have huge pinnate leaves, the longest of any plant in the world. The leaflets are often armed with fine spines. The inflorescences, which either arise in the upper leaf axils or terminate a stem, carry unisexual flowers of both sexes. The large fruit are covered with rows of overlapping scales. After fruiting the stem dies, this being the whole plant in the case of solitary species. (For a treatment of the species in West Africa see T.A. Russell, 'The *Raphia* Palms of West Africa', *Kew Bulletin* 19 (2) (1965):173–95; for a recent revisionary treatment see M.O. Otedoh, 'Revision of the genus *Raphia* Beauv. (Palmae)', *Journal of the Nigerian Institute for Oil Palm Research* 6(22) (1982): (145–89).

Raphia palms are of major significance to local people in Africa and their distribution may have been influenced by cultural practices. The leaf stems are used for the construction of dwellings, furniture and other purposes and the leaves for an excellent thatch. Wine can be made by tapping the emerging inflorescence or fermenting the fruit. Raffia fibre is made from the developing leaflets of unexpanded sword leaves or leaf-bases and used for twine, basketware and weaving. At one time this fibre was exported widely to western countries. The cabbage may be eaten and the fleshy layer around the seed used as a source of oil for cooking. These palms are so important to local economies that there is considerable interest in establishing commercial plantations of various species, especially in Nigeria. They also have the considerable advantage of growing in swampy areas which are not used for other forms of agriculture.

Cultivation: Massive, handsome palms which must have plenty of room to develop and reach their potential. They are not suitable for home gardens and are best planted in parks and on large acreages. They grow very well in wet soils but can also be induced to grow on well-drained sites if watered, mulched and fertilised regularly. All species need a sunny position from an early age. Single plants are capable of producing fertile seeds.

Propagation: Seed germinates readily but can be sporadic, with seedlings appearing from one month to six months after sowing. Each fruit contains a single seed.

Raphia australis

(southern)

Kosi Palm, Southern Raphia

SOUTHERN AFRICA

This species occurs in Mozambique and north-eastern South Africa close to the border with Mozambique. It forms loose groups in freshwater swamps and the base of each plant is surrounded by erect roots which take in oxygen from the atmosphere. It is an attractive, solitary palm which produces a massive terminal inflorescence and dies after fruiting, this process usually taking place after 25–35 years. The long leaves (to 18 m) are held stiffly erect with the attractive, reddish petioles contrasting with the blue-green to dark green leaflets. These leaflets have spiny margins and a reddish midrib which may also be spiny. The huge terminal inflorescence (to 3 m long) has brownish flowers which are followed by shiny brown, egg-sized scaly fruit. The leaves are used locally for thatch, house construction and as a source of raffia fibre. A colony established at Mtunzini further south in South Africa, has thrived and is now a national monument. Although this palm thrives in wet conditions, plants will also grow in well-drained soils if watered regularly.

Raphia farinifera

(yielding flour)

Raffia Palm

MADAGASCAR—EAST AFRICA

A spectacular Madagascan and east African palm which colonises wet areas and spreads by suckers. Each trunk may grow to 10 m tall and bears an impressive crown of stiffly upright, feathery fronds. These are among the largest leaves produced by any palm and indeed are close to the largest leaf in the plant kingdom, with individuals measuring over 20 m long (see also *R. regalis*). Each trunk is monocarpic but the colonies spread by the continual production of new basal suckers. The inflorescence is unusual and is similar to that described for *R. vinifera*. The fruit are most attractive, being about 8 cm long and covered with overlapping, shiny brown scales. Raffia fibre, which today has decreased in importance, was made from the leaf-bases of this palm. This palm is highly regarded for dwelling construction and would appear to have been widely planted in Africa for this purpose.

Because of their size, cultivated raffia palms are restricted to parks and large gardens. They succeed best in the tropics but have also been successfully grown in the subtropics. This species has also been known as *R. ruffia* and *R. kirkii*.

Raphia hookeri

(after J.D. Hooker)

Wine Palm

WESTERN AND CENTRAL AFRICA

A widespread species in West Africa which is distributed from Guinea to Cameroon, Togo and Gabon. It grows naturally in swamps and is also commonly planted in many West African countries for its numerous valuable products (see genus

Raphia farinifera young plant.

introduction). These include piassava fibre or 'nduvui', which is a stiff fibre obtained from the leaf-bases and used for brooms, mats, baskets, etc. The species is also being considered as a source of fibre for paper production. It is a solitary or sparsely suckering species with a trunk to 10 m tall and shiny, dark green leaves to 12 m long which have rigid leaflets. The long (to 2.5 m) inflorescences are pendulous and after flowering bear top-shaped or ellipsoid, brown fruit to 12 cm x 5 cm, each with a stout beak.

Raphia palmapinus

(pine-palm, in reference to the scaly fruit)

WESTERN AFRICA

This species is distinguishable by its dull, pale yellowish-green fronds and the leaflets which have a prominent, sharply angled midvein obvious on the upper surface. The brown, ovoid fruit, to 9.5 cm x 4 cm, have a short, sharp apical beak. It is native to West Africa where it is widely distributed from Senegal to Ghana. It grows naturally in swamps (including brackish swamps adjacent to mangroves) and suckers to form dense thickets. Many useful products are obtained from this palm (including piassava fibre—see *R. hookeri*) and the species may be planted specifically for these products.

Raphia regalis

(regal, royal)

WESTERN AFRICA

This species has the longest leaves of any palm and indeed the longest leaves of any plant in the world. A specimen collected in the Congo was measured at 25.11 m or about 75 feet. Of this, about one-third consisted of the petiole and the rest was the rachis bearing the leaflets. This palm is unusual in the genus in that it grows in well-drained soils on forested slopes. It occurs naturally in eastern Nigeria, Gabon, Cameroon and the Congo. Plants are almost stemless or have a short, stout trunk, each with few leaves. The much-branched, fan-shaped inflorescence is very distinctive in being erect and having gracefully curved branches. The fruit, which have a long, sharp apical beak, are dark brown and shiny. This species appears to be rarely, if at all, cultivated.

Raphia sudanica

(from the Sudan)

Northern Raphia Palm

WESTERN AND NORTHERN AFRICA

In its native state this palm forms extensive, impenetrable thickets. Like most members of the genus it grows in swamps, but this species occurs in a drier climatic zone than others, being distributed from Senegal eastwards to northern Nigeria and southern Sudan. The palm is used locally for a range of purposes and plantations of it have been established in Nigeria. In this species the young fronds have a prominent yellow to orange rachis and the leaflets are densely armed with spines on the veins and margins. The large top-shaped fruit are dark brown when ripe.

Raphia taedigera

(torch bearing)

Yolillo Palm

CENTRAL AND SOUTH AMERICA

This, the only species of *Raphia* to occur outside Africa and Madagascar, is found in Colombia, Central America—Nicaragua, Costa Rica, Panama—and is also widespread around the mouth of the Amazon River in Brazil. It grows in extensive swamp forests where it dominates the vegetation to the exclusion of other species. Plants sucker profusely and develop trunks to 9 m tall and with arching fronds from 12–20 m long. The ground between plants is covered with pneumatophores and these specialised roots also occur on the trunk up to about 1 m above the ground. The shiny brown fruit, to 7 cm x 4 cm, is oblong in shape. This species is cultivated but only on a limited scale.

Raphia vinifera

(wine yielding)

African Bamboo Palm

WESTERN AFRICA

An interesting palm readily recognised by its upright crown of tall, feathery leaves with long leaflets, which are dark green and shiny above, waxy and glaucous beneath. The trunks are usually short and stout (to 5 m tall) and sucker from the base and hence the plant eventually develops into a clump. Each trunk dies after flowering. The inflorescence is a most unusual structure best described as a thick, sausage-like growth. When it emerges it is erect but later becomes pendulous and may reach more than 3 m long. In Nigeria, where this species occurs naturally, the inflorescences are cut off as they emerge and the collected juice is fermented into wine. The species also occurs in Benin and Bioko (Equatorial Guinea). *R. vinifera* grows far too big for the home garden and can only be grown on acreages where the plants can spread. They are tropical in their requirements but can also be successfully grown in the subtropics. Plants need to be grown where they can receive plenty of water.

RAVENEA

(after Louis Ravene, French consular official)

A genus of ten species of palm, eight being endemic to Madagascar and two on the islands of Comoros. They grow in rainforests, often forming communities. They are solitary, unarmed feather-leaved palms which lack a crownshaft, with the long fronds often forming a large, impressive crown. The inflorescences arise among the leaves and bear unisexual flowers. The genus includes the only known aquatic palm.
Cultivation: Handsome palms which deserve to become widely grown. At least one species has become popular for its ornamental qualities and others have good prospects. Male and female trees are essential for fertile seed production.
Propagation: Fresh seeds germinate readily two to four months after sowing. Each fruit contains a single seed.

Ravenea hildebrandtii

(after J.M. Hildebrandt, original collector)

COMOROS

This species is from Comoros off the coast of Madagascar where it grows in basalt soil among rocks. It is a large palm with a trunk in excess of 15 m tall and an impressive crown of obliquely erect to spreading, dark green fronds with a contrasting white rachis. These fronds have numerous narrow, sword-shaped leaflets to 45 cm x 2 cm. Long-stalked inflorescences hang below the crown and at fruiting carry colourful clusters of brilliant orange fruit, each about 1.5 cm across. This handsome palm was cultivated to a limited extent in the nineteenth century and in 1965 was reintroduced from seed collected on Grande Comore.

Ravenea musicalis

(musical, from the melodious sound of falling fruit)

MADAGASCAR

An amazing palm from Madagascar which is the only true aquatic plant in the whole palm family. It was discovered as recently as 1992 and described in 1993. It is restricted to a single stream growing in water to 2.5 m deep, with the palms either erect or bent over by the force of the water. The plants develop a swollen trunk to 10 m tall and have an attractive crown of arching fronds with orange sheathing bases. Ripe fruit are orange, to 23 mm x 19 mm and the seeds germinate within the fruit. After falling, the fruit float initially but then the flesh rots and the seeds sink, become anchored on the bed of the stream and grow. The seedlings have soft pendulous leaves and wave about in the current, with the older leaves eventually emerging above the water surface. Recently, significant quantities of seed of this remarkable palm have been distributed to various botanic gardens around the world, where it is hoped they can become established as a novelty item.

Ravenea rivularis

(growing beside streams)

Majestic Palm

MADAGASCAR

A magnificent palm which has become deservedly popular in cultivation. It has proved to be fast growing and adaptable, with a highly ornamental appearance. Young plants have a crown of arching, bright green fronds whereas in older plants the crowns are rounded, with a large proportion of the fronds being drooping or pendulous. It is a tall palm with a tapered, white trunk to 30 m and a large crown (up to 25 leaves). Young leaves have a whitish appearance and the rachises and petioles are covered with cottony fibres. Small, red, globose fruit, 7–8 mm across, are borne on large, much-branched infructescenses. An excellent specimen plant but also well suited to planting in groups. Small plants look highly decorative in a container and perform well as an indoor plant. Plants will grow in tropical, subtropical and warm–temperate

The Majestic Palm *Ravenea rivularis* young plant.

regions. The species is native to Madagascar where it grows in moist areas near streams and swamps from sea level to about 600 m altitude.

Ravenea robustior

(very robust)

MADAGASCAR

This strong-growing palm has an erect, shuttlecock-like crown of fronds which gives the whole plant a feather-duster appearance. Plants have twelve to fifteen dark green leaves, in a crowded crown atop a stout, grey to brown trunk to 25 m tall. Large clusters of small, brilliant-red fruit provide an eye-catching splash of colour. The species grows in moist forests of Madagascar from sea-level to about 1000 m altitude. Although it is apparently rare in cultivation, this species has good prospects for parks or similar large sites in the subtropics and tropics.

REINHARDTIA

(Possibly after Johannes C.H. Reinhardt, one-time professor of zoology, University of Copenhagen)

A small genus of six species of palm mainly distributed from Mexico to Panama, with one extending to Colombia and another recently described from the Dominican Republic. They are understorey palms found in moist forests and rainforests. Four species have distinctive, broad, pleated leaves and those of the other two are regularly pinnate. At least two species have a unique pattern of small slits or openings near the base of each pinna and this has earned them the name of 'Window Pane Palms'. They are small to moderately sized, solitary or clumping, unarmed, entire or feather-leaved palms which lack a crownshaft. The inflorescence arises among the leaves and bears unisexual flowers of both sexes. The genus *Malortiea* is synonymous.
Cultivation: Dainty palms which are excellent for sheltered gardens and pots. They need a protected situation in good, well-drained soil. Some species have proved to be adaptable and will grow successfully in the subtropics. Single plants are capable of producing fertile seed.
Propagation: Clumping species can be propagated by division but these are slow to establish. Fresh seed germinates readily although it may be slow and sporadic. Each fruit contains a single seed.

Reinhardtia gracilis in fruit.

Reinhardtia elegans

(elegant, graceful)

MEXICO

A solitary species which has a very slender, ringed stem to 6 m tall and a crown of ten to twelve widely spreading fronds. These have about eighty narrow, dark green leaflets, to 40 cm x 1.5 cm, which are deeply notched at the apex. Ellipsoid to obovoid fruit, to 18 mm x 10 mm, are dark purple when mature. This palm, essentially a collector's item, is native to southern Mexico where it grows in moist forests at about 1000 m altitude.

Reinhardtia gracilis var. gracilis

(slender, graceful)

Window Pane Palm

CENTRAL AMERICA

This dainty, small palm grows in dense rainforests from Mexico to Central America, always in protected, shady positions. The plants grow in small clumps with thin stems up to about 1.5 m tall. The small, irregularly pinnate leaves have prominent veins and small gaps or 'windows' between the main veins at the base of each leaflet. These 'windows' create interest in the species and it has become a popular small palm to be grown mainly in subtropical regions. It makes an excellent container plant or, if to be grown in the ground, it likes a shady, protected position in good soil. Plants are fairly slow growing and can be propagated from seed which germinates readily or by suckers.
R. gracilis is an extremely variable species and four varieties have been named, based on leaflet numbers and the number of stamens in the flowers. The typical variety (var. *gracilis*) has 16–22 stamens in the male flowers and large leaves with 14–22 veins on each side of the rachis. It occurs naturally in British Honduras, Guatemala and Honduras.

Reinhardtia gracilis var. tenuissima

(very slender)

This variety has 16–22 stamens in the male flowers and small leaves with 8–9 veins on each side of the rachis. It is restricted to Oaxaca, Mexico, growing at 1000–1500 m altitude.

Reinhardtia gracilis var. *gracilior*

(more slender than another)

This variety has 8–10 stamens in the male flowers, fruit with a low-pointed crown and leaves with 8–11 veins on each side of the rachis. It is distributed from southern Mexico to Honduras.

Reinhardtia gracilis var. *rostrata*

(beaked)

This variety has 8–10 stamens in the male flowers, fruit with a swollen, truncate boss and leaves with 11–15 veins on each side of the rachis. It is found in Nicaragua and Costa Rica.

Reinhardtia koschnyana

(after T. Koschny, original collector)

CENTRAL AMERICA

A delightful palmlet from shady rainforests of Costa Rica, Panama and Colombia. Plants have stems less than 1 cm across and up to 4 m tall, although they are often decumbent on the ground. The leaves are a very interesting feature, being undivided but with a narrow, deep, apical notch and coarse, sharply toothed lobes on the margins. They are dark green with prominent veins and are carried on slender, stiff petioles. The inflorescence is a simple spike and the fruit are purplish-black when mature. This gem is extremely rare in cultivation, if it is grown at all.

Reinhardtia latisecta

(with broad segments or divisions)

Giant Window Pane Palm

CENTRAL AMERICA

This species is the giant of the window pane palms, with trunks that may grow to nearly 8 m tall. These are slender (about 6 cm across) and carry the attractive, arching to drooping leaves, scattered up the stems. Each leaf may be up to 1.3 m long and is divided into broad, conspicuously pleated pinnae which have slits or openings at the base. The leaflet tips are toothed and the terminal leaflets are broadly united with a deeply notched apex. Black, ovoid to obovoid fruit about 1.7 cm x 1 cm contrast with

the bright red rachillae of the infructescence. Native to Belize and perhaps Guatemala, this attractive palm forms sparse clumps in very shady situations. It is essentially a collector's item requiring a shady position in the tropics.

Reinhardtia paiewonskiana

(after B. Paiewonsky)

DOMINICAN REPUBLIC

A recently described (1987) species which is endemic to the Dominican Republic where it grows in moist, broadleaf forests at about 800 m altitude.

Reinharditia gracilis var. *gracilior* published as *Malortiea gracilis* (*Botanical Magazine*, Vol. XXXIII of the 3rd series, plate 5291)

It is the largest member of the genus, growing to about 12 m tall and with a solitary trunk to 4 cm across. It has a rounded crown of about twelve graceful pinnate fronds to 3 m long. Each frond has about 110 narrow leaflets to 70 cm x 3.5 cm. The ovoid to nearly globose fruit, about 2 cm long, are purplish-black when ripe. This species, which grows in inaccessible areas, is rarely collected and may not be cultivated.

Reinhardtia simplex

(simple, undivided)

CENTRAL AMERICA

Widely distributed from Honduras to Panama, this small palm grows in shady situations on the forest floor. The plants form a sparse cluster of slender stems to about 1 m tall and have simple, deep green leaves which are variously lobed and with prominently toothed margins, but lack an apical notch. The leaves are generally small (10–15 cm long) and held at right angles to the stems. The obovoid, black fruit are about 1 cm long. Unlike the commonly grown members of this genus, the leaves of this species lack the slits or windows in their fronds. Essentially a collector's palm, this attractive small species favours a protected, shady position in the tropics.

RHAPIDOPHYLLUM

(from the Greek *rhapis*, needle and *phyllon*, leaf, in reference to the needle-like leaf tips)

A monotypic genus of palm which is endemic to the south-eastern United States of America. It grows in woodlands in humus-rich, sandy or calcareous soils, usually in shady situations. It is a clumping, formidably armed fan-leaved palm. The inflorescences arise among the leaves and may bear unisexual flowers of one sex, unisexual flowers of both sexes or bisexual flowers. (For a comprehensive treatment of the species see A.G. Shuey and R.P. Wunderlin, 'The Needle Palm: *Rhapidophyllum hystrix*', *Principes* 2 (1977): 47–59.)
Cultivation: A hardy, very adaptable palm which is unfortunately very slow growing. It can be successfully grown from the tropics to temperate regions. Single plants are capable of producing fertile seed.
Propagation: Clumps can be divided successfully with the divisions taking readily but being slow to

establish. Seeds germinate sporadically taking from six months to two years to appear. Each fruit usually contains a single seed.

Rhapidophyllum hystrix

(resembling a hedgehog or porcupine)

Needle Palm, Hedgehog Palm, Blue Palmetto

NORTH AMERICA

A very slow-growing palm which forms clumps made impenetrable by the thick fibre and long, black spines on the leaf sheaths and upper trunk. Some plants may remain sparse whereas others sucker profusely to form dense clumps. The trunks often lean away from each other and in time may become separate plants when the connecting tissue rots. The large palmate leaves (about 1 m across) are deeply divided into narrow, stiff, dark green segments. The inflorescences are hidden within the crown and the fruit often become caught in the petioles, fibre and needles and remain there until they decay. An interesting, hardy and adaptable palm which will tolerate very heavy frosts (-12°C). Plants will grow in a shady situation and will also happily tolerate full sun if well watered and mulched. The species is native to the American states of Mississippi, Alabama, Georgia, South Carolina and Florida, but has been greatly reduced in the wild by commercial exploitation and is now rare and threatened. In the 1800s and early 1900s, whole crowns of this palm were cut and used for indoor decoration. More recently the plants have been poached for nurseries and landscaping.

RHAPIS

(from the Greek *rhapis*, a needle, an apparent reference to the leaf segments)

Lady Palms

A small genus of about twelve species of palm distributed mainly in southern China, Laos, Vietnam and Thailand, with a disjunct undescribed species occurring in northern Sumatra. They grow as understorey palms in relatively dry forests, apparently often on limestone. They are dwarf to small, clumping fan-leaved palms with the slender stems covered in adherent fibres and the leaves deeply divided into narrow segments diverging like the fingers of a hand. The short, much-branched inflorescence arises among the leaves and bears either unisexual flowers or bisexual flowers.

Cultivation: Highly prized palms and of significant economic importance to the horticultural industry, since they are among the best palms for indoor decoration. Although only three species have become widely cultivated, they are very easy to grow and have proved to be adaptable to a range of situations. They can be grown in the ground and are also unsurpassed as subjects for containers. For the commonly grown species, male and female plants are essential for fertile seed production. Some of the other species may produce bisexual flowers.

Propagation: Rhapis palms are commonly propagated by division and this is usually straightforward and successful. Some dwarf cultivars may be more difficult and it is then advisable to wait until

Rhapis excelsa in flower.

the sucker has a good root system before it is severed. Fresh seed germinates readily, usually within three months of germination. Each fruit contains one to three seeds.

Rhapis excelsa

(high, lofty, elevated)

Lady Palm, Bamboo Palm

SOUTHERN CHINA

A multi-stemmed fan palm from southern China that forms dense clumps or thickets that are leafy to ground level. The leaves have five to eight widely divergent segments that are stiff and spreading or held erect and with no tendency to droop. The leaves are light green and generally take on a yellowish hue if the plants are starved or grown in full sun. The stems are very slender and covered with woven brown fibre. *R. excelsa* is one of the best garden palms but because of its very slow growth, plants tend to be expensive and hence the species is not as widely planted. It is very cold-tolerant and will grow well in temperate areas. Plants can also be induced to grow in the tropics but in a shady position. In temperate regions the species will tolerate considerable exposure to sun, even full sun, but the leaves bleach badly and may even burn, especially if not watered regularly. For best appearance this species should be grown in a semi-protected position where it only receives partial sunlight during the day. It can be successfully established under large trees and will make satisfactory growth providing it is mulched and kept moist. As a tub plant it is unsurpassed and will remain in the same container for years without the need for repotting. Indoors the plants are very decorative and will last for long periods without

Rhapis excelsa.

the need for spelling. In fact it is one of the most durable of indoor palms. Most propagation is by division of the clumps. Seedling-grown plants are rarely available.

A number of cultivars with a dwarf habit or variegated leaves are known and these are mainly a choice collector's item, although they are becoming more widely grown. Most have apparently originated in Japan where more than 100 cultivars are known, with each having an individual growth habit and leaflet shape and development. They must be propagated by division. Most of these are extremely slow growing and well-established specimens are highly valued items. Their slow growth is exacerbated by using some bonsai techniques, that is by growing the plants in small pots, thus greatly restricting their root system and potting into well-drained but nutritionally poor potting mixes. Some of these novelties, which include clones with variegated leaves, have been imported from other countries and there has also been limited local selection from seedling variants. Variegated cultivars include

Rhapis excelsa cream banded variegation.

Rhapis excelsa cream flushed variegation.

the prized selections which are evenly striped and those of lesser value which have random patterns.

Cultivars include:

'Ayanishiki'—a choice variegated cultivar with very even, creamy-white stripes.

'Chiyodazuru'—a variegated cultivar with narrow whitish stripes; plants maintain good colour in low to medium light.

'Daruma'—grows to about 1 m tall; very compact with narrow leaflets giving a lacy impression.

'Gyokuho'—grows to about 30 cm tall; suckers sparsely, has broad leaflets.

'Koban'—grows to about 60 cm tall; a compact grower with broad leaflets.

'Kodaruma'—grows to about 50 cm tall; very compact with broad, thick leaflets.

'Kotobuki'—a variegated cultivar with prominent white stripes which may brown off in intense light.

'Tenzan'—grows to about 2 m tall; relatively open habit with broad leaflets.

'Zuikonishiki'—a vigorous variegated cultivar which suckers freely; leaflets are variable in their creamy-white stripings.

'Zuiko-Lutino'—a choice variegated cultivar with very broad, creamy-white stripes.

Rhapis filiformis

(in the form of a thread)

SOUTHERN CHINA

A compact species which grows to about 1.5 m tall. The dark green leaves have three pointed segments. Inflorescences are much branched and very hairy. This poorly known species originates in southern areas of China (Kwangsi Province) and would be suitable for cultivation in temperate and subtropical zones.

Rhapis humilis large clumps.

Rhapis gracilis

(slender)

SOUTHERN CHINA

A poorly known species from southern areas of China. It has been introduced into cultivation but remains essentially a collector's item. Plants have very slender stems covered in brown fibre and the leaves have three or four widely divergent segments, with the tip irregularly and sharply toothed. The globular, purplish-brown fruit are about 8 mm across. This species should succeed in subtropical and temperate regions.

Rhapis humilis

(of low growth)

Slender Lady Palm

SOUTHERN CHINA

A lovely garden palm that was very popular early this century and was widely planted in private gardens as well as botanic gardens and other municipal gardens. It is native to southern China and forms a spreading clump that may extend to more than 2 m across and 4 m high. The clumps consist of numerous, slender stems densely covered with closely woven, brown fibres and attractive palmate leaves scattered up their length. The leaves are deeply divided into numerous, dark green, drooping segments and impart a very ferny appearance to the palm. In contrast to those of the Lady Palm, the clumps tend to be taller and more open at the base and the leaves much deeper green. In the garden they like a well-drained soil with plenty of organic matter and regular watering. Plants will tolerate some sun but for best appearance like a moist position in deep shade. They grow very well in temperate and subtropical regions and are an excellent palm for mingling with or providing protection to ferns. Clumps also grow very well in tubs and seem to live for years without the need for repotting. They are also excellent indoors, tolerating considerable darkness and neglect. Female plants of this species are unknown and all cultivated plants in the world have resulted from vegetative propagation by division.

Rhapis laosensis

(from Laos)

Laos Lady Palm

LAOS—VIETNAM

A species which is similar in habit and general appearance to *R. subtilis*, but which is readily identified by the leaves, these having only three segments which are more or less joined at the base.

These segments are generally lighter green and much longer than those of *R. subtilis* (to 25 cm x 6 cm). Both species have pointed tips to their leaflets but those of *R. laosensis* are more drawn out and tend to droop. *R. laosensis* is still a novelty item but reports indicate that it grows well in tropical and subtropical regions. The species is native to the valley of the Mekong River in Laos and may extend into Vietnam. At this stage it is only propagated by division since all cultivated plants are reported to be female.

Rhapis micrantha

(with small flowers)

VIETNAM

A little-known species from Vietnam which forms spreading clumps of slender (8 mm thick) stems to about 2 m tall. The leaves are deeply divided into six to ten, widely divergent segments (to 25 cm x 4 cm) which are separated to the base. The apex of each is drawn out into a point. The stems are covered in closely appressed fibrous sheaths. Although the flowers of this species are reputed to be smaller

Rhapis multifida.

than those of *R. excelsa*, the globose, white fruit are larger (8–9 mm across). The species is most suitable for tropical and subtropical regions.

Rhapis multifida

(divided into many segments)

Finger Palm

SOUTHERN CHINA

The leaves of mature plants of this species are divided into about twelve segments, whereas those of young plants commonly have five segments. The segments, to 18 cm x 2 cm, radiate widely in an attractive pattern. An ornamental clumping palm which grows to about 2 m tall and has very thin stems covered with brown fibres. It is native to the Kwangsi Province of southern China and has recently become well established in cultivation. Although generally similar to *R. humilis*, it shows tremendous promise because plants will also grow in temperate regions and seed is available for propagation.

Rhapis robusta

(strong, vigorous, robust)

SOUTHERN CHINA

This species has slender canes (about 1 cm across or less) covered with fine, brown fibres and leaves which are deeply divided into three or four segments. These segments are relatively short and

Rhapis laosensis.

broad (about 5 cm x 2–2.5 cm) with minutely toothed margins and a pointed apex, and diverge widely from the point of attachment. They are scattered along the stems, which may grow to 2 m tall. The inflorescences are much branched and the rachillae are clothed with fine, brown hairs. This species, which is apparently not cultivated outside of China, is native to southern China in the Kwangsi Province.

Rhapis subtilis
(fine, thin, slender)

Dwarf Lady Palm, Thailand Lady Palm

THAILAND

This palm has become widely cultivated in many countries and in recent years has become a familiar nursery plant, especially in subtropical districts. It has been confused with *R. excelsa* but can be distinguished by the fewer leaf segments (three to six) which are widely spaced, of irregular width on a single leaf and often elliptical in shape with a pointed apex. It also has a different inflorescence. Plants do not grow much more than 1 m tall and remain compact, with the new growths arising close together at the base of the plant. They devel-

op into handsome clumps with dark green, somewhat glossy leaf segments which are thinly textured but tough. This is an excellent palm for tub culture and indoor decoration. It is native to Thailand where it grows in shaded forests. The species grows very well in tropical and subtropical regions but is too cold-sensitive for temperate zones.

RHOPALOBLASTE

(from the Greek *rhopalon*, club and *blasto*, embryo or shoot, literally with club-like shoots)

A genus of six species of palm occurring in Malaysia, the Nicobar Islands, the Moluccas, New Guinea and the Solomon Islands. They grow as understorey plants in rainforest. They are solitary or clumping, unarmed feather-leaved palms with a crownshaft. Plants range from small-growing species to tall and robust. The leaflets have a cushion-like pulvinus at the base. The inflorescence arises below the crownshaft and bears unisexual flowers of both sexes. Some species were previously included in the genus *Ptychoraphis*. (For a review of the genus see H.E. Moore, 'The Genus *Rhopaloblaste* (Palmae)', *Principes* 14 (3) (1970): 75–92.)

Rhapis subtilis.

Cultivation: Popular palms, some species of which are widely grown. They require shelter from excess sun and wind and succeed best in the tropics. Single plants are capable of producing fertile seed.
Propagation: Some species can be propagated by division, though seed is the usual method. Fresh seeds germinate readily, usually within three to four months of sowing. Each fruit contains a single seed.

Rhopaloblaste augusta

(exalted, majestic)

NICOBAR ISLANDS

A solitary, robust palm with a trunk to 30 m tall and 20–30 cm across. Leaves spread widely in a delightful, rounded crown, with the long, narrow (to 70 cm x 3 cm) leaflets hanging gracefully. The leaflets taper to a long, drawn-out point and are dark green and shiny on the upper surface with scattered brown scales beneath. Green to whitish flowers are followed by large showy, clusters of orange to red fruit which have a prominent apical beak. This species, native to the Nicobar Islands (India), is a truly delightful palm for tropical and perhaps subtropical regions and deserves to become more widely cultivated.

Rhopaloblaste brassii

(after L.J. Brass, original collector)

NEW GUINEA

A little-known species from Papua New Guinea and West Irian which grows on ridges and moist alluvial flats in rainforest at low to moderate altitudes. Plants have a slender, solitary, ringed stem to 3 m or more tall and a purplish-green crownshaft. The leaves, to 4 m long, form a graceful crown and have widely spreading, narrow leaflets. The ovoid to obovoid fruit, to 2 cm long, are red when ripe. An attractive palm which would make an ideal garden plant for the tropics.

Rhopaloblaste ceramica

(from Ceram)

MOLUCCAS

This species has been introduced into cultivation and has proved to be an ornamental palm suitable for the tropics. It is native to the Moluccas where it grows in lowland rainforests. Plants have a solitary trunk to about 15 m tall and a crown of spreading fronds with long, pendulous leaflets (to 1 m x 2.5 cm). Green flowers are followed by clusters of large, showy, scarlet fruit (to 3.5 cm x 1.8 cm). This species may be cultivated as *R. hexandra*.

Rhopaloblaste dyscrita

(doubtful, hard to determine)

NEW GUINEA

A poorly known species which was originally collected from the Morobe District in Papua New Guinea. It is apparently a clumping palm which has soft, white, woolly scales on the young leaves and narrow, rigid, spreading leaflets. It is not known to be cultivated but, like other members of this genus, has good potential.

Rhopaloblaste elegans

(elegant, graceful)

GUADALCANAL

A well-named, delightful palm which has a very slender (about 15 cm across), light grey trunk to 12 m tall and a light green to pinkish, densely scaly crownshaft. The leaves spread in a graceful, open crown and the narrow, light green, stiff, papery leaflets are bent downwards forming an angle of about 45° with the rachis. The fruit show colourful changes during the ripening process, maturing through yellow and orange to crimson. Each fruit is globose to obovate in shape and covered with a glaucous bloom when fresh. This species, which originates on the island of Guadalcanal (Solomon Islands), grows well in tropical regions.

Rhopaloblaste singaporensis

(from Singapore)

Kerinting

MALAYSIA—SINGAPORE

When dried, lacquered and polished, the slender stems of this palm make handsome walking sticks. Native to Singapore and Malaysia, it is a very common species of rainforests and moist, open forests. Clumps usually consist of two or three mature stems which may reach 3 m tall. These stems are supple and whippy and usually less than 2.5 cm thick. Leaves grow to 1.6 m long and have numerous stiff, narrow, spreading leaflets which are 35–45 cm long. Clusters of fruit are very colourful, each fruit being about 1.4 cm long, fleshy and brilliant-orange to red when ripe. All in all, this is a very attractive and ornamental, slender palm for a shady or semi-protected position in a tropical or subtropical garden. It was previously known as *Ptychoraphis singaporensis*.

RHOPALOSTYLIS

(from the Greek *rhopalon*, club and *stylos*, pillar or pole (the style), literally with a club-like style)

A genus of three species (considered by some authors to be two species and two varieties) of palm distributed in Norfolk Island, New Zealand, Chatham Island and Raoul Island. They grow in dense, moist to wet forests mostly in lowland regions. They are solitary, unarmed feather-leaved palms with a prominent, inflated crownshaft and obliquely erect fronds which impart a distinct silhouette. The inflorescences are carried below the crownshaft and bear unisexual flowers of both sexes.

Cultivation: Interesting palms which are moderately popular in cultivation, but valuable because of their suitability for temperate regions. Plants require shelter when young but will tolerate sun when well established. They are moisture-loving palms which should not dry out. They respond strongly to the use of fertilisers. Single plants are capable of producing fertile seed.

Propagation: Fresh seed germinates readily one to three months after sowing. Seedlings are best grown in deep pots to accomodate the growing point, which becomes deeply buried in the soil.

Rhopalostylis baueri

(after F.L. and F.A. Bauer, botanical artists)

Norfolk Island Palm

NORFOLK ISLAND

This 'feather duster' palm resembles the Nikau Palm of New Zealand but is generally thicker and more robust in all its parts. The two can be readily distinguished when growing side by side because the crownshaft of the Norfolk Island Palm is stout, bulging and very prominent and the trunk and petioles thicker. Seedlings and young plants have distinct reddish tonings in the leaves and the leaflets are notched at the tips. The species is native to Norfolk Island, Australia, where it is still common in remnant vegetation. In cultivation it is quite a fast-growing and handsome palm, but for best appearance should be protected from the wind as the fronds readily become shredded. It is most successful when grown in moist, shady conditions and is an ideal palm for mixing with ferns. It succeeds quite admirably in warm–temperate regions and can also be grown in the cooler subtropics, but its tolerance to tropical conditions is doubtful (see under *R. cheesemanii* for differences between the two taxa).

Rhopalostylis cheesemanii

(after T.F. Cheeseman, prominent New Zealand botanist)

Kermadec Nikau Palm

KERMADEC ISLANDS

This species is endemic to Raoul Island (also called Sunday Island) in the Kermadec Group off the east coast of New Zealand. Plants are very

Rhopalostylis baueri.

Rhopalostylis baueri published as *Areca baueri* (*Botanical magazine*, Vol. XXIV of the 3rd series, plate 5735)

Rhopalostylis sapida.

west coast of the South Island where it grows on forested slopes facing the Tasman sea. It commonly grows in wet, lowland forests, often in dense, dark situations and is distributed from the coast to ranges at about 700 m altitude. Plants are often retained in cleared paddocks. The Maoris made considerable use of this palm, eating the young inflorescence and the heart and plaiting the leaves into baskets and other utilities. The leaves were used to thatch huts known as 'whares' and when properly constructed, these were waterproof. The fruit are extremely hard and it is recorded that they were used as bullets when shot was scarce. This palm tolerates light frosts and grows well in temperate regions. Young plants need shady, moist conditions and protection from direct sun for the first five or six years. Even mature plants look best in a sheltered position as the fronds are easily damaged by wind. Plants are generally slow growing and do not form a trunk before they are fifteen years old, nor flower until about thirty years of age. They need well-drained soil and respond to watering during dry periods, mulching and fertilisers.

R. sapida is somewhat variable (for an interesting account of this variation see K. Boyer, *Palms and Cycads Beyond the Tropics*, Palm and Cycad Societies of Australia, 1992). Plants from the South Island of New Zealand are more robust than those from the North Island and with a more open crown and broader leaflets. Those from the Chatham Islands are more robust again, with pinkish petioles, broader, more crowded leaflets and larger fruit. These differences would seem to be genetic and suggest that this population is worthy of taxonomic recognition. Unfortunately it has suffered badly from clearing and grazing and is now severely threatened and perhaps endangered. Other variants have been noted on islands to the north-east of New Zealand.

similar to those of R. baueri from Norfolk Island, but the fruit are larger (12–16 mm across), distinctly globose (ovoid to ellipsoid in R. baueri) and glossy red when ripe (dull red-brown in R. baueri). Some authorities include this species as a variety of R. baueri (R. baueri var. cheesemanii) but it is certainly distinctive and requires recognition at more than varietal rank. Seedlings of this species are reddish and have notched leaflets. This species is rarely cultivated but has requirements similar to those of R. baueri. It will grow in warm–temperate and subtropical regions.

Rhopalostylis sapida

(pleasant to taste, a reference to the edible inflorescence)

Nikau Palm, Feather Duster Palm, Shaving Brush Palm

CHATHAM ISLANDS

This New Zealand palm is the southernmost naturally occurring palm in the world, extending as it does to Pitt Island of the Chatham group at a latitude of 44°18' south. In New Zealand, it is widely distributed on the North Island and on the

ROSCHERIA

(after Albrecht Roscher)

A monotypic genus of palm which is confined to Silhouette and Mahé Island in the Seychelles. Plants grow in the understorey of rainforests at moderate altitudes (500–750 m), sometimes in pure stands. It is a solitary, spiny, feather-leaved palm which has a prominent crownshaft. The inflorescences arise among the leaves and hang below the crownshaft after leaf fall. They carry unisexual flowers of both sexes.

Cultivation: An elegant palm which seems to be

Roystonea borinquena group planted.

uncommon to rare in cultivation. Some reports suggest that the species has specific requirements and may be difficult to maintain. Single plants are capable of producing fertile seed.

Propagation: Fresh seed germinates two to four months after sowing. Each fruit contains a single seed.

Roscheria melanochaetes

(with black bristles)

Latanier Hauban

SEYCHELLES

Young plants of this palm have entire leaves whereas mature plants have irregularly pinnate leaves, with the leaflets being of uneven width and often splitting irregularly near the apex. The stems of young plants are also quite spiny, whereas on mature plants the stems are unarmed or with a few, weak spines. The crownshaft and petioles are also spiny. Clusters of small, globular fruit (about 7 mm across) are red when ripe. A handsome palm which likes rich, loamy soils and suffers badly if allowed to dry out. Warm, humid conditions and a sheltered position when young are also essential requirements. In the

late 1800s and early 1900s, this species was grown as a pot plant in European conservatories.

ROYSTONEA

(after General Roy Stone)

Royal Palms

A genus of about twelve species of tall, majestic palm centred on the Caribbean region and adjacent areas in Central America, northern South America, Florida and Mexico. Some species grow in well-drained soils in forests and woodlands, others are found in the wet soils of swamps, bogs and morasses. They have a very distinctive appearance, with a stout, tall, solitary trunk; prominent, long crown-shaft beset with phallic spadices; and a graceful crown of feathery fronds, with most species having the leaflets arranged in several ranks. The inflorescences arise below the crownshaft and bear unisexual flowers of both sexes. Most species in this genus are basically similar in appearance, with identification often relying on features of the leaflets, inflorescence, flowers and fruit.

Cultivation: Highly ornamental palms which are commonly planted in suitable regions around the world. They are unexcelled for planting in rows and are frequently used to line avenues and driveways where they provide an imposing spectacle. They are also excellent for planting in parks and, despite their size, are very popular in home gardens. They respond strongly to cultural practices such as heavy mulching, regular fertiliser dressings and watering during dry periods. These palms are often still sold in nurseries under the old generic name *Oreodoxa*. Single plants are capable of producing fertile seed. Palms of this genus are purported to hybridise freely but supporting evidence is lacking and growers familiar with the group say that hybridisation is rare.

Propagation: Fresh seed germinates readily within one to four months of sowing. Each fruit contains a single seed.

Roystonea borinquena

(from *Borinquen*, a native name for Puerto Rico)

Puerto Rican Royal Palm

PUERTO RICO

Native to Puerto Rico, this Royal Palm is basically very similar to *R. regia* but has a trunk that is prominently bulging above the middle and pale brown rather than purple fruit. It can also be distinguished by its densely crowded flowers (well separated in *R. regia*) and the leaflets being glossy green on the upper surface. Its dark green fronds have a feathery appearance due to the arrangement of the leaflets in many ranks. This species is uncommonly grown, seems to be slightly cold sensitive and is best suited to tropical conditions.

Roystonea dunlapiana

(after Dr R.V.C. Dunlap)

HONDURAS

An attractive palm known from Honduras where it grows in open groves on hills and as isolated specimens along tidal estuaries, often with its roots in water. It is distinctive for its slender trunk (30–40 cm across) and the leaflets being arranged prominently in two ranks, the rows standing at about 30° to each other. Plants have an open, rounded crown with the lower fronds prominently drooping. The crownshaft is prominent and the fruit are reverse

Roystonea elata in flower.

pear-shaped (about 14 mm x 10 mm) with a constricted base. This species is apparently not grown.

Roystonea elata

(tall, lofty)

Florida Royal Palm

FLORIDA

In its natural state this palm occurs in the low, swampy ground of Florida, USA and although its populations have been reduced by clearing, it is still present in the Everglades. It is often confused with *R. regia* but the two are quite distinct. *R. elata* is a tall palm with a slender, (bulging in *R. regia*), ash-grey (white in *R. regia*) trunk and a large crown of dark green leaves above a prominent, bright green crownshaft. Its inflorescences are much longer than wide (about as wide as the length in *R. regia*). The species has been widely planted and is especially popular in Florida. Plants will tolerate some cold but are not as hardy as *R. regia*. They can be grown in subtropical and tropical regions.

Roystonea hispaniolana

(from Hispaniola)

Hispaniolan Royal Palm

HISPANIOLA

This species is common on the island of Hispaniola, growing in valleys and on low hills, sometimes in extensive stands. It is also reported to be planted in other places in the Caribbean region. Plants are similar to *R. borinquena* but the leaflets are not glossy on the upper surface and the fruit are broad throughout and are not narrowed to the base.

Roystonea jenmanii

(after G.S. Jenman, original collector)

GUYANA

This species was described from cultivated plants growing in Georgetown, Guyana. It is a handsome palm, having a trunk slightly swollen near the middle and a rounded crown of dark green leaves. The leaflets are arranged in several ranks imparting a plumose appearance. This species is apparently still grown in Georgetown but is apparently uncommon elsewhere.

Roystonea lenis

(soft, smooth)

CUBA

A little-known species from Cuba which is common in the Guantanamo Province on the eastern end of the island, often in mountainous areas. It is a tall, robust palm (to 40 m) similar in general appearance to *R. regia* but with kidney-shaped sepals on the flowers and larger, spherical fruit (to 15 mm across) which have a very hard endocarp. Seed of this species has recently been introduced into cultivation. *R. regia* var. *pinguis* may be synonymous with this species

Roystonea oleracea

(used as a vegetable)

West Indian Royal Palm, Feathery Cabbage Palm

WEST INDIES

This is the tallest of the royal palms, with trunks being recorded at more than 30 m high. They are smooth, of uniform width throughout, but usually with a conspicuous bulge at the base and topped by a bright, shiny green crownshaft and a large, flat-bottomed crown of dark green, spreading fronds. These fronds appear flat because of the placement of the leaflets in one plane and in this feature and the flexuose rachillae, the species differs from all other royal palms. A very popular palm which can be grown in tropical and subtropical regions and also, with care, in warm–temperate districts. The species originates in the West Indian islands of Trinidad and Barbados.

Roystonea peregrina

(strange, foreign)

Round-fruited Royal Palm

WEST INDIES

This species is apparently distinguished from other royal palms by its large, globose fruits. Plants have a relatively slender, columnar trunk to 10 m tall and an attractive, rounded crown of dark green leaves, with the leaflets arranged in several planes. It is recorded that this poorly-known species is grown in the West Indies and was described from plants cultivated on the island of Guadeloupe, but of unknown native origin.

Roystonea oleracea.

Roystonea regia var. *regia*

(regal, royal)

Cuban Royal Palm

CUBA

This majestic palm is the national tree of Cuba where it is still abundant, growing on fertile soils to about 1000 m altitude. The bulging, concrete-white trunks are a useful guide to its identity. The bulge is usually present in the middle but may occur anywhere along its length. Its plumose leaves carried in a graceful, rounded crown and the oblong fruit, which are flattened or compressed on one side are also distinctive. In its native country the trunks are cut for timber, the leaf-bases are used as a waterproof covering for bales of tobacco and the fruits (called 'palmiche') are used to feed pigs. Cuban Royal Palm is a familiar sight as it is very commonly planted throughout the tropics and to a lesser extent, the subtropics. It is frequently planted in rows beside driveways, roads and avenues and makes a uniform and stately palm for this purpose.

343

A sunny position in well-drained soil is essential for success, although plants may grow rapidly in wet soils where the water is not stagnant. Plants respond vigorously to heavy applications of fertiliser rich in nitrogen. Two distinctive varieties are recognised.

Roystonea regia var. *hondurensis*

(from Honduras)

HONDURAS

Differs from the typical in being more robust (trunk to 35 m x 75 cm) and with oddly shaped fruit—these being reverse pear-shaped with a constricted base, each 12 mm x 8 mm. It is native to Honduras.

Roystonea regia var. *maisiana*

(from Cape Maisi, eastern end of Cuba)

CUBA

Differs from the typical by having a narrower, hardly bulging trunk, smaller crown and oblong–globose fruit. An elegant palm which has recently been introduced into cultivation.

Roystonea regia.

Roystonea venezuelana

(from Venezuela)

Venezuelan Royal Palm

VENEZUELA

In the original description of this species H.E. Moore stated, 'I had seen this tree in native stands from a plane and even then thought it to be distinct'. *R. venezuelana* is a relatively slender species in this genus of stout palms, with its trunk —which may reach 30 m tall—only being about 40 cm across. Its crown is not distinctly rounded as in other species, owing to its erect and horizontal leaves. The leaflets, which are arranged in several ranks, are dull on the upper surface and ash-grey beneath. This species is native to eastern Venezuela where it grows in lowland rainforests. It is a fine palm which is widely planted in tropical America and grown to a lesser extent in other tropical and subtropical countries.

Roystonea ventricosa

(swollen near the middle)

GUYANA

This species was described in the genus *Euterpe* in 1906 and later transferred to *Roystonea* when its characters became apparent. As with *R. jenmanii*, it was described from plants cultivated in Georgetown, Guyana, this time from plants growing in the botanic gardens. Plants have a narrower trunk than *R. jenmanii* (which is distinctly swollen near the middle) and smaller fruit. It is apparently still grown in Georgetown but is uncommon elsewhere.

Roystonea violacea

(violet-blue)

Blue Palm

CUBA

The male flowers of this species can be immediately recognised by their violet-blue to purple colouration at the base of the petals and filaments. The palm is distinctive in other ways, having a reddish to brownish, cylindrical trunk (reddish fibres inside as well) and oblong fruit (about 14 mm x 8 mm) which are flattened on one side towards the base. The crown is rounded, with the dark green leaflets arranged in several ranks. A very decorative palm, native to eastern Cuba,

Sabal bermudana.

which is uncommonly grown. Seed has been recently introduced into cultivation.

SABAL

(probably from an American native name)

Palmetto Palms, Hat Palms

A genus of fifteen species of fan palm found in the south-eastern USA, north-eastern Mexico, various Caribbean islands, Panama and northern South America. They mostly grow in open habitats, including savanna, low hills, coastal dunes and swamps. They are unarmed solitary palms with costapalmate leaves, unarmed petioles and split leaf-bases. Trunk growth ranges from subterranean to tall and lofty. The inflorescence arises among the leaves and bears bisexual flowers. Species of *Sabal* are notoriously difficult to identify, particularly those of cultivated origin, because hybridism can be frequent if species are planted in close proximity. (For a treatment of cultivated species see H.E. Moore, 'Notes on *Sabal* in Cultivation', *Principes* 15 (1971): 69–73. A complete detailed study has been prepared, see S. Zona, 'A Monograph of *Sabal*

(Arecaceae: Coryphoideae)', *Aliso* 12 (4) (1990): 583–660.)

Cultivation: A group of hardy palms which are widely grown in many countries. They are sun-loving palms which demand excellent drainage. Some species are well-suited to group planting, while others make impressive specimen plants and can be used for avenues. Seed should be collected with caution from mixed plantings because of hybridisation. Single plants are capable of producing fertile seed.

Propagation: Fresh seed germinates readily, although seedlings of some species may be slow and erratic to appear. Seedlings are generally slow growing. Each fruit contains one to three seeds.

Sabal bermudana

(from Bermuda)

Bermuda Fan Palm

BERMUDA

A fan palm from Bermuda which has proved to be adaptable in cultivation, with mature specimens thriving in temperate regions as well as the tropics. Plants develop a stout, grey trunk to 7 m tall and 35 cm across, with a large crown of green leaves. The fruit are distinctly pear-shaped, to 18 mm across and black when ripe. They have a sweet flesh which attracts birds and animals. Occasionally massive plants of this palm are encountered in cultivation. They are often labelled as *S. princeps* but this name has no botanical standing. *S. beccariana*, also described from a cultivated plant, is another synonym.

Sabal causiarum

(pertaining to hats, in reference to a use of the leaves)

Puerto Rican Hat Palm

ANEGADA—PUERTO RICO—HISPANIOLA

The hallmarks of this species are a very stocky, grey trunk which may rise to more than 10 m tall and be more than 50 cm across and a dense, heavy crown of fan leaves which always seems to be too small for the massive trunk. Although of ungainly appearance, the palm is nevertheless impressive and an excellent choice to line a driveway or scatter in open parkland. Native to Puerto Rico, Anegada and Hispaniola, it is now widely planted

Sabal causiarum.

where it grows in the interior, from the lowlands up to about 1000 m altitude. It has also recently been recorded from the eastern part of Cuba where the leaves are used for thatch. Plants are very similar overall to *S. causiarum* but have larger, pear-shaped fruit (11.5–14 mm across). They develop a massive, smooth, grey trunk and a large crown of green leaves. Because of confusion with *S. causiarum* it is uncertain whether *C. domingensis* is in cultivation. It is certainly a handsome palm worth growing if sufficient space is available. Plants probably have similar requirements to those of *S. causiarum*.

Sabal etonia

(from Etonia Scrub, Florida, the type locality)

Scrub Palmetto

FLORIDA

This is one of the easier *Sabal* species to identify with certainty, since most plants have a subterranean trunk (some may be emergent to 3 m tall) and three to five strongly costapalmate, yellowish-green leaves which have many threads hanging from the segments near the division. It also has an erect, bushy inflorescence and brownish-black fruit, 1–1.5 cm across. It is native to central and south-eastern Florida, USA, where it grows in deep sand ridges clothed with pine-oak woodland. The species is still common in reserves but much of its habitat has been destroyed for agriculture and urban development. *S. etonia* is an interesting palm which is relatively uncommon in cultivation. It grows well in tropical areas but its adaptability to colder climates is largely unknown.

Sabal gretheriae

(after R. Grether)

MEXICO

A recently described (1991) species from the Yucutan Peninsula in Mexico. It is similar to *S. mexicana* but can be distinguished by its broader leaf segments (6–7 cm versus 3.2–5.3 cm in *S. mexicana*), different seeds and it flowers in May to August (January to April for *S. mexicana*). This species probably has similar cultural requirements to those of *S. mexicana*.

in many countries of the world. It will grow well in tropical and temperate regions and also thrives in the drier atmosphere of inland districts. Its deeply divided leaves are usually dull green but may be bluish and after flowering the long inflorescences carry masses of small, black fruits that hang among the outer canopy of the leaves. The spherical to oblate fruit, to 1 cm across, are black when ripe. In cultivation this palm needs very well drained soil and responds to added fertiliser. Once established, the plants are very hardy to dry conditions.

Sabal domingensis

(from Santo Domingo)

Palmetto Palm

HISPANIOLA

A poorly known species which is native to Hispaniola (both Dominican Republic and Haiti),

Sabal mauritiiformis.

Sabal guatemalensis

(from Guatemala)

CENTRAL AMERICA

A palm which is very similar to *S. mexicana* but differing in features of the flowers and fruit. The flowers have an urn-shaped calyx (cup-shaped in *S. mexicana*) and the fruit are pear-shaped and 10–14 mm across (spherical and 15–19 mm across in *S. mexicana*). The species occurs naturally in Mexico and Guatemala.

Sabal maritima

(growing on the seashore)

Bull Thatch Palm, Cuban Palmetto

CUBA—JAMAICA

A common Cuban palm found growing in sandy or calcareous soils throughout the island and also occurring in southern and western Jamaica. It is a tall species to 15 m with a grey trunk to 40 cm across and a large crown of green, strongly costapalmate, thread-bearing leaves. The inflorescence is about as long as the leaves and the oblate to pear-shaped black fruit are about 14 mm across. This is a very hardy palm which grows well in coastal situations and also thrives in hot, dry climates further inland. It can be successfully grown from tropical to temperate regions. *Sabal floridana* and *S. jamaicensis* are synonymous.

Sabal mauritiiformis

(resembling a *Mauritia*)

CENTRAL AND SOUTH AMERICA—TRINIDAD

A widely distributed species which is found in disjunct populations from southern Mexico to northern coastal areas of Colombia and Venezuela and also Trinidad. It commonly grows on calcareous soils derived from limestone and ranges from sea-level to about 1000 m altitude. It is a distinctive species, with a slender (15–20 cm across) grey-brown trunk and green to glaucous, weakly costapalmate leaves. Each leaf has numerous slender segments (to 2 m long) which are united in groups of two or three, with the long, deeply lobed apex of each segment often drooping. The erect to arching inflorescence extends beyond the leaves and the spherical to pear-shaped, black fruit are about 1 cm across. This palm is best suited to tropical and subtropical regions and prefers some protection when young. *S. glaucescens*, *S. nematoclada*, *S. allenii* and *S. morrisiana* are synonymous with this species.

Sabal mexicana

(from Mexico)

Rio Grande Palmetto, Texas Palmetto

TEXAS—MEXICO—EL SALVADOR

A common species which dominates the lowland vegetation of tropical parts of Mexico. It also occurs in southern Texas along the valley of the Rio Grande River and has been collected once in El Salvador. Plants are cultivated in plantations for thatch in the Yucatan Peninsula of Mexico. It is a robust species, having a grey trunk to 15 m tall and 35 cm across and a large crown of green, strongly costapalmate, thread-bearing leaves. The inflores-

Sabal etonia.

Sabal minor in fruit.

Sabal mexicana.

cence is about as long as the leaves and the spherical to oblate black fruit are 15–19 mm across. This is a very hardy palm which thrives in hot, dry climates and is excellent for inland districts. It can be successfully grown from tropical to temperate regions. *S. texana* and *S. exul* are synonymous.

Sabal minor

(smaller, lesser)

Dwarf Palmetto, Bush Palmetto, Swamp Palmetto

NORTH AMERICA

This, the most northerly member of the genus, is widely distributed in south-eastern USA. It grows in alluvial sites along floodplains and in swamps in broadleaf deciduous forests. Plants usually have a subterranean trunk and the leaves arise in a crown at the surface, although some plants may have an erect trunk 1–2 m tall. The crown consists of four to ten, weakly costapalmate leaves which may be dark green or bluish–green and lack threads. The segments are stiff and the apex is usually not divided. Plants are generally precocious and flower while quite young, producing a succession of inflorescences during summer. The inflorescence is usually slender and arches out from the rosette of leaves. Small, white, fragrant flowers are followed by shiny, black, spherical fruit 7–10 mm across. This palm is an excellent garden plant which can be grown from the tropics to temperate zones. *S. lousiana* and *S. deeringiana* are synonyms.

Sabal palmetto

(a small palm)

Palmetto Palm

NORTH AMERICA—BAHAMAS—CUBA

A wide-ranging, weedy and highly variable palm which is commonly cultivated in the tropics and subtropics. It is widely distributed in south-eastern USA (south from North Carolina) to various islands in the Bahamas and Cuba. It grows in a wide variety of habitats, from coastal dunes and tidal flats to seasonally flooded savannas, swamp and stream margins. Plants have a trunk to 20 m tall and 35 cm across with a large crown of green to yellow-green, strongly costapalmate, thread-bearing leaves. The leaf segments are joined for about one-third of their length and the apex is deeply notched. The inflorescence is about as long

Sabal palmetto young mature plants.

Sabal palmetto mature plant.

as the leaves and the spherical black fruit 8–14 mm across. This palm has proved to be extremely easy to grow. It succeeds very well in sandy soil, tolerating some coastal exposure and brackish water inundation. Plants grow very strongly where their roots can tap ground water. This palm looks good when planted in groups and is particularly successful in tropical and subtropical regions. Plants are fast growing and generally flower and fruit when quite young. *S. viatoris*, *S. jamesiana* and *S. parviflora* are synonyms.

Sabal pumos

(a native name for the fruit)

MEXICO

A slender, graceful palm which has a trunk to 15 m tall and 35 cm across and a large, rounded crown of green, strongly costapalmate, thread-bearing leaves. The arching to nodding inflorescence is shorter than the leaf petioles and the oblate to spheroidal, greenish-brown to black fruit are among the largest in the genus, being 18–27 mm across. *Sabal pumos* is endemic to Mexico where it is locally common, growing in sandy soils at 600–1300 m altitude. Its leaves are used for thatch and the fruit are eaten. Seed of this species has been recently introduced into cultivation. *S. dugesii* is a synonym.

Sabal rosei.

Sabal rosei

(after J.N. Rose, original collector)

MEXICO

A slender species to 15 m tall with a grey, smooth trunk to 30 cm across and a large, open crown of green, strongly costapalmate, thread-bearing leaves. The arching to nodding inflorescence is about as long as the leaves and the greenish-brown to black fruit are 15–22 mm across. This species is common in western Mexico, growing in tropical deciduous forests from sea-level to 600 m altitude. It grows readily in cultivation, particularly in tropical and subtropical zones.

Sabal uresana

(from the locality of Ures in Mexico)

Sonoran Palmetto

MEXICO

Young plants of this species have deep bluish-green leaves but this is not so noticeable on mature plants, although some specimens are markedly

glaucous. Mature plants develop a large trunk, to 20 m tall and 40 cm across and also have a large, rounded crown of strongly costapalmate leaves. The arching inflorescence is about as long as the leaves and the oblate, spherical or pear-shaped fruit are 13–18 mm across. This species is locally common in western Mexico and has a wide altitudinal range, extending from sea-level to about 1500 m. It is a hardy palm which grows well in subtropical and warm–temperate regions.

Sabal yapa

(from the Cuban name of *cana japa*)

Thatch Palm

CUBA—CENTRAL AMERICA

This species is of similar general appearance to *S. mauritiiformis*, especially in the finely divided leaves with conspicuously drooping segments. It can, however, be immediately distinguished by the slender trunk and flowers which have a bell-shaped calyx and fleshy petals. It occurs naturally in Belize, western Cuba and the Yucatan Peninsula of Mexico, growing on well-drained calcareous soils derived from limestone. In cultivation this species has proved to be slow growing. It is tropical in its requirements. *S. mayarum*, *S. peregrina* and *S. yucatanica* are synonymous.

SALACCA

(from the Malay native name, *salac*)

Salak Palms

A group of about eighteen species of palm, distributed in Yunnan (China), Burma, Thailand, Malaysia, the Philippines, Borneo and Java, with the centres of diversity in Malaysia and Borneo. Some species have edible fruit and the leaves are used for a wide variety of purposes, consequently plants are often left during clearing operations. They grow in a range of habitats, from well-drained slopes to swamps, with some species forming thickets. They are unique, very spiny palms ranging from small to robust, with very short, branching stems which are either decumbent and subterranean or shortly erect. The erect leaves form a tuft and may be entire with a notched apex, or pinnate and have stout spines, either variously scattered or grouped in whorls. The simple or branched inflorescences arise among the leaves (breaking through a slit in the leaf sheath) and bear unisexual flowers.

The fruit are covered with irregular rows of scales. *S. conferta* is now placed in the genus *Eleiodoxa*. The genus *Salacca* is being actively studied. (See for example, J.P. Mogea, 'The Flabellate-leaved Species of *Salacca* (Palmae)', *Reinwardtia* 9(4) (1980): 461–79 and 'Notes on *Salacca wallichiana*', *Principes* 25 (1981): 120–3.)

Cultivation: The genus includes some imposing palms which, although highly spiny, have ornamental qualities. Some species are grown for their edible fruit which have commercial potential. Most species are very sensitive to cold and will only succeed well in the tropics. Plants will tolerate exposure to sun but should be mulched and watered regularly if fully exposed. Male and female plants are essential for fertile seed production, although some plants may produce occasional bisexual flowers.

Propagation: Clumping species can be divided, although it is more usual to sever a developing sucker after it has formed roots. Fresh seeds may germinate erratically, appearing three to twelve months after sowing. Each fruit contains one to three seeds.

Sabal yapa.

Salacca affinis var. affinis

(similar to another)

MALAYSIA—SUMATRA

A prickly, clumping palm from Malaysia and Sumatra where it grows as scattered individuals in shady forests. Leaves extend to about 2.5 m long and are held erect, arising in rosettes from the short, prostrate trunks. The smooth, relatively broad, green, leaflets (to 45 cm x 10 cm) are borne in clusters of two to four on each side of the rachis and are held in one plane. Most parts are well armed with long spines (3–6 cm long). The fruit grow to about 4 cm long and are scaly, with a sweet, edible pulp and a short, nipple-like apex.

Salacca affinis var. borneensis

(from Borneo)

This variety differs in having eighteen vertical rows of scales in the fruit (22–26 rows in var. *affinis*).

Salacca clemensiana

(after Mary Strong Clemens, 19th & 20th century botanical collector)

PHILIPPINES—SABAH

A clumping species with subterranean or shortly erect trunks and erect fronds to about 4 m long. These are densely spiny on the leaf sheaths and petioles and the leaflets are dark green on the upper surface and chalky white beneath. The terminal leaflets are united. The female inflorescence is short and the flowers have strongly reflexed, crimson petals. The male inflorescence is arching to pendulous and has colourful pink bracts. The scales of the fruit have reflexed spiny tips. An interesting palm known from Mindanao island in the Philippines and recently discovered in Sabah.

Salacca dolicholepis

(with long scales)

BORNEO

This species has spiny fronds to about 2 m tall with a triangular rachis and clustered leaflets arranged irregularly along the rachis. The leaflets, to 40 cm x 3.5 cm, are more or less linear in shape, curved at the base, dark shiny green on the upper surface and whitish beneath. The fruit are rounded to top-shaped and about 4 cm long. An interesting palm which is native to Mount Kinabalu in Borneo.

Salacca dransfieldiana

(after John Dransfield, noted contemporary palm botanist)

BORNEO

This palm is known only from a single population of about thirty plants. It is endemic to Borneo where it grows on an alluvial flat in lowland forest. Plants have simple notched leaves similar to those of *S. flabellata*, but the male inflorescence is very short, erect and not whiplike. Female plants and fruits are unknown and it is surmised that all plants of *S. dransfieldiana* may be clonal, having arisen by vegetative reproduction.

Salacca flabellata

(fan-like)

MALAYSIA

This species multiplies in a novel way with the tip of the slender, whip-like male inflorescence often producing a vegetative shoot which takes root and develops into a separate plant. These inflorescences are 1–3 m long and by this means, the species builds up into localised colonies of male plants. The inflorescence remains viable for some time and it is recorded that flowers may be formed even after the terminal plantlet develops. The female inflorescences do not produce new plantlets. This palm has simple leaves which are deeply notched at the apex and carried on slender, spiny petioles to 1 m long. Young leaves are greyish-pink, mature leaves are glossy green above and grey or silvery beneath. A novelty item for a shady tropical garden. It is native to lowland areas of Malaysia where it grows in forests near streams.

Salacca glabrescens

(becoming glabrous or nearly glabrous)

MALAYSIA

A Malaysian species which grows on forested slopes. Plants have a subterranean trunk and clumps of large, spiny leaves to 5 m long. The shallowly curved, green leaflets, which are arranged in groups of two or three along the rachis, have

small spines along the margins. The inflorescences are very sparsely hairy and the pear-shaped fruit, to 5 cm x 4 cm, are abruptly narrowed into a beak. The fruit has edible flesh and this species is cultivated on a limited scale in the tropics for these fruit.

Salacca graciliflora

(with slender flowers)

MALAYSIA

A dwarf salak with very short (20 cm long), clumping stems and pinnate leaves 1–1.25 m long. These have up to thirty leaflets, to 24 cm x 2.5 cm, which are alternately arranged and held in one plane. The apical leaflets are united into a fishtail-like structure. The leaf sheath and petioles are spiny but the rachis and leaflets are not. The male inflorescences are slender, whip-like structures which lie on the ground. Details of the female inflorescence are unknown. This species is known from the State of Johore, Malaysia where it grows

Salacca dransfieldiana. G. Cuffe

on forested slopes at about 300 m altitude. It apparently reproduces vegetatively by forming plantlets on the inflorescence.

Salacca magnifica

(magnificent)

BORNEO

This species has very large, simple, notched leaves which are bright green and glossy on the upper surface and paler green beneath. The leaves grow in tufts, to 6 m tall, with the blade to about 5 m long and 70 cm wide. The fruit are reported to be deep pink. Although rarely cultivated, this species would have great potential because of its impressive leaves. It is native to Borneo where it grows in soaks at the base of scree slopes.

Salacca minuta

(small, tiny)

MALAYSIA

Apparently this species is the smallest in the genus, with plants being only about 1 m tall. It can be distinguished by its entire leaves which have a wedge-shaped to triangular blade, which is deeply notched at the apex, the margins of this notch being lobed or toothed. The upper surface is bright green and glossy and the lower surface has a coating of brown hairs. The leaf sheaths, petioles and rachis are spiny. The male inflorescences, which are slender and whip-like, to 1.5 m long, lie prostrate on the ground and produce a plantlet from near the apex. This attractive small palm would appear to have excellent potential for cultivation. It is native to the State of Johore, Malaysia.

Salacca multiflora

(with many flowers)

MALAYSIA

A trunkless palm which has long, simple leaves which erupt out of the forest litter in a tussock. Each leaf, which may be up to 3 m long, has a short spiny petiole and a narrowly triangular to wedge-shaped blade which is deeply notched at the apex, with the margins of the notch toothed where a vein ends. The blade is stiff-textured, to 2 m x 28 cm, and is bright green and glossy on the upper

surface and with a few brown dots beneath. The male inflorescences are branched and the small, globose fruit are covered with upturned, brown scales. This highly ornamental species is known only from a small area in Malaysia.

Salacca ramosiana

(after M. Ramos, a plant collector in the Philippines and Sabah.)

PHILIPPINES—SABAH

This species occurs on the Philippine islands of Sulu and Palawan where it grows on the inland margin of mangrove swamps and in small freshwater swamps. It also occurs in Sabah. It is unique in the genus in that the leaflets have deeply lacerated or lobed tips. Plants have subterranean stems and clusters of leaves 3–5.5 m long which are densely spiny and have pale brown hairs. The leaflets tend to be held upright and are clustered along the rachis in groups of three or four. The ellipsoid fruit, about 6 cm x 3 cm, are bright reddish-brown and have a prom-inent apical beak. This species, which is not known to be cultivated, would be an interesting ornamental palm for the tropics.

Salacca rupicola

(growing in rocky places)

BORNEO

This is the only species of the genus to be associated with limestone. It grows in small clumps in crevices of limestone cliffs and hills in Borneo. It is a clumping species, with narrow, bluish-green leaves about 2 m long and curved leaflets ending in long, slender drip tips. The apical leaflets are united into a flabellum, with the drip tips of the united leaflets protruding from the margin. An interesting palm which is unknown in cultivation but would merit trialling.

Salacca sarawakensis

(from Sarawak)

BORNEO

This species is known only from female plants. It has simple, notched leaves with the blade to 120 cm long and 66 cm wide, with pale yellowish spines. Female inflorescences grow to about 20 cm long but details of the male inflorescence and fruit are unknown. Although not cultivated, this Bornean species would be an attractive novelty item for the tropics.

Salacca sumatrana

(from Sumatra)

SUMATRA

A robust clumping palm with leaves to 3 m or more long, the petiole and lower parts of the rachis being liberally armed with formidable spines to 5 cm long. The rigid leaflets, to 75 cm x 7.5 cm, are regularly arranged in one plane and are dark green and glossy on the upper surface, ash-grey beneath. The fruit are top-shaped, to 7 cm x 4.5 cm and covered with dark, chestnut-brown, glossy scales. This palm is native to south-western Sumatra, growing in forests at about 350 m altitude.

Salacca wallichiana

(after Nathaniel Wallich)

SOUTH-EAST ASIA

A robust species which may have either a creeping trunk, branching on the older parts, or be shortly erect. Large leaves (to 7.5 m long), which are borne in erect, dense tufts, have densely spiny petioles and clustered leaflets which spread widely from the rachis and give the impression of being whorled. The slender male inflorescences are pendulous and sometimes produce a plantlet at the apex in the manner of *S. flabellata*. The fruit are orange-brown, edible, but usually cooked before eating. The species is widespread, growing in swampy forests in Burma, Thailand and Malaysia. Three plants of an unusual, spineless variant have been found in Thailand.

Salacca zalacca

(from a native name)

Salak

INDONESIA

The large, pear-shaped, brown fruit of this spiny palm have an edible layer of pale yellow (or pink), succulent flesh surrounding the seeds. In parts of Asia these fruit are sold in the marketplace and are a popular item. The palm itself forms huge

clumps, spreading by branching subterranean trunks. Each trunk, where it emerges, carries a crown of long, erect pinnate fronds (to 6 m tall) with whorled leaflets (to 60 cm x 5 cm) that are white beneath. Numerous sharp spines clothe most of the above-ground parts. The palm is native to the Indonesian islands of Java and Sumatra where it grows in thickets in wet, swampy soil. It has been introduced into other countries where it has become commonly cultivated for its edible fruit. Although most populations of this species are dioecious it has been observed that plants from Bali (known as Bali Salak) may produce polygamous flowers, although some researchers believe that apomixis may be involved rather than monoecy. The species is very tropical in its requirements and succeeds best in lowland tropical areas. The plants like plenty of water and it has been observed that they respond to regular dressings of fertiliser.

Schippia concolor.

Flower production is enhanced by regularly thinning clumps and removing debris and senescing fronds. Plants will tolerate full sun from a very early age. This species was previously known as *S. edulis*.

SATAKENTIA

(after Toshihiko Satake and *Kentia* another genus of palm)

A monotypic genus endemic to the Ryukyu Islands to the south-west of Okinawa where it grows on the slopes of low hills. It is a solitary, unarmed feather-leaved palm with a prominent crownshaft. The inflorescences arise below the crownshaft and bear unisexual flowers of both sexes.
Cultivation: An attractive palm which grows well in cultivation and which has become popular in some areas. Single plants are capable of producing fertile seed.
Propagation: Fresh seed germinates readily about three months after sowing. Each fruit contains a single seed.

Satakentia liukiuensis

(from Liukiu, the original site of collection)

Satake Palm

RYUKYU ISLANDS

A handsome palm with a general appearance reminiscent of a Coconut. It is becoming popular for cultivation in the southern states of the USA but at this stage is rarely cultivated elsewhere except in its native country. It is a clean-cut palm with a neat, ornamental appearance. Cultivated plants probably will not grow much more than 10 m tall and have a crown of spreading, deep-green, graceful fronds. Newly emerged inflorescences are a colourful pink to purple. The small fruit, 6–7 mm across, ripen black. Although sensitive to severe cold, this species has proved to be adaptable and will succeed in a warm position in the subtropics.

SCHIPPIA

(after W.A. Schipp, collector in Belize, Guatemala and Mexico)

A monotypic genus which is endemic to Belize where it grows as an understorey palm in pine forest. It is a slender, solitary, unarmed fan palm with long petioles and split leaf-bases. The short inflorescence arises among the leaves and bears bisexual flowers or a mixture of bisexual and unisexual flowers.
Cultivation: An attractive, elegant palm which deserves to become widely grown. Cultivated plants are slow but have proved to be reliable.
Propagation: Fresh seed takes six months or longer to germinate. Each fruit contains a single seed.

Schippia concolor

(of one colour)

Silver Pimento Palm, Mountain Pimento Palm

BELIZE

This palm must surely be one of the most graceful of all the fan palms. It has a slender trunk to about 10 m tall and an open crown of delicately segmented leaves. These are carried on long, slender, smooth petioles which are divided at the base. The leaves themselves are dark, glossy green above, paler beneath and deeply divided into slender, arching segments. Fruit are about 2.5 cm across, rounded and white when ripe. The species grows well in tropical and warm subtropical regions. Plants grow happily in semi-shade but will also tolerate full sun. Drainage must be unimpeded. Like many of the smaller fan palms, this species is slow growing.

SERENOA

(after Sereno Watson, 19th century American botanist)

A monotypic genus of palm which is endemic in the south-eastern USA, growing in colonies in woodlands and coastal dunes. It is a densely clumping, prickly fan palm with very short trunks and thorny petioles. The leaves are deeply divided into narrow, stiff segments. The inflorescence arises among the leaves and bears bisexual flowers.
Cultivation: Although regarded as a pest in the wild, this hardy palm is frequently planted in the USA and to a lesser extent elsewhere. It is an excellent barrier plant and has proved to be adaptable to a range of soils and climates. A form with glaucous leaves is more popular than the normal green-leaved type. Single plants are capable of producing fertile seeds.
Propagation: Plants can be divided, but this is a

Serenoa repens.

Serenoa repens inflorescence.

difficult task unless they are small. Fresh seed germinates readily three to six months after sowing. Each fruit contains a single seed.

Serenoa repens

(creeping)

Saw Palmetto

NORTH AMERICA

Native to various states of the south-eastern part of the USA, the Saw Palmetto grows in huge colonies and is most common on the coastal plain. It is a fan palm with a branching trunk that is usually subterranean but sometimes emergent and reaching to 3 m tall. The leaves are extremely variable in colour, often being green or yellowish but in some forms being a splendid silvery-white. They are held stiffly erect on toothed petioles and form an impassable barrier. Saw Palmettos are an attractive palm for cultivation and succeed extremely well in coastal districts, tolerating considerable exposure to salt-laden winds. They are rather cold-sensitive however and grow best in tropical or warm–subtropical regions. Plants need a sunny aspect and will withstand full sunshine even when quite small. Once established, they are very hardy, especially if their roots can tap ground water.

SIPHOKENTIA

(from the Greek *siphon*, a tube and *Kentia*, another genus of palm)

A small genus of two species of palm from islands in the north-west Moluccas. They are understorey plants growing in lowland rainforest on ultrabasic rock. They are solitary unarmed feather-leaved palms with a slender, ringed trunk and a prominent crownshaft. The inflorescence arises below the leaves and carries unisexual flowers of both sexes. There is some doubt among botanists about whether the two species are distinct from each other. Single plants are capable of producing fertile seeds.
Cultivation: Handsome palms, one species of which is well established in cultivation but still mainly the province of enthusiasts. It is adaptable and of easy culture. Plants grow readily in tropical and warm–subtropical regions. Single plants are capable of producing fertile seed.
Propagation: Fresh seed germinates readily two to four months after sowing. Each fruit contains a single seed.

Siphokentia beguinii

(after V.M.A. Beguin, Dutch collector in Indonesia)
MOLUCCAS

Seed of this species was distributed in 1940 after being collected from Kahatola Island in the Moluccas. Since then the palm has become well established in cultivation and is now grown in many tropical and subtropical countries. It is a slender palm with a shiny green, prominently ringed trunk about 10 cm thick and a crown of arching to drooping, shiny green fronds, each about 2 m long. These are divided rather irregularly into both narrow and broad leaflets, with the terminal part undivided and prominent. The inflorescence is spreading, with creamy-white flowers followed by red, cylindrical fruit about 2 cm long. This species is an ideal garden palm, with a slender, manageable habit and attractive appearance. Plants grow well in a sheltered position and need regular watering and mulches. They can flower when quite small. For several years, young plants have broad leaves (entire or nearly so) with an attractive net venation and make very decorative pot subjects. Flowering occurs at an early age.

Above: *Siphokentia beguinii.*
Below: *Siphokentia beguinii* inflorescence.

SYAGRUS

(an early Latin name for a kind of palm tree used by Pliny)

A complex genus of thirty-two species of palm distributed from Venezuela to Argentina, with the centre of diversity in Brazil. Some species grow in rainforest but most are found in drier open habitats such as savanna and semi-arid scrublands. The genus has suffered severely from habitat destruction, with at least three species becoming extinct (*S. leptospatha*, *S. lilliputiana* and *S. macrocarpa*) and others reduced to rarity. They are small to robust, solitary or clustered, unarmed or spiny feather-leaved palms. One distinctive group is apparently acaulescent (the trunk is actually sub-terranean), with fronds arranged in erect, often dense tufts. The inflorescence arises among the leaves and bears unisexual flowers of both sexes. (For a recent treatment of the genus see S.F. Glassman, 'Revisions of the palm genus *Syagrus* Mart. and other selected genera in the *Cocos* alliance'. *Illinois Biological Monographs* 56: (1987): 1–230.) The genera *Arecastrum* and *Arikuryroba* are included in *Syagrus*.

Cultivation: Some species are highly ornamental, others are of horticultural interest because of unusual growth features. Relatively few species are cultivated, however, primarily because of the unavailability of seed. The species that have been grown have proved to be adaptable and easy. Many are relatively cold-tolerant and will grow in warm–temperate and subtropical regions. As a group these palms are sun-loving species which need free drainage. Single plants are capable of producing fertile seed.

Propagation: Fresh seed germinates readily two to four months after sowing. It has been reported that seedlings of the acaulescent species are sensitive to damage of their fine, much-branched root system. Each fruit contains one or two seeds.

Syagrus amara

(bitter)

Overtop Palm, Moca Palm

WEST INDIES

This *Syagrus*, native to various islands of the West Indies including Trinidad, is the only species of the genus not found on the continent of South America. It grows in dry forests close to the coast. The immature seeds contain a bitter fluid that is fermented into a drink. Plants are tall, with a trunk to 20 m and often only 10 cm across. The crown is somewhat reminiscent of that of a Coconut but with drooping leaflets. Clusters of fruit, each to 7 cm x 3.5 cm and resembling a small coconut, are orange when ripe. This palm is widely grown as an ornamental, succeeding in tropical and subtropical regions. Plants are well suited to coastal conditions, tolerate sun from an early age and are moderately fast growing. They require excellent drainage and respond to fertilisers, mulching and watering during dry periods. This species is placed in the segregate genus *Rhyticocos* (as *R. amara*) by some authors.

Syagrus botryophora

(bearing clusters like grapes)

BRAZIL

This species has had a confused botanical history, for many years being wrongly regarded as a robust variant of *S. romanzoffiana*. It is, however, quite distinct from that species, being readily distinguished by the leaflets which are regularly distributed in two ranks along the rachis (clustered and arranged in several ranks in *S. romanzoffiana*). Plants grow to about 18 m tall, with a trunk to

Syagrus amara.

Syagrus comosa.

Syagrus comosa

(with hairy tufts)

BRAZIL

A common Brazilian species which grows in woodlands. It is cultivated on a limited scale, growing well in the tropics and perhaps also successful in the subtropics. Plants have a slender trunk less than 10 cm across, which can grow to 5 m tall. Arching fronds are green to greyish and have the leaflets clustered along the rachis. The ovoid to globose fruit are about 3 cm across.

Syagrus coronata

(crowned or wreathed)

Licury Palm, Ouricury Palm

BRAZIL

The trunk of this species is covered by persistent leaf-bases which are arranged more or less in five spiral rows. Plants can grow to about 10 m tall and have a crown of graceful, arching, bluish-green fronds in which the leathery leaflets are arranged along the rachis in groups of three or four. Yellow flowers are followed by clusters of ovoid, orange fruit each about 2.6 cm long. These fruit have an edible flesh and the endosperm of the seed is rich in oil and is the basis for a commercial industry.

Syagrus coronata.

25 cm across and a graceful, rounded crown of dark green leaves. This is a very attractive palm which grows in coastal woodland of Brazil. Plants are easily grown in a range of situations but are not as popular as *S. romanzoffiana. Arecastrum romanzoffianum* var. *botryophum* is a synonym.

Syagrus cocoides

(resembling a *Cocos*)

BRAZIL

This species, the type of the genus *Syagrus*, is native to Brazil, growing in moist to wet forests and woodlands. Plants have a slender trunk (about 7 cm across) and crown of erect to arching fronds with the leaflets arranged in clusters along the rachis. The petiole and rachis are covered with white, mealy powder. The fruit are large, being about 6 cm long. An attractive palm, cultivated on a limited scale, which is suitable for tropical and subtropical gardens.

Syagrus oleracea infructescence.

A commercially important wax can also be extracted from the leaves. This highly ornamental palm is popular in cultivation and widely grown. It is adaptable and will grow in calcareous soils. Plants are best in a sunny situation and will grow in the tropics and subtropics. *S. treubiana* is a synonym.

Syagrus flexuosa

(wavy, zigzagged)

BRAZIL

An interesting palm which may have a solitary trunk but more usually grows in clumps with several trunks prominent. Plants grow 2–4 m tall and have an attractive crown of arching grey to glaucous fronds. The narrow leaflets are arranged in groups of two to four along the rachis. The fruit have a prominent beak. This species is widely distributed in Brazil and apparently often colonises disturbed sites. It grows readily but is not well known. Success has been achieved in tropical and subtropical regions.

Syagrus macrocarpa

(with large fruit)

SOUTH AMERICA

This palm, which is known from a few cultivated specimens, is apparently extinct in the wild since no plants have been collected since 1877. It is a very distinctive species, having a superficial resemblance to *S. romanzoffiana* but with a much more slender trunk and clustered leaflets arranged in several ranks. Plants grow to about 10 m tall and the leaves arch in a graceful, rounded crown. The petiole and rachis are covered with white woolly hairs, a feature prominent on young fronds.

Cultivation provides the only hope for the survival of this species. Plants are apparently easy to grow and succeed in tropical and subtropical regions. The species is native to southern coastal areas of Brazil and to Argentina.

Syagrus oleracea

(used as a vegetable)

SOUTH AMERICA

A tall palm with a trunk to more than 15 m tall and a crown of graceful, arching green to glaucous fronds to about 3 m long. The leaflets are clustered and have long-pointed tips. Clusters of fruit are colourful, being bright orange, with each fruit rounded and about 4.5 cm across. The species is native to Brazil and Paraguay where it grows in woodland. It is cultivated in many tropical countries and has proved to be adaptable in a variety of situations. *S. picrophylla* and *S. gomesii* are synonyms.

Syagrus pseudococos

(false *Cocos*)

Coco Verde

BRAZIL

A Brazilian palm which is now very rare in the wild owing to clearing of its habitat. It is a tall palm with a trunk reaching in excess of 20 m and 25 cm across. The crown is dense and rounded, with the dark green leaves arching gracefully and the leaflets irregularly clustered in groups of two to four. The large fruit, to 7 cm x 4 cm, have a sharply pointed apex, as does the seed. This palm deserves to be more widely grown, although it apparently does succeed well in alkaline soils. It has been grown in tropical regions and may be worth trialling in the subtropics. The species is placed in the segregate genus *Barbosa* (as *Barbosa pseudococos*) by some authors and has also been known as *Langsdorffia pseudococos*. *S. mikania* is another synonym.

Syagrus romanzoffiana

(after Count N.P. Romanzoff, Russian nobleman)

Queen Palm, Giriba Palm

SOUTH AMERICA

A familiar sight in tropical and subtropical countries where it is planted in the thousands, this palm has proved to be surprisingly adaptable, as mature

A clump of the popular Queen Palm *Syagrus romanzoffiana*.

specimens are also known from temperate zones. In the tropics it will grow in inland areas and also thrives on the coast, withstanding quite a deal of salt-laden wind. It is planted in parks and in streets, as well as home gardens, for its decorative value. This palm is fast growing and is especially responsive to nitrogenous fertilisers. The grey trunk is fairly slender and grows up to 15 m tall and is crowned by spreading, feathery-looking fronds. The species is widely distributed in Brazil, Paraguay, Argentina and Uruguay where it grows in forests and woodlands. It is often sold in the nursery trade as *Cocos plumosus*. This palm has a couple of drawbacks, notably the retention of untidy dead fronds and the large mass of fruits which, when ripe, attract fruit bats and cockroaches. Large plants can be shifted readily, often with little setback and the species is widely used in landscaping and commonly grown to an advanced stage by nurseries. The species is somewhat variable, with some plants being extremely robust and developing massive crowns. The typical variety (var. *romanzoffiana*) has plump fruit 2–2.5 cm long while in the var. *australe* they are quite slender and up to 3 cm long. *S. romanzoffiana* var. *botryophora* is a distinct species (see *S. botryophora*). *S. romanzoffiana* may hybridise with other species of *Syagrus* and the following natural hybrids have been named: *S. X camposportoana* (*S. romanzoffiana* x *S. coronata*); *S. X teixeiriana* (*S. romanzoffiana* x *S. oleracea*). *S. romanzoffiana* was previously well known as *Arecastrum romanzoffianum*. Cultivated plants may hybridise with species of *Butia* (see *X Butiagrus*).

Syagrus ruschiana

(after A. Rusch, collector in Brazil)

BRAZIL

An uncommon to rare palm which is native to Brazil, where it grows in sparse habitats associated with columnar cacti in soils derived from black gneissic rocks. Plants have a clumping habit with slender trunks to 10 m tall and arching fronds. New and recently mature fronds are glaucous, contrasting with the older fronds which age to green. The leaf petioles lack spines. Fruit are about 2.5 cm x 2 cm, with a short apical beak. A hardy, sun-loving palm suited to tropical and subtropical regions. It has also been known as *Arikuryroba ruschiana*.

Syagrus sancona

(a native name)

SOUTH AMERICA

A widely distributed species which occurs in Ecuador, Venezuela, Peru and Colombia, growing in rainforest. Much of the habitat of this palm has now been cleared, especially in Colombia. Plants grow to about 30 m tall, with a trunk to 35 cm across and a graceful crown of arching fronds. The leaflets, to 1 m x 4.5 cm, are arranged in groups of two to four along the rachis. The fruit, which are about 3 cm long, have a short beak. This species, which is grown in a few countries, appreciates tropical conditions.

Syagrus schizophylla

(with split leaves or leaflets)

Arikury Palm

BRAZIL

Although mainly a collector's palm, this species has proved to be fairly adaptable and more cold-tolerant than its Brazilian habitat would suggest. It is a solitary pinnate palm, the most conspicuous feature of which are the long, black leaf-bases which cover the trunk. These are very spiny near the base. Its crown consists of numerous, dark green, pinnate leaves which have very slender petioles and arch

Syagrus schizophylla.

pleasantly, sometimes with a slight twist. The species can be grown in tropical and subtropical regions and needs protection from direct sun at least when small. Young plants make attractive tub specimens. Seed germinates readily within two months of sowing. Previously known as *Arikuryroba schizophylla.*

Syagrus smithii

(after Dr E.E. Smith, Forestry Adviser, Peru)

PERU

An attractive palm from Peru which is now very rare owing to clearing of its habitat. It was formerly placed in its own genus, *Chrysallidosperma* (as *C. smithii*), being distinguished by the distinctive seed, which is triangular in cross-section and abruptly cut off at each end. Plants have a solitary trunk to about 10 m tall and a rounded crown of arching, dark green leaves. The flowers are reported to be sweetly scented and the ovoid fruit, to 7 cm x 4 cm, are crowded in heavy clusters. Although an attractive palm, this species appears to be rare in cultivation.

Syagrus vagans

(scattered, spread)

BRAZIL

An apparently trunkless palm with a tuft of fronds arising from the ground and growing to about 2 m tall. The petioles have short, marginal teeth and the narrow, glaucous leaflets are arranged regularly along the rachis. The ovoid to ellipsoid fruit, to 3.7 cm x 1.8 cm, have a prominent apical beak and are borne on an inflorescence about as long as the leaves. This interesting palm is native to Brazil growing in woodland, often in association with *S. coronata*. It is rarely cultivated but has been successfully grown in Florida.

Syagrus werdermannii

(after E. Werdermann, the original collector)

BRAZIL

A clumping trunkless palm which forms dense tufts of leaves arising directly from the soil surface. The leaves have about forty narrow, glaucous leaflets variably arranged in loose or tight clusters along

Left: Syagrus vagans.
Below: Syagrus vagans inflorescence.

the rachis. Young leaves are strongly glaucous and contrast with the older, darker leaves. An interesting small palm from Brazil which would be suited to tropical regions.

SYNECHANTHUS

(from the Greek *syneches*, holding together, joined, *anthos*, a flower)

A small genus of two or three species of palm occurring in southern Mexico, Central America and northern South America. They grow in rainforests of the lowlands and mountains. They are small to medium-sized, solitary or clumping, unarmed feather-leaved palms which lack a crownshaft. The inflorescence arises among the leaves and bears unisexual flowers of both sexes.

Cultivation: Highly ornamental palms, with one species being widely cultivated. Plants have prominently ringed trunks, pleasant foliage and colourful fruit. They are shade-loving palms which grow well in tropical and subtropical regions. Single plants are capable of producing fertile seed.

Propagation: Fresh seed germinates readily two to four months after sowing. Each fruit contains a single seed.

Synechanthus fibrosus
(fibre-like)

Palmilla, Jelly Bean Palm

CENTRAL AMERICA

In general appearance, this species could easily be mistaken for a *Chamaedorea*. *S. fibrosus* is a solitary palm with a slender, shiny green, ringed trunk to 5 m tall (2–3 cm across) and a crown of arching pinnate leaves to 2 m long. These have thin, flexible, dark green, glossy leaflets with the terminal pair united and fishtail-like. The leaflets, have a single vein and are irregularly arranged, usually in loose groups of two to four. The inflorescence is a small panicle which emerges from among the lower leaves. Fruit are globose to ellipsoid, about 2 cm long and ripen through yellow and orange to scarlet. In cultivation this palm likes shady conditions in deep, organically rich soil. It is rather sensitive to dryness and should be liberally watered. The species is native to southern Mexico, Belize, Guatemala, Honduras and Costa Rica. It is common in wet lowland forests but in some areas it extends up to 1200 m altitude. This palm can flower when quite small, often when a trunk is just developing.

Synechanthus warscewiczianus

(after J. Warscewicz, Polish collector)

CENTRAL & SOUTH AMERICA

A clumping palm from shady forests of Costa Rica, Panama, Colombia and Ecuador. It is an attractive species with tall, slender, dark green stems (to 6 m tall and 5 cm across) and an open crown of oblique-ly erect to arching fronds. The apical leaflets are united into a broad fishtail segment with the other leaflets of various widths and variously arranged. The leaflets have one to several veins. Clusters of ellipsoid to nearly globose fruit, which ripen through yellow and orange to bright red, provide an attractive display. This species grows readily in tropical regions.

THRINAX

(from the Greek *thrinax*, a trident)

Thatch Palms, Peaberry Palms

A small genus of seven species of palm which occur in Mexico and various islands of the Caribbean region, particularly Jamaica and Cuba. They may also occur in coastal areas of adjacent mainland countries. They grow only on alkaline soils, particu-larly those derived from coral near sea-level, but also extend to about 1200 m altitude and grow on limestone outcrops, ultrabasic soils and serpentinic soils. They are slender, solitary, unarmed fan palms with a split leaf-base and slender petioles. The inflo-rescences arise among the leaves and bear bi- sexual flowers followed by small white fruit. (For a study of the genus see R.W. Read, 'The genus *Thrinax* (Palmae: Coryphoideae)', *Smithsonian Contributions to Botany* 19 (1975): 1–98.) The genus *Hemithrinax* is synonymous.

Cultivation: Appealing, slender palms which have much to offer horticulture but plants are seldom grown. They are excellent in coastal regions and grow well on calcareous soils. The major drawback is that plants are very slow growing. They are sun-

Synechanthus fibrosus.

loving palms which must have excellent drainage. These palms are not amenable to transplanting. Single plants are capable of producing fertile seed. **Propagation:** Fresh seed germinates readily within two to four months of sowing but some seedlings may appear sporadically over twelve months or longer. Seedlings are very slow growing. Each fruit contains a single seed.

Thrinax compacta

(compact, dense)

CUBA

A palm of compact growth which originates in the savannah vegetation of Cuba. Plants have a slender trunk to about 3 m tall and a dense, com-pact crown of fronds which have a bright, shiny green (almost varnished) upper surface, whereas the lower surface is glaucous to greyish. Inflores-cences shorter than the leaves carry white flowers followed by clusters of white, spherical fruit each about 6.5 mm across. This is a poorly known species which has good prospects for cultivation.

Thrinax excelsa

Jamaican Thatch Palm, Broad Thatch, Silver Thatch

(high, lofty, elevated)

JAMAICA

This slender palm is endemic to Jamaica where it grows in exposed, rocky situations, always on limestone, between 350 m and 750 m altitude. It can be distinguished from other species of this genus by the large, impressive leaves, which are silvery white beneath from a dense covering of fringed scales. The leaves are very striking, nearly circular in outline, sometimes to more than 3 m across and with stiffly pointed segment tips. The inflorescence branches and flowers are colourful, being pink to purple, with the flowers having an alluring smell of rum and spice. The globular, white fruit are 8–11 mm across. Although rarely grown, this is a

Thrinax morrisii.

handsome palm and is deserving of much wider recognition. It is best suited to the tropics and requires a semi-protected to open position in well-drained soil. *T. rex* is a synonym.

Thrinax morrisii

(after Sir D. Morris, original collector and one-time Assistant Director of Kew Gardens)

Brittle Thatch Palm, Buffalo-top, Buffalo Thatch, Pimetta

CARIBBEAN

A palm which occurs throughout islands of the northern Caribbean, the Bahamas, Virgin Islands, Anguilla, Florida Keys and islands off the coast of Mexico. Although basically similar to other *Thrinax* species, *T. morrisii* can usually be distinguished by the presence of rows of numerous small, white dots on the underside of the leaves. There is also a tendency for the leaves to be blue-green to grey beneath. The inflorescences arch well out from the leaves and are longer than in other species. Small, white fruit (4–8 mm across) are borne in dense clusters and are most attractive. The plants grow well in coastal conditions and are best suited to tropical and subtropical regions. The species has had a confused botanical history and has been known by a variety of names, including *T. microcarpa*, *T. keyensis*, *T. ponceana*, *T. praeceps*, *T. bahamensis*, *T. drudei*, *T. punctulata* and *T. ekmanii*.

Thrinax parviflora ssp. *parviflora*

(with small flowers)

Mountain Thatch Palm, Thatch Pole, Iron Thatch

JAMAICA

A variable palm endemic to mountainous areas of western Jamaica where it grows on exposed bluffs, ridges and cliffs in soils derived from limestone, from sea-level to about 500 m altitude. It has the heaviest-textured leaves of the genus and these are of similar colouration on both surfaces and often have a corrugated appearance. The flowers are cream to yellowish and are followed by clusters of smooth, white fruit, each 6–7 mm across. The typical subspecies has a very sparse, open crown of small leaves with conspicuously twisted and curled segments, a short hastula (to 1.2 cm long) and an erect, glabrous or sparsely pubescent inflorescence. As the common names suggest, the leaves of this palm were used for thatching the roofs of dwellings in the mountainous region where it occurs naturally. It is an elegant fan palm, useful in a variety of situations and tolerating infertile, sandy soil. Plants are moderately frost-sensitive and best suited to warm–subtropical and tropical regions. They need well-drained soils and will tolerate full sun when quite small. *T. harrisiana* and *T. tessellata* are synonyms.

Thrinax parviflora ssp. puberula

(shortly hairy)

This subspecies differs from the typical (ssp. *parviflora*) by its denser crown of leaves with flat to curved or slightly twisted segments, a hastula to 4.4 cm long and an inflorescence about as long as the petioles, which has densely pubescent branches. It also is native to Jamaica.

Thrinax radiata.

Thrinax radiata

(radiating from the centre)

Florida Thatch Palm, Silk-top Thatch, Sea Thatch

GULF OF MEXICO—BAHAMAS

In nature this palm always grows within the range of salt-laden winds in near-coastal areas. It is widely distributed in the Caribbean region of the Bahamas, Mexico and Belize and also extends to southern Florida. It often grows in dense stands, dominating the vegetation. It is an excellent palm for cultivation in exposed coastal districts and although restricted to calcareous soils in nature, cultivated plants will succeed in a wide range of soil types. The species is best suited to subtropical and tropical regions but can be induced to grow in a warm–temperate climate. Plants are very intolerant of wet soils and will happily take exposure to full sun from a very early age. They also make very attractive and long-lived tub plants. The species has also been known as *T. floridana*, *T. wendlandiana* and *T. martii*.

Thrinax rivularis

(growing by streams)

CUBA

A little-known palm which is endemic to Cuba where it grows in forests close to streams and in swamps. It is a slender palm to 6 m tall with a dense crown of yellowish-green leaves. The old leaves persist as a skirt beneath the crown. This species is not known to be in cultivation.

TRACHYCARPUS

(from the Greek *trachys* rough, *carpos* a fruit)

Windmill Palms

A small genus of about nine species (including three undescribed species and two other doubtful species named from cultivation) of palm occurring in northern India, Burma, Nepal, northern Thailand and China. They are very hardy palms from forests at high altitudes (to 2400 m). They are dwarf to medium-sized, solitary or clumping fan-leaved palms. The trunk is usually obscured by a dense covering of coarse fibre. The inflorescence arises among the leaves and bears unisexual (rarely bisexual) flowers. The genus is currently under study.
Cultivation: Easy and rewarding palms for cultivation,

Trachycarpus fortunei inflorescence.

with all species being especially useful in temperate climates. They have a pleasant appearance and are popular for their symmetrical, radiating crown of fan-shaped leaves. All species need good drainage and will tolerate sun from an early age. Male and female trees are required for fertile seed production. **Propagation:** Seeds germinate readily within three or four months or longer of sowing but the seedlings are slow growing. Each fruit contains a single seed.

Trachycarpus fortunei
(after Robert Fortune, collector in China)

Chinese Windmill Palm, Chusan Palm

CHINA

This species, one the most cold-hardy of all palms, is a familiar sight in temperate zones. It is readily recognised by the fan-shaped leaves and slender trunk covered with persistent but loosely arranged, dark brown fibre. It grows up to 10 m tall and the dark green leaves are green to glaucous beneath with the blades unevenly divided more than half-way to the petiole. Some older leaves often persist and clothe the upper part of the trunk as a skirt. This palm is native to China where it grows in cold, mountainous regions. The fibre can be stripped from the trunk in pads and provides a very useful lining material for hanging baskets. In China, these fibres are used to make fine waterproof cloaks, brooms, brushes and door mats. The inflorescence bears yellow flowers which are quite showy. Solitary specimens of this palm rarely produce seeds. This hardy palm thrives in temperate regions even in southern latitudes (Tasmania, South Island of New Zealand, the British Isles) but is very difficult to grow in the tropics. It has also been known as *T. excelsus* and is frequently wrongly sold in nurseries as *Chamaerops excelsa*. A variant having strongly glaucous leaves is often confused with *T. martianus*. The palm known as *T. caespitosus* was described from plants cultivated in California and has not been found in the wild. It is merely a cultural variant of *T. fortunei* which supposedly has smaller leaves with rigid leaflets and suckers from the base to produce multi-trunked clumps. It is an interesting variant which performs well in cultivation.

Trachycarpus fortunei.

Trachycarpus martianus in habitat, Khasia Hills, India.
M. Gibbons and T. Spanner

Trachycarpus martianus

(after C.F.P. von Maritius, father of palm botany)

Martius' Windmill Palm

HIMALAYAS

A very distinctive species which is uncommon
to rare in cultivation. It can be recognised by
the slender trunk, with the fibres being closely
appressed (not loose and woolly) and confined to
about 60 cm below the crown, with the trunk bare
and ringed below that. The slender trunk tends to
impart a heavier aspect to the crown than is noticeable in *T. fortunei*. The leaves are large (more than
1 m across), dark green and shiny, with a glaucous
undersurface and up to 75 segments. The blade is
divided about halfway to the base, with the palman
being very even. The inflorescence is drooping and
bears yellow flowers. The fruit are oblong, ripening
through yellow to black. The seed is distinctive in
having a prominent longitudinal groove, much like
a date seed. An attractive, slow-growing palm well
suited to subtropical and temperate regions. Plants

like an open, sunny aspect in well-drained acidic
soil. They are tolerant of moderate to heavy frosts.
The species is native to the Himalayan region of
north-eastern India, Nepal and Burma growing at
altitudes up to 2000 m on acid soil.

Trachycarpus nanus

(dwarf)

Yunnan Dwarf Palm

SOUTH-WESTERN CHINA

A dwarf palm which usually appears trunkless but
which in fact has a curved, subterranean trunk
about 5 cm across and up to 2 m long. Most plants
consist of a clump of leaves arising directly from
the ground, but rare specimens have a short,
fibrous, above-ground trunk to 30 cm high. Plants
have a crown of five to twenty leaves with the
blades divided nearly to the bases into narrow,
deeply folded, stiff segments which have shortly
notched tips. Leaf colouration ranges from light
green to silvery blue, with the underside prominently glaucous (unique in *Trachycarpus*). The
inflorescence is erect, about as long as the petioles
and after flowering produces dense clusters of
kidney-shaped fruit. This hardy, but poorly known
palm, grows in south-western China on shrubby
slopes and among grass at about 2000 m altitude,
where it is under extreme threat from grazing by
goats. Soils range from acidic, red clay to neutral,
stony loam. Seed of this species has recently been
introduced into cultivation. (See M. Gibbons and
T. Spanner, 'In search of *Trachycarpus nanus*',
Principes 37 (2) (1993): 64–72.)

Trachycarpus martianus in habitat, Khasia Hills, India.
M. Gibbons and T. Spanner

Trachycarpus nanus in habitat, China. M. Gibbons and T. Spanner

Trachycarpus takil

(from Mount Takil or Thakil)

HIMALAYAS

This species was introduced into cultivation in 1887 when the famed palm botanist Odoardo Beccari planted one specimen in his garden in Florence, Italy. The plant thrived and it is recorded that this male plant flowered in 1902 and in 1904 and had reached about 4 m tall. The species is native to the western end of the Himalayas at altitudes of 2000–2400 m. It grows in cool, north-facing, narrow wooded valleys where the plants are covered with snow during winter. Young plants have a creeping to oblique stem which then becomes erect. Plants grow to about 12 m tall and the trunk is covered with closely appressed, brown fibres (not woolly as in *T. fortunei*). The leaf-blades are about 1 m or more across and have 45–50 segments (sometimes up to 60) with a slightly irregular palman. The blades are shiny green on the upper surface and glaucous beneath. The fruit and seeds are distinctly kidney-shaped. Although rarely cultivated, this species would make an attractive palm for temperate climates and is extremely cold-tolerant. This species has been greatly reduced on Mount Thakil (now Mount Thalkedar) due to young plants being cut off at ground level for fibre. Doubt has also been expressed about the distinctiveness of this taxon. (See M. Gibbons, 'Trekking on the *Trachycarpus* trail', *Principes* 37 (1) (1993): 19–25.)

Trachycarpus takil in habitat, Himalayas, India. M. Gibbons and T. Spanner

Trachycarpus wagnerianus

(after a grower by the name of Wagner)

This species was described from cultivated plants and is not known in the wild. It is probably just a variant of *T. fortunei* since plants of the two species are basically similar, but those of *T. wagnerianus* have smaller leaves with more rigid segments. Plants of this species are very slow growing when

young but increase in growth rate after a few years and may eventually reach 5 m in height. They are sometimes wrongly sold in the nursery trade as *T. takil* or *T. nanus*.

TRITHRINAX

(from the Latin or Greek *tri*, three and *thrinax*, trident, probably in reference to the leaf division)

A small genus of about five species of palm occurring in Bolivia, Brazil, Paraguay, Uruguay and Argentina. Most species grow in open habitats, forming dense clumps in calcareous and saline soils. They are solitary or clumping, spiny, fan-leaved palms with the trunk mostly covered by an intricate layer of fibres and spines. The leaf-blades are deeply divided into stiff spreading segments. The inflorescence arises among the leaves and bears bisexual flowers.

Cultivation: Very hardy palms which are tolerant of cold, dryness and calcareous soils and some species will grow in saline soils. They are generally regarded as being only moderately ornamental and are rarely encountered in cultivation. They are sun-loving palms which will not tolerate water logging and must have excellent drainage. Single plants are capable of producing fertile seed.

Propagation: Fresh seed is erratic to germinate, with seedlings appearing three to eight months or more after germination. Each fruit contains a single seed.

Trithrinax acanthocoma

(with spiny tufts of hairs)

Spiny Fibre Palm

SOUTH AMERICA

An uncommonly grown palm which can be readily recognised by the intricate webbing of brown fibre which covers the trunk. This fibre is attached to the leaf-bases which adhere to the trunk long after the leaves fall. Unfortunately the leaf-bases also bear numerous long, slender spines and these detract somewhat from the palm's popularity. The combination of fibre and spines, however, does give the species a very distinctive, if formidable appearance. The leaves are fan-shaped and grey-green above, with stiffly spreading segments. Creamy-white flowers are borne densely in a stout inflorescence which arches out from the lowest leaf. The globose, white fruit are about 18 mm across. A hardy, distinctive palm deserving of wider recognition. The species is native to Brazil and Argentina and will grow in tropical, subtropical and temperate regions tolerating temperatures to about –6°C. Plants often flower when quite young.

Trithrinax biflabellata

(with two fans)

SOUTH AMERICA

A clumping palm with trunks 2–5 m tall and the fibrous sheaths covered with thick spines. The leathery leaves are divided into about twenty-five stiff, rigid segments, with the central ones (about 40 cm long) longer than the lateral ones (about 24 cm long). All segments are divided to the base. Clusters of white fruit are showy, each fruit being about 1 cm across with a granular surface. A hardy palm from open habitats in Argentina and Paraguay which deserves to be introduced into cultivation.

Trithrinax acanthocoma.

Trithrinax brasiliensis

(from Brazil)

SOUTH AMERICA

This species grows in open habitats, often in thickets, in dry, calcareous soils. Plants may be clumping or have a solitary trunk which is covered with coarse fibres and spines. The leaves are green to grey-green, greyish beneath and deeply cleft into stiff, spreading segments. Flowering is spectacular with the dense inflorescences supported by prominent, snowy-white bracts and the flowers being yellow. The ovoid to globose, black fruit are about 1 cm across. As the specific name suggests, the species is native to Brazil and also Argentina. It has similar cultural requirements to *T. acanthocoma*.

Trithrinax campestris

(growing in the fields)

SOUTH AMERICA

A little-known palm which forms a compact clump of stout trunks to about 2 m tall. Each trunk has a dense crown of stiff, spreading leaves that are woolly white on the upper surface and glossy green beneath (sometimes partly woolly). The trunks are covered with a woven webbing in which are embedded spreading spines. This species is slow growing and quite showy when in flower and fruit, with the dense white flowers being followed by heavy clusters of white, globose fruit each about 2.5 cm across. The species is native to Argentina and Uruguay, growing in semi-arid habitats, often in thickets. Some plants are single trunked and do not appear to sucker. An interesting, hardy species for subtropical and temperate regions.

Trithrinax schizophylla

(with split leaves or leaflets)

SOUTH AMERICA

A species from western areas of Brazil, Argentina and eastern Bolivia. It is a clumping species with slender trunks to 5 m tall and 20 cm across, each covered with fibrous sheaths and spines. The crown consists of large (blades 80 cm across), rigid, grey-green leaves which have acuminate, almost pungent, leathery segments. Clusters of white flowers are followed by greenish to yellowish fruit each about 8 mm across. The fruit are apparently inedible. An interesting species which appears to be rare in cultivation if it is grown at all.

VEITCHIA

(after James and John Veitch, 19th century English nurserymen)

A genus of eighteen species of palm distributed in the Philippines, New Hebrides and Fiji (where there are ten species), all endemic. They grow in forests and rainforests from sea-level to moderate altitudes. They are solitary, unarmed feather-leaved palms with a prominent crownshaft. The inflorescence arises below the crownshaft and bears unisexual flowers of both sexes. The fruit are produced in dense, colourful clusters. (For a study into the genus see H.E. Moore, 'Synopses of various genera of Arecoideae: 21. *Veitchia*', *Gentes Herbarium* 8 (7) (1957): 483–536; and for the Fijian species see *Flora Vitiensis Nova*, Pacific Tropical Garden,

Trithrinax campestris.

Veitchia macdanielsii.

Lawaii, Kauai, Hawaii (1979): 412–19.)

Cultivation: Clean-cut, handsome palms which are excellent subjects for landscaping. Most species are cold-sensitive and achieve their best appearance in the tropics. Single plants are capable of producing fertile seeds. There has been much confusion as to the identity of cultivated palms of this genus.

Propagation: Fresh seed germinates readily within two to four months of sowing. Each fruit contains a single seed.

Veitchia arecina

(like an *Areca*)

VANUATU

A graceful palm from various islands of Vanuatu, growing in coastal and lowland forests. Plants have a slender, grey trunk which can grow to more than 15 m tall, a prominent whitish crownshaft and a flattish crown of fronds. The leaflets spread widely or droop towards the apex. The large, whitish

inflorescence is prominent, as also are the clusters of globose fruit, each about 2.5 cm long. This palm grows well in coastal districts of the tropics.

Veitchia joannis

(after John Gould Veitch)

FIJI

An elegant palm which in nature may attain heights of more than 30 m. Its slender, grey trunk is crowned with graceful arching, shiny, dark green fronds more than 3 m long and which have prominently drooping leaflets, each with an unequal apex. Large clusters of sharply beaked, red fruit (each 5–6 cm long) are borne at the base of the bright green crownshaft and are most decorative. This palm, which is common on many Fijian islands growing in rainforest, is now widely grown throughout the tropics. It is a fast-growing palm and has proved to be relatively cold-tolerant, surviving in frost-free, temperate areas. When young, it prefers a semi-shaded position, but mature plants thrive in full sunshine. They like an organically rich, well-drained soil and respond to side dressings of nitrogenous fertilisers.

Veitchia macdanielsii

(after Dr L.H. MacDaniels, American academic and horticulturist)

VANUATU

An excellent palm for the tropics which is gaining popularity for its ease of culture and graceful appearance. It is native to the island of Espiritu Santo, Vanuatu, where it grows in coastal and lowland forests. It is a tall palm (to 25 m) with a slender trunk and rounded crown of arching to drooping fronds. The leaflets are drooping and densely scaly beneath. Clusters of ellipsoid to oblong fruit (each about 3 cm long) are bright red and showy.

Veitchia merrillii

(after Elmer D. Merrill)

Manila Palm, Christmas Palm, Dwarf Royal Palm, Adonidia Palm

PHILIPPINES

A familiar tropical species which is native to the Philippine islands. It is becoming widely planted, especially in tropical areas and is favoured because of its neat habit and compact crown of arching, bright green, feathery fronds atop a green crownshaft and a slender, ringed trunk. The crowded leaflets are held erect and have drooping tips. The lower segments retain prominent slender reins (lorae). Its ornamental appeal is enhanced by large clusters of bright, glossy-red fruit (each about 3 cm long) which in the USA ripen at about Christmas time, hence the alternative common name. It is an ideal species to plant in the lawn of the average suburban garden of the tropics. Plants will grow happily in a sunny position and can be planted out when small. They are fast growing and may flower when quite young. Although they will survive in subtropical areas, they are very cold-sensitive and have their best appearance in the warm tropics and are excellent for coastal districts. Potted plants are reputed to be good for indoor decoration. *Adonidia merrillii* is a synonym.

Veitchia montgomeryana

(after Colonel R.H. Montgomery)

VANUATU

This species has become well established in cultivation and is very popular in Florida. It is an excellent palm for coastal districts in the tropics as it withstands considerable exposure to buffeting, salt-laden winds. It is a tall palm (to about 30 m) with a green crownshaft and a crown of stiffly arching to horizontal fronds. The leaflets spread widely and the unequal apex appears as if obliquely cut or bitten off. The stout, yellowish inflorescence bears green flowers which are followed by bright red, oblong fruit about 4.5 cm long. An attractive, useful palm which is endemic to the island of Efate, Vanuatu. The cabbage is avidly eaten and the species has been reduced to rarity by excessive collection for this purpose.

Veitchia petiolata

(with a petiole)

FIJI

A little-known palm from the islands of Vanua Levu and Taveuni, Fiji, where it grows on forested slopes at about 500 m altitude. It has recently been introduced into cultivation but its response remains largely unknown. Plants develop a slender trunk to 10 m tall and only about 10 cm thick. The crownshaft is white from a dense covering of scales and the fronds form a rounded crown. White flowers are followed by orange to red fruit each about 2.2 cm long.

Veitchia sessilifolia

(with sessile leaves or leaflets)

FIJI

This species is grown on a limited scale in Florida and also in Cuba where a plant flowered in 1952 at a height of about 2 m. It is native to the island of Vanua Levu, Fiji where it grows in moist, forested valleys. It is a graceful palm with a relatively small crown (about seven fronds) of erect to obliquely erect fronds which have strongly arching tips. The petiole and crownshaft is covered with white, fluffy scales and the leaflets are dark green and glossy on the upper surface, dull and paler beneath. Prominent white flowers are followed by clusters of red fruit (each about 2.2 cm long).

Veitchia simulans

(resembling another)

FIJI

This palm grows in dense, wet forests on the island of Taveuni, Fiji, at about 500 m altitude. It appears to be rarely cultivated but has ornamental appeal. Plants grow to about 7 m tall, with a slender trunk and a dark green crownshaft. The leaves are held horizontally and have dark green, glossy, pendulous leaflets. Clusters of white flowers and orange-red fruit (each about 2 cm long) are additional attractive features. *V. subglobosa* is a synonym.

Veitchia vitiensis

(from Viti Levu, Fiji)

FIJI

A variable palm which can be recognised by its relatively short, broad, widely spaced leaflets which appear as if they have been abruptly cut or bitten off at the apex. Plants often appear to have a sparse crown (usually of about eight fronds) because of the wide leaflet spacing. The fronds arch gracefully on a slender grey trunk to about 15 m tall and 20 cm across. The crownshaft is green and the

Veitchia winin.

greenish flowers are followed by clusters of red, ellipsoid fruit each about 2.2 cm long. An attractive palm from the Fiji islands of Viti Levu, Kandavu and Ovalau, which appears to be rarely cultivated. *V. smithii* is a synonym.

Veitchia winin

(from a native name)

VANUATU

This species adapts well to cultivation in tropical regions. Plants are fast growing and flower when quite young. They have proved to be adaptable and are successful in a warm position in the subtropics. Plants grow to more than 15 m tall, have a very slender trunk (less than 20 cm across), a pale greenish to whitish crownshaft and an arching crown of stiffish, dark green fronds. Showy inflorescences bearing masses of white flowers are followed by bunches of bright red fruit (each 1.5–1.8 mm long). An attractive palm which originates from various islands in Vanuatu.

VERSCHAFFELTIA

(after A.C.A. Verschaffelt, 19th century Belgian nurseryman)

A monotypic genus endemic to the Seychelles (Mahé, Silhouette and Praslin) where it grows on steep slopes and gorges from sea-level to about

600 m altitude. It is a unique, spiny solitary palm with prominent stilt roots. The leaves are entire with a prominently notched apex, but usually become irregularly split by the wind. The inflorescence arises among the leaves and bears unisexual flowers of both sexes.

Cultivation: An extremely ornamental and interesting palm which is widely grown in the tropics. Solitary plants are capable of producing fertile seed.

Propagation: Fresh seed germinates readily within two to four months of sowing. Each fruit contains a single seed.

Verschaffeltia splendida

(splendid)

Seychelles Stilt Palm, Latanier Latte

SEYCHELLES

An extremely ornamental palm prized by collectors for both its beauty and its interesting growth characteristics. Its leaves are simple and undivided and in young specimens are quite majestic, being bright green, broad and pleated, with conspicuously red rachises. Unfortunately, in mature palms the entire leaves split readily and appear as if pinnate. They are still, nevertheless, unusually broad and form an attractive crown. Other interesting features of this palm are the rings of long, black spines on the trunk and the cluster of long, stout, aerial roots which support the base of the trunk and can arise about 1 m above ground level. Young specimens are extremely decorative and make fantastic plants for indoor or glasshouse use. In the garden they need shelter at least while young and are best protected from strong winds. Warm, humid, tropical conditions with gentle air movement seem to suit them best.

WALLICHIA

(after Nathaniel Wallich)

Wallich Palms

A genus of seven species of palm occurring in the Himalayan region of northern India, adjacent areas of Burma, southern China to Thailand. Most species are understorey palms found in humid, tropical forests from near sea-level to the mountains. They are dwarf to tall, solitary or clumping, unarmed palms with pinnate fronds, the leaflets of which are folded upwards. The leaflets are generally broad with ragged margins and may resemble

Verschaffeltia splendida. G. Cuffe

those of *Caryota* species, to which the genus is related. The inflorescences arise among the leaves (they burst through a leaf sheath) and bear unisexual flowers sometimes of both sexes. Solitary species die after the lowermost inflorescence has matured its fruit; in clumping species the whole stem dies after fruiting.

Cultivation: Intriguing and attractive palms well suited to tropical, subtropical and warm–temperate regions. They are uncommonly grown even though some species have proved to be easy and adaptable. Some species require partial shelter, others are tolerant of full exposure. Plants of some species can produce fertile seeds by themselves, others may need to be planted in groups.

Propagation: Clumping species can be divided readily providing the divisions retain some roots. Fresh seed germinates irregularly, with seedlings appearing sporadically from three months to twelve months after sowing. Each fruit contains one to three seeds. The fruit pulp contains irritant stinging crystals and should be handled with care.

Wallichia caryotoides

(resembling a *Caryota*)

HIMALAYAS

An interesting palm which forms dense clusters (about 3 m tall and spreading a similar distance). The trunks are virtually subterranean and bear at their apex a cluster of erect or arching fronds which have very irregularly shaped leaflets scattered along their length. These vary from oblong to fishtail-shaped and are bright green above and silvery beneath. The inflorescences are borne on slender stalks a bit shorter than the leaves and after flowering carry a dense cluster of purplish-black fruit each about 1.5 cm long. In cultivation this palm has proved to be adaptable and ornamental. It is hardy once established and is suitable for tropical and subtropical regions. A semi-shaded position is favoured, with the plants receiving exposure to some sun for part of the day. The species is native to shady, moist valleys in the Himalayas, growing to about 1200 m altitude and sometimes forming thickets.

Wallichia chinensis

(from China)

SOUTHERN CHINA

A poorly known palm from China which has large, broadly wedge-shaped to triangular leaflets, each to 45 cm x 15 cm. They are dark green on the upper surface and paler beneath, with prominent veins. The flowers are yellowish-white and the light green fruit are about 18 mm x 10 mm at maturity. The species is native to the Kwangsi Province of southern China and would be an excellent palm for temperate regions.

Wallichia densiflora

(with densely clustered flowers)

HIMALAYAS

The most outstanding feature of this small palm is its cluster of long, arching or erect fronds with unusual leaflets not unlike those of a fishtail palm but bright green above and silvery-white beneath. As these fronds wave lazily in the breeze, the contrast between the upper and lower surfaces is quite apparent. Each clump consists of a few short, broad and very crowded trunks with the leaves

Wallichia densiflora.

arising from near the end. Male flowers are yellow and the female flowers are purplish. Dull purple, oblong fruit about 1.2 cm long are carried in narrow clusters. This species is native to the Himalayas of northern India and adjacent high areas of Burma, where it grows in shady, moist gullies. Plants are relatively cold-tolerant and are well suited to warm–temperate and subtropical areas. In cultivation it likes a shady aspect in well-drained, organically rich soil.

Wallichia disticha

(in two rows)

HIMALAYAS

In Latin '*distichus*' means arranged in two ranks and this epithet describes to perfection the leaves of this unusual palm, which are placed in two vertical rows each on opposite sides of the trunk. This combination results in a very flat profile when the plant is viewed from the side. Native to the Himalayas where it grows up to 600 m altitude, this intriguing palm is unfortunately both rare in cultivation and short-lived. In cultivation the plants may reach 7 m tall in ten to fifteen years, at which stage they begin to flower and then die four or five years later when they have matured all their clusters of dull red fruit. Despite their curtailed life the plants are well worth growing. The leaves arch stiffly from the trunk and have long leaflets which are grouped in bundles. These are dark green above and greyish beneath, with a few short lobes and teeth near the apex. The trunk is invisible, being covered with persistent leaf-bases and their associated layers of fibre. Plants are best suited to tropical regions, although they may be induced to grow in warm situations in the subtropics. Seedlings are fairly sensitive to cold. This species is native to Sikkim and the Himalayas of northern India, growing in deep valleys up to about 1200 m altitude. It is often locally common, growing in colonies and the local people make sago from the stem and use the leaves for thatch.

Wallichia gracilis

(slender)

SOUTHERN CHINA—VIETNAM

This species forms a dense clump with slender stems to 1.5 m tall and wedge-shaped leaves to 40 cm x 10 cm. These are stiff, papery-textured, green on the upper surface and silvery white beneath. Ovate to elliptical fruit, about 13 mm long, are borne in dense clusters. The species is native to northern Vietnam and perhaps southern China, growing in forests at about 800 m altitude.

Wallichia mooreana

(after Dr. H.E. Moore, noted American palm botanist)

SOUTHERN CHINA

Described in 1982, this palm honours Dr Harold E. Moore for his remarkable contribution to our knowledge of palms. A native of southern China, this species grows within the canopies of humid forests. It is a clumping palm, with the trunks growing to less than 2 m tall and the leaves less than 1 m long. The dark green leaflets, which are usually borne in groups of three, taper to the base and have the outer margins deeply toothed and notched. Male inflorescences, which bear yellow flowers, are pendulous and much branched whereas the female inflorescences are erect and have fewer branches. The small, reddish fruit each contain two seeds. This poorly known palm remains rare in cultivation but probably has similar cultural requirements to other species of *Wallichia*. Some authorities believe this species to be synonymous with *W. caryotoides*.

Above: *Wallichia disticha* showing leaf-bases and fibres.
Below: *Wallichia disticha*.

Wallichia siamensis

(from Siam)

THAILAND

A clumping species which has a slender trunk about 0.5 m long and fronds about 1.5 m long. These have large, stiff, almost papery leaflets (to 50 cm long) which are more or less wedge-shaped, with the upper surface dark green and the lower silvery white. The outer margins are wavy and have numerous sharp teeth and lobes. Short inflorescences arise among the leaves. An interesting palm which is native to northern Thailand where it grows in forests near streams. It originates at about 900 m altitude and has excellent prospects for cultivation in subtropical gardens.

Wallichia triandra

(with three stamens)

HIMALAYAS

An ornamental palm which forms sparse clumps to 3 m tall. The slender, cane-like stems are densely clothed with dark fibres and the irregularly wedge-shaped leaflets (to 40 cm x 10 cm) are dull green above and silvery white beneath, with the apex

looking as if it has been irregularly browsed by a herbivore. The fruit are oblong, to 13 mm x 7 mm and dark red or purple when ripe. This palm is native to northern India, in the region of the Arunchal Pradesh Himalayas, growing in dense forest at altitudes between 1000–2000 m. It has much to offer horticulture. At one time it was placed in its own genus *Asraoa* (as *A. triandra*).

WASHINGTONIA

(after George Washington, first President of the USA)

Cotton Palms

A small genus of two species of palm native to the west coast of the USA, Baja California and Sonora, Mexico. They grow in semi-arid, desert regions, usually forming colonies near water, often in gorges and canyons. They are tall-growing, solitary, spiny fan-leaved palms, the old leaves of which clothe the trunk with a persistent petticoat. The inflorescences, which arise among the leaves, often extend beyond the canopy and bear bisexual flowers.

Cultivation: Imposing, stately palms which are very popular in cultivation. They grow very well in semi-arid climates but have also proved to be highly adaptable, growing well in coastal districts, mountainous regions and from the tropics to cool–temperate climates. They are excellent for avenue planting in rows. They are often sold as indoor plants but are generally unsuitable for this purpose. Single plants are capable of producing fertile seed.

Propagation: Fresh seed germinates readily two to four months after sowing. Each fruit contains a single seed.

Washingtonia filifera

(thread-bearing)

American Cotton Palm, California Palm

NORTH AMERICA

A widely grown and familiar palm that will succeed admirably in a range of soils and climates, from the

Washingtonia filifera in habitat, Palm Springs, California.

tropics to southern temperate regions as well as inland and coastal districts. It is a tall, distinctive palm with a fat, heavy, barrel-shaped, grey trunk and a spreading crown of grey-green, fan-shaped leaves in an open crown. The old leaves persist as a petticoat covering the trunk and this may be clothed right to ground level. The name Cotton Palm arises from the white, cottony threads found on and between the leaf segments. The species originates in the USA where it forms colonies in canyons of south-eastern California and western Arizona, also extending over the border to Baja California. Plants grow to 16 m tall and have an imposing crown. This palm has proved to be very hardy in many countries and is a fine palm for street, avenue and parkland planting. Once established, the plants are very drought resistant. They prefer a sunny position and they respond strongly to the use of fertilisers. Plants will tolerate exposure to direct sun even when quite small.

Washingtonia robusta

(robust, vigorous)

Washington Palm, Skyduster, Mexican Fan palm

MEXICO

This species is frequently confused with W. *filifera* but grows taller, with a much thinner trunk flared widely at the base and a compact crown of fronds. The leaves are also brighter green, with prominent, red-brown basal sheaths and the blades of mature palms lack the adornment of the cottony threads. Like the former, this species is also a very hardy and adaptable palm. It grows very well in cool, temperate regions and also in the subtropics and tropics. Literature records that this species is not as cold-tolerant as W. *filifera*. It makes an excellent palm for avenues, driveways and in parklands, although it is often also planted in gardens. It is native to north-western Mexico and Baja California, often growing near the sea.

WODYETIA

(after Wodyeti, last male Melville Range Aborigine)

A monotypic genus of palm endemic to north-eastern Australia where it grows in low scrub among boulders. It is a solitary, unarmed feather-leaved palm with a prominent crownshaft. The leaflets present a densely plumose appearance, with the central primary pinnae each divided into many secondary linear pinnae, which are ribbed on both margins. The inflorescence arises at the base of the crownshaft and bears unisexual flowers of both sexes.

Cultivation: Although described only in 1983, this palm has become extremely popular in cultivation. Indeed it has become so popular that illegal poaching of seed from natural populations has become an ecological problem. Plants grow readily in cultivation and have proved to be adaptable to a range of soils and situations. Single plants are capable of producing fertile seeds.

Propagation: Fresh seed germinates readily two to three months from sowing, with odd batches still appearing after seven to twelve months. Each fruit contains a single seed.

Washingtonia robusta.

Wodyetia bifurcata

(twice forked, in reference to the pinnae and fibres in the fruit)

Foxtail Palm

AUSTRALIA

A delightful palm which has exploded in popularity and is now widely planted in many countries. Plants have proved to be fast growing and adaptable. Best growth is achieved in drier tropical regions but success has also been gained in the subtropics. Plants demand good drainage but seedlings will take hot sun from an early age. Ornamental features include a slender, closely ringed, columnar to slightly bottle-shaped trunk, graceful, arching, densely plumose fronds, a slender crownshaft and colourful clusters of large, orange-red fruit. The species occurs naturally in the Melville Range, near Bathurst Bay on Cape York Peninsula, north-eastern Queensland. It grows in loose, granitic, sandy soils among huge granite boulders. The climate is tropical, with a prolonged dry season.

ZOMBIA

(from *latanier zombi*, a Haitian native name)

A monotypic palm endemic to Hispaniola where it grows in sparse scrub on dry hills. It is a medium-sized, clumping fan-leaved palm. The roots around the trunk bear erect, spinelike, breathing roots (pneumatophores), 5–8 cm long above the soil and the trunk is covered with overlapping, fibrous, spiny sheaths. The leaf petioles do not split. The inflorescence arises among the leaves and bears bisexual flowers.

Wodyetia bifurcata.

Zombia antillarum.

Cultivation: An intriguing palm which attracts attention because of its appearance and curious features. It is rare in cultivation and performs well in tropical and subtropical regions. Single plants are capable of producing fertile seed.

Propagation: Fresh seeds germinate within two to four months of sowing. Each fruit contains a single seed.

Zombia antillarum

(from the Antilles)

Zombi Palm

HISPANIOLA

Where it grows naturally, this strange palm obviously influenced the local people, for they called it *latanier zombi*. The first word is applied generally to fan-leaved palms and *zombi* is the name applied to the dead who have returned to life. This palm certainly has some unusual growth features. Its trunks bear rings of spreading to downward-pointing spines enmeshed at the base in a woven mat of fibres which completely envelopes the trunk. If a trunk dies, the central core may rot away leaving a hollow fibrous column with its projecting spines. The leaves, which tend to droop (and look wilted) are deeply divided into narrow segments which are dull green above and silvery beneath. Small white flowers on short inflorescences are followed by attractive clusters of starkly white, globose fruit

Zombia antillarum showing characteristic fibres and spines.

each about 2 cm across. This species, which is rare in cultivation, grows readily in a sunny position. Plants must have excellent drainage but otherwise are not demanding in their requirements.

GLOSSARY

abaxial On the side of a lateral organ away from the supporting axis; the lower side of a leaf or petiole

abscission Shedding of plant parts such as leaves either through old age or prematurely as a result of stress

acaulescent Without a trunk; often used for palms which have a subterranean trunk

accessory roots Lateral roots developing from the base of the trunk as opposed to those arising from the seed root system

acropetal Developing upwards, towards the apex

aculeate Bearing short, sharp prickles or spines

acuminate Tapering into a long, drawn-out point

acute Bearing a short, sharp point

adaxial On the side of a lateral organ next to the supporting axis: the upper side of a leaf or petiole

adnate Fused together tightly

aerial roots Adventitious roots arising on stems and growing in the air

aff. or affinity A botanical reference used to denote an undescribed species closely related to an already described species

after-ripening The changes that occur in a dormant seed and render it capable of germinating

albumen An old term used for the endosperm of seeds

alternate Organs borne at different levels in a straight line or spiral—for example, leaves

androecium The male parts of a flower (that is, the stamens)

annular rings Prominent ring-scars left on the trunk of certain palms after leaf fall—for example, *Archontophoenix*

annulate Bearing annular rings on the trunk

anomalous An abnormal or freak form

anther The pollen-bearing part of a stamen

apical dominance The dominance of the apical growing shoot which produces hormones and prevents lateral buds or suckers developing while it is still growing actively—for example, the growth habit of *Linospadix minor*

apiculate With a short, pointed tip or beak

apocole The cotyledonary sheath of germinating palm seedlings

apocarpous With free carpels

appendage A small growth attached to an organ

arborescent With a tree-like growth habit

arcuate Arched, as in the fronds of *Howea belmoreana*

aril A fleshy outgrowth of the stalk of the ovule which covers or partially covers the seed—for example, Lepidocaryoid palms

armed Bearing spines or prickles

arundinaceous With slender reed-like or cane-like stems

asexual reproduction Reproduction by vegetative means without the fusion of sexual cells

attenuated Drawn out

auricle An ear-like appendage at the base of the leaflets of some species—for example *Arenga*

axil Angle formed between a leaf petiole and a trunk

axillary Borne in an axil

barbed Bearing sharp, backward-sloping hooks as in *Calamus* spp.

basipetal Developing downwards, towards the base

berry A simple, fleshy, one- to many-seeded fruit which does not split when ripe

bifid Deeply notched for more than half its length

bilobed Two-lobed

binding Threads joining young leaflet tips together before the frond has expanded

bipinnate Twice pinnately divided as in fronds of *Caryota* spp.

bisexual Both male and female sexes present

blade The expanded part of a leaf

bole The trunk of a tree or palm

bottom heat A propagation term used to denote the application of artificial heat around a seed or cutting

bract A leaf-like structure which subtends a flower stem or part thereof

bracteole A small, leaf-like structure which subtends a single flower

bulbil In palms a vegetative aerial growth arising from a modified inflorescence and sometimes caused by damage to the growing apex of a stem

bulbous Bulb-shaped or swollen

caducous Falling off prematurely

caespitose Growing in a clump, as in suckering or clumping palms

calcareous An excess of lime in a soil

callus Growth of undifferentiated cells that develop on a wound or in tissue culture

calyx All of the sepals of a flower

cambium The growing tissue lying just beneath the bark (absent in palms)

cane A reed-like plant stem, as in species of *Calamus* and *Chamaedorea*

canopy The cover of foliage

capitate Enlarged and head-like, as in the female inflorescence of *Elaeis* and *Nypa*

carpel Female reproductive organ

catkin A dense, pendulous spike of unisexual flowers

caudex A trunk-like growth axis, as in the trunks of palms

chartaceous Thin, stiff and papery

ciliate With a fringe of hairs

cirrate Bearing a cirrus

cirrus A whip-like organ bearing recurved hooks used as an aid for climbing: it arises as an extension of the leaf rachis and is present in some species of *Calamus*

clavate Club-shaped

clone A group of vegetatively propagated plants with a common ancestry—for example, commercial cultivars of the Date Palm which are all propagated by suckers

clustered Clumping, with several stems

colonial A term sometimes used for palms which branch basally by rhizomes—for example, *Calamus* spp.

coma A term sometimes used for the crown of a palm

compound leaf A leaf with two or more separate leaflets or divisions (both pinnate and palmate leaves are compound)

compressed Flattened laterally

confluent Leaflets remaining united and not separating

congested Crowded close together

contracted Narrowed

cordate Heart-shaped

coriaceus Leathery in texture

costa A main vein; often used for the rib of a costapalmate leaf

costapalmate Palmate leaves with a well-developed rib which is an extension of the petiole into the blade and is the equivalent of the rachis

cotyledon The seed leaf of a plant, in palms there is a single cotyledon which may be greatly modified

crenulate With shallowly toothed margins

cross Offspring or hybrid

cross-pollination Transfer of pollen from flower to flower

crown The head of foliage of a palm

crownshaft A series of tightly packed, specialised tubular leaf-bases which terminate the trunk of some pinnate-leaved palms

cupule A bowl or cup-shaped calyx developed at the base of the fruit of some palms

cultivar A horticultural variety—for example, variegated *Rhapis* palms

cymbiform Boat-shaped, sometimes used to describe the spathes of some palms

deciduous Falling off or shedding of any plant part

decumbent Reclining on the ground with the apex ascending, as in the trunk of *Elaeis oleifera*

decurrent Running downward beyond a junction, as in the leaflets of many pinnate palms

deflexed Abruptly turning downwards

dentate Toothed

denticulate Finely toothed

depauperate A weak plant or one imperfectly developed

depressed Flattened at one end, as in the fruit of some palms

determinate With the definite cessation of growth in the main axis, as in monocarpic palms

dichotomous Regular forking into equal branches, as in the branching of the trunks of *Hyphaene* spp. and *Nypa fruticans*

dicotyledons A section of the Angiosperms bearing two seed leaves in the seeding stage

diffuse Of widely spreading and open growth

digitate Spreading like the fingers of a hand from one point

dimorphic Existing in two different forms—the juvenile and seedling leaves of most palms are dimorphic: sucker leaves may be different from leaves on mature stems; species of *Nengella* and *Heterospathe humilis* have simple and pinnate-leaved plants present in the same population

dioecious Bearing male and female flowers on separate plants

dissected Deeply divided into segments

distal Farthest from the point of attachment

distichous Alternate leaves or leaflets arranged along a stem or rachis in two opposite rows in the one plane—for example, *Wallichia disticha*

divaricate Widely spreading

divided Separated to the base

drupe A fleshy, indehiscent fruit with seed(s) enclosed in a stony endocarp

ebracteate Without bracts

ecology The study of the interaction of plants and animals within their natural environment

effuse Very open and spreading, usually referring to the growth habit

elliptic Oval and flat and narrowed to each end which is rounded

elongate Drawn-out in length

emarginate Having a notch at the apex

embryo Dormant plant contained within a seed

endemic Restricted to a particular country, region or area

endocarp A woody layer surrounding a seed in a fleshy fruit; the innermost fruit wall

endosperm Tissue rich in nutrients which surrounds the embryo in seeds

ensiform Sword-shaped, as in the leaflets of many pinnate palms

entire Whole, not toothed, or divided in any way

eophyll The first green leaf produced by a palm seedling

epicarp The outermost layer of the fruit wall

epigeal A term used for roots which grow above ground—for example, *Catostigma* spp.

equable A term used to describe the endosperm of a seed when it is smooth and uniform

erect Upright

exocarp The outermost layer of the fruit wall

exotic A plant introduced from overseas

exserted Protruding beyond the surrounding parts as in the terminal inflorescence of *Corypha* and *Metroxylon*

extrafoliar Said of an inflorescence arising from the stem below the leaf-bases

falcate Sickle-shaped

farinaceous Containing starch, as in the trunks of palms used for sago; also appearing as if dusted or coated with flour

fasciculate Arranged in clusters as in the flowers of some palms; also refers to the arrangement of leaflets of some pinnate palms—for example, *Calyptrocalyx* spp.

ferruginous Rusty brown colour

fertile bract Bracts on a palm inflorescence which subtend branches and rachises (cf. sterile bracts)

fibrose, fibrous Containing fibres, as on the leaf-bases and trunks of many palms and outer covering of some fruit—for example, coconut

filament The stalk of the stamen supporting the anther

fimbria The fine, hair-like fringes of a scale

fimbriate Fringed with fine hairs

flabellate Fan-shaped

flabellum A term sometimes applied to the united pair of terminal leaflets of a pinnate leaf—for example, *Hydriastele*, *Pinanga*

flaccid Soft, limp, lax

flagellum A whip-like organ that bears curved hooks and is used as an aid to climbing; it is a modified inflorescence and arises in a leaf axil—for example, *Calamus*

flexuose Having a zigzag form, as in the rachises of some palms

floccose Having tufts of woolly hairs, as on the spathes of *Pritchardia*

floriferous Bearing numerous flowers

foetid Having an offensive odour

forked Divided into equal or nearly equal parts

form A botanical division below a species

free Not joined to any other part

frond Leaf of a palm or fern

fruit The seed-bearing organ

fused Joined or growing together

fusiform Spindle-shaped, swollen in the middle and narrowed to each end, as in the trunk of *Hyophorbe verschaffeltii*

geniculate Bent like a knee

genus A taxonomic group of closely related species

germination The active growth of an embryo resulting in the development of a young plant

glabrous Smooth, hairless

glaucous Covered with a bloom giving a bluish lustre

globose Globular, almost spherical

growth split A vertical crack or split that develops in the trunk of fast-growing palms

gynoecium Collectively the female parts of a flower

habit The environment in which a plant grows

hapaxanthic A term describing clumping palms, individual stems of which die after flowering, cf. monocarpic, pleonanthic

haustorium In palms this term refers to the apex of the cotyledon which is embedded in the endosperm and on germination of the seed, enlarges, secretes enzymes and transports soluble materials to the developing seedling

hastula A collar-like extension of the petiole in certain palmate palms—for example, *Pritchardia* spp. and *Sabal* spp.

head A composite cluster of flowers or fruit—for example, *Elaeis guineensis, Nypa fruticans*

hermaphrodite Bearing both male and female sex organs in the same flower

hilum The scar left on the seed from its point of attachment with the seed stalk

hirsute Covered with short white hairs or wool giving the surface a greyish appearance

homogenous Of uniform texture

hybrid Progeny resulting from the cross-fertilisation of parents either within the same genus—for example, *Phoenix* spp.—or between different genera—for example, *Arecastrum* X *Butia*

imbricate Overlapping, as in the scales covering the fruit of such palms as *Calamus* spp. and *Metroxylon* spp.

imparipinnate Pinnate leaves bearing a single terminal leaflet which extends from the end of the rachis—for example, *Phoenix* spp.

incurved Curved inwards

indehiscent Not opening on maturity

indeterminate Growing on without termination as in the trunks of most palms

indigenous Native to a country, region or area but not necessarily restricted there

indumentum A collective term describing the hairs or scales found on the surface of an organ

induplicate Leaflets folded longitudinally with the 'V' opened upwards—that is, the margins upwards

inflorescence The flowering structure of a plant

infrafoliar Below the foliage; a term used for palms which bear their inflorescences below the

leaves—for example, *Archontophoenix* spp.

infructescence A term used to describe a fruiting inflorescence

interfoliar Between the leaves; term used for palms which bear their inflorescences between the leaves in the crown—for example, *Sabal* spp.

internode The part of a stem between two nodes

involute Rolled inwards

jointed Bearing joints or nodes, used in reference to the cane-like stems of *Chamaedorea* spp.

juvenile The young stage of growth before the plant is capable of flowering

kerangas A specialised heath forest developed in Borneo

lacerate Irregularly cut or torn into narrow segments

laciniate Cut into narrow segments

lamina The expanded part of a leaf

lanceolate Lance-shaped, tapering to each end especially the apex

lateral Arising at the side of the main axis

lax Open and loose

leaf-base Specialised expanded and sheathing part of the petiole where it joins the trunk

leaf-spine A term sometimes used for the spine-like basal leaflets of *Phoenix* leaves; may also be used to refer to spines on leaves

leaflet Strictly a segment of a bipinnate leaf, as in *Caryota* spp. (also pinnule) but generally also used loosely for pinnae

lepidote Dotted with persistent, small, scurfy, peltate scales

liana A large, woody climber

ligule An extension at the junction of leaf sheath and petiole: in palms often incorrectly referring to the hastula; in seedlings of some palm species it is a tubular structure developed around the cotyledonary sheath

linear Long and narrow with parallel sides

littoral Growing in communities near the sea

loricate Covered with scales, as in the fruits of lepidocaryoid palms—for example, *Calamus*

mangrove A specialised plant growing in brackish or sea-water—for example, *Nypa fruticans*

marcescent Withering while still attached to the plant

maritime Growing near the sea

mealy Covered with fine, flour-like powder

membranous Thin-textured

meristem The apical growing point which is an area of active cell division

mesocarp The middle layer of a fruit wall

midrib The main vein that runs the full length of a leaflet on segment

monocarpic A term describing plants which flower once then die after fruiting. In palms this term is now applied to solitary trunked species while clumping palms are described as hapaxanthic

monocotyledons The section of Angiosperms to which palms belong and characterised by bearing a single seed leaf

monoecious Bearing separate male and female flowers on the same plant

monopodial A term used to describe a growth habit with unlimited apical growth—for example, all solitary, polycarpic palms

monotypic A genus with a single species

mucronate With a short, sharp point

nerves The fine veins which transverse the leaf blade

node A point on the stem where leaves or bracts arise

obcordate Ovate with the broadest part above the middle

obovoid Ovoid but widest above the middle

obtuse Blunt or rounded at the apex

ochrea A swollen or membranous appendage at the opening of the leaf-sheath, found in some climbing palms—for example, *Korthalsia* spp. In some species ants inhabit this organ

offset A growth arising from the base of a plant

orbicular Nearly circular

oval Rounded but longer than wide

ovary Part of the gynoecium that encloses the ovules

ovate Egg-shaped in a flat plane

ovoid Egg-shaped in a solid plane

ovule The structure within the ovary which becomes the seed after fertilisation

palman That region of a palmate leaf lamina where the segments are fused together at their margins

palmate In palms this refers to a circular or semi-circular leaf with the segments radiating from a common point

palmetum A collection of planted palms

palmito Palm hearts harvested and prepared for eating

panicle A much-branched racemose inflorescence

paniculate Arranged in a panicle

paripinnate Compound pinnate leaves lacking a terminal leaflet

parthenocarpic Fruit developing without fertilisation and seed formation, such seedless fruit are known in *Bactris gasipaes*

patent Spreading out

pedicel The stalk of a flower

peduncle The main axis of an inflorescence

pellicle The membrane surrounding a palm seed

peltate Circular with the stalk attached in the middle on the underside: leaves of *Licuala* spp. appear as if peltate but are not strictly so; the scales of some palms are peltate

perianth Collectively the sepals and petals

pericarp The hardened ovary wall that surrounds a seed

petal The innermost whorl of the corolla

petiole The stalk of a leaf

petiolule The stalk of a pinna or leaflet

petticoat A collective term for the persistent, hanging, dead leaves of some palms—for example, species of *Copernicia* and *Washingtonia*

pinna A primary segment of a pinnate leaf (see also segment), commonly called a leaflet, a term strictly applied to a pinnule

pinnate Once divided with the divisions extending to the midrib, usually referring to leaves

pinnule The segment of a compound leaf divided more than once (also leaflet)

pistil The female reproductive part of a flower

pistillate Bearing a pistil; a term used for female flowers

pistillode A sterile pistil, often found in male flowers

pleonanthic A term describing palms which flower regularly each year after reaching maturity, cf. hapaxanthic

plicate Folded or pleated longitudinally, as in the leaves of many fan palms

plumose In palm leaves, referring to pinnae orientated in different directions—for example, *Syagrus romanzoffiana*—as opposed to leaflets in one flat plane

pneumatophore Specialised upright roots carrying oxygen to the plant, breathing roots—for example, *Raphia* spp.

pollination The transference of the pollen from the anther to the stigma of a flower

polycarpic Flowering over many years, as opposed to monocarpic

polygamous Having mixed unisexual and bisexual flowers together

praemorse As though bitten off; often used in reference to the leaflet tips of palms such as species of *Hydriastele* and *Ptychosperma*

proliferous Bearing offshoot and other vegetative propagation structures

prophyll The first (or outer) sheathing bract of a palm inflorescence

prostrate Lying flat

proximal Closest to the point of attachment

pubescent Covered with short, downy hairs

pulvinus A cushion-like growth of inflated cells at the base of leaflets, spines or inflorescence branches

punctations A term to describe the presence of small, rounded scales

puncticulations As above but describing very small scales

pyriform Pear-shaped

rachilla A small rachis, the secondary and lesser axes of a compound inflorescence; the branch that bears the flowers

rachis The main axis of an entire or compound leaf of a palm extending from the petiole to the end of the lamina

radical Arranged in a basal rosette

radicle The undeveloped root of the embryo; also the primary root of a seedling

ramenta Elongated scales, often ragged, attached at one end only

raphe A ridge or depression on the surface of the seed

raphides Needle-shaped crystals of *Calcium oxalate*; stinging crystals

recurved Curved backwards

reduplicate Leaflets folded with the 'V' opened downwards (margins downward)

reflexed Bent backwards and downwards

reins Leaf fibres of palms, in particular referring to marginal strips that tie the developing leaflets together

repent Creeping; may be used to describe palms with a spreading growth habit

revolute With the margins rolled backwards

rheophyte A plant adapted to growing close to streams and surviving regular submergence

rhizome An underground stem which produces growths at intervals—for example, species of *Calamus* and other climbing palms

rib The section of the petiole of a costapalmate leaf that extends into the blade—for example, *Sabal palmetto*

ring-scars Annular rings

root spines Thin, spiny projections on the trunk of some palms (they may actually be primordial aerial roots)

rostrate With a beak, as in the fruit of some palms

rugose Wrinkled

ruminate Folded like a rumen, often with darkish markings where folded

saccate Pouch or sac-like, as in the spathes of some palms—for example, *Astrocaryum mexicanum*

sagittate Arrowhead-shaped

scabrous Rough to the touch

scale A dry, flattened, papery body, found on the young fronds and petioles of some palms

scandent Climbing

scarious Thin, dry and membranous, as in the bracts found on the inflorescence of many palms

scurfy Bearing small, flattened, papery scales

secondary thickening The increase in trunk diameter as the result of cambial growth (absent in palms)

secund With all parts directed to one side

seed A mature ovule, consisting of an embryo, endosperm and protective coat

seedcoat The protective covering of a seed, also called testa

seedling A young plant raised from seed

segment A subdivision or part of an organ; in palms the term is generally used for the divisions of a palmate leaf

sepal A segment of the calyx or outer whorl of the perianth

serrate Toothed

sessile Without a stalk

shag A collective term for the persistent, hanging, dead leaves of some palms—for example, *Washingtonia* spp.

simple Undivided as in the leaves of *Phoenicophorium*

sinker A term for the long shoot of some palm

species which emerges from the seed prior to leaf development

sinuous Wavy, like a snake

sinus A junction; the joint where the segments of a palmate leaf meet

soboliferous Bearing creeping, rooting stems, although sometimes interpreted as bearing suckers and applied to clumping palms (also caespitose)

solitary Describing a palm with a single stem or trunk

spadix Used for a palm inflorescence but misleading as applied differently in other plant families—for example, Araceae

spathe A large sheathing bract which encloses the young inflorescence, often very prominent in palms

spathel A small sheathing bract subtending a secondary or lesser rachilla

spathulate Spoon-shaped, spatula-shaped

spear-leaf The erect, unopened young leaf of a palm

species A taxonomic group of closely related plants, all possessing a common set of characteristics which sets then apart from another species

spicate Arranged like a spike

spike A simple, unbranched inflorescence with sessile flowers; found in palms such as *Howea belmoreana*, *Linospadix* spp. and *Raphia* spp.

spine A sharp, rigid projection

spinous Modified or resembling a spine

spinule A weak spine

stamen A male part of a flower producing pollen, consisting of an anther and a filament

staminate flowers Male flowers

staminode A sterile stamen

stem clasping Enfolding a stem, as in the leaf sheath of palms

sterile bracts Bracts on the palm inflorescence between the prophyll and the first branch (cf. fertile bracts)

stigma The expanded area of the style that is receptive to the pollen

stipitate Stalked

stolon A basal stem growing just below the ground surface and rooting at intervals

stoloniferous Spreading by stolons

strain An improved selection within a variety; also cultivar

stratification The technique of burying seed in moist, coarse sand to expose it to periods of low temperature or to soften the seed coat

striate Marked with narrow lines

style Part of the gynoecium connecting the stigma with the ovary

subspecies A taxonomic subgroup within a species used to differentiate geographically isolated variants

subulate Narrow and drawn out to a fine point

succulent Fleshy or juicy

sucker A shoot arising from the roots or trunk below ground level

sulcate Grooved or furrowed

sympodial Used to describe a growth habit that branches from the base—for example, *Calamus* spp., *Chamaedorea microspadix*

syncarpous With united carpels

taxon A term used to describe any taxonomic group—for example, genus, species

taxonomy The classification of plants or animals

ternate Divided or arranged in threes

terminal The apex or end

terete Slender and cylindrical

testa The outer covering of the seed, the seedcoat

tomentose Densely covered with short, matted soft hairs or scales

tortuose Twisted, with irregular bending, as in the rachillae of some palm inflorescences

transpiration The loss of water vapour to the atmosphere through openings in the leaves

triad A group of three; in some palms the flowers are arranged in triads

tribe A taxonomic group of related genera within a family or subfamily

trichomes A term used to describe outgrowth of the epidermis such as scales or hairs

truncate Ending abruptly, as if cut off

trunkless Without a trunk, in palms apparently trunkless species usually have a subterranean trunk

tuberculate With knobby or warty projection, as on the fruit of *Pelagodoxa*

tuberous Swollen and fleshy

tufted Growing in small, erect clumps

turgid Swollen or bloated

unarmed Smooth, without spines, hooks or thorns

undulate With wavy margins

unisexual Of one sex only

united Joined together, wholly or partially

variety A taxonomic subgroup within a species used to differentiate variable populations

vegetative Asexual development or propagation

vein The conducting tissue of leaves

venation The pattern formed by veins

ventricose Swollen or inflated as in the trunks of some palms—for example, *Hyophorbe verschaffeltii*

verrucose Rough and warty

verticillate Arranged in whorls as in the spines on the stems of some *Calamus* species

viable Alive and able to germinate

viviparous Germinating while still attached to the parent plant—for example, seeds of *Nypa fruticans*

whip Used generally for climbing organs

xerophytic Adapted to growing in dry conditions

BIBLIOGRAPHY

Allen, P.H. (1952), 'Distribution and Variation in *Roystonea*', *Ceiba* 3 (1), 1-18.

Allen, P.H. (1965), 'Palms in Middle America', *Principes* 9, 44-7.

Allen, P.H. (1965), 'Rain Forest Palms of Golfo Dulce', *Principes* 9, 48-62.

Anderson, A.B. & M.J. Balick (1988), 'Taxonomy of the Babassu Complex (*Orbignya* spp.: Palmae)', *Systematic Botany* 13 (1), 32-50.

Backer, C.A. & Bakhuizen van den Brink (1968), *Flora of Java: Arecaceae* Vol 3, 165-99.

Bailey, L.H. (1938), 'Article 6. *Thrinax*—The Peaberry Palms', *Gentes Herbarium* 4(4), 129-49.

Bailey, L.H. (1939), 'Article 11. Species of *Rhapis* in cultivation—The Lady Palms', *Gentes Herbarium* 4(6), 199-208.

Bailey, L.H. (1939), 'Article 14. *Coccothrinax* of Florida', *Gentes Herbarium* 4(6), 220-25.

Bailey, L.H. (1939), 'Article 16. New Haitian Genus of Palms', *Gentes Herbarium* 4(7), 239-46.

Bailey, L.H. (1939), 'Article 17. *Coccothrinax* in the Southern Greater Antilles', *Gentes Herbarium* 4(7), 247-59.

Bailey, L.H. (1939), 'Article 20. The Royal Palm of Hispaniola', *Gentes Herbarium* 4(7), 266-70.

Bailey, L.H. (1939), 'Article 21. The Sabals of Hispaniola', *Gentes Herbarium* 4(7), 271-5.

Bailey, L.H. (1940), 'Article 33. *Acoelorraphe* vs. *Paurotis*—Silver-Saw Palm', *Gentes Herbarium* 4(10), 361-5.

Bailey, L.H. (1942), 'Article 1. Palms of the Seychelles Islands', *Gentes Herbarium* 6(1), 3-48.

Bailey, L.H. (1942), 'Article 2. Palms of the Mascarenes', *Gentes Herbarium* 6(2), 51-104.

Bailey, L.H. (1943), 'Article 4. Brahea, and an *Erythea*', *Gentes Herbarium* 6(4), 177-97.

Bailey, L.H. (1944), 'A Revision of the American Palmettoes', *Gentes Herbarium* 6(7), 366-459.

Balick, M.J. (1986), 'Systematics and Economic Botany of the *Oenocarpus-Jessenia* (Palmae) Complex', *Advances in Economic Botany* 3, 1-140.

Balick, M.J., A.B. Anderson & J.T. De Medeiros-Costa (1987), 'Hybridization in the Babassu Palm Complex. II. *Attalea compta* X *Orbignya oleifera* (Palmae)', *Brittonia* 39(1), 26-36.

Balslev, H. & A. Barfod (1987), 'Ecuadorean Palms—an overview', *Opera Botanica* 92, 17-35.

Barbier, C. (1985), 'Further Notes on *Livistona carinensis* in Somalia', *Principes* 29, 151-5.

Basu, S.K. (1987), '*Corypha* Palms in India', *Journal of Economic & Taxonomic Botany* 11(2), 477-86.

Beccari, O. (1909), 'New or Little-known Philippine Palms', *Leaflets of Philippine Botany* 2, 639-50.

Beccari, O. (1909), 'Notes on Philippine Palms, II', *The Philippine Journal of Science* 9, 601-37.

Beccari, O. (1931), 'Asiatic Palms—Corypheae, revised and edited by V. Martelli', *Annals of Royal Botanic Gardens, Calcutta* 13.

Beccari, O. & J.F. Rock (1921), 'A Monographic Study of the Genus *Pritchardia*', *Memoirs of the Bernice Bishop Museum* 8.

Beach, J.H. (1984), 'The Reproductive Biology of the Peach or "Peijibaye" Palm (*Bactris gasipaes*) and a Wild Congener (*B. porschiana*) in the Atlan-Atlantic Lowlands of Costa Rica', *Principes* 28, 107-19.

Bernal, R.G., G. Galeano & A. Henderson (1991), 'Notes on *Oenocarpus* (Palmae) in the Colombian Amazon', *Brittonia* 43(3), 154-64.

Blatter, E. (1926), *The Palms of British India and Ceylon*, Oxford University Press, London.

Bodley, J.H. & F.C. Benson (1980), 'Stilt-Root Walking by an Iriarteoid Palm in the Peruvian Amazon', *Biotropica* 12(1), 67-71.

Bonde, S.D. (1988), 'Palms of Pune Area', *Journal of Economic & Taxonomic Botany* 12(1), 93-101.

Borchsenius, F.,R.G. Bernal & M. Ruiz (1989), '*Aiphanes tricuspidata* (Palmae), A New Species from Colombia and Ecuador', *Brittonia* 41(2), 156-9.

Borchsenius, F. & H. Balslev (1989), 'Three new species of *Aiphanes* (Palmae) with notes on the genus in Eucuador', *Nordic Journal of Botany* 9(4), 383-93.

Borhidi, A. & O. Muniz (1971), 'New Plants in Cuba I', *Acta Botanica Academiae Scientiarum Hungaricae Tomus* 17(1-2), 1-5.

Boyer, K. (1992), *Palms & Cycads Beyond the Tropics*, Publication Fund: Palm & Cycad Societies of Australia.

Braun, A. (1968), 'Cultivated Palms of Venezuela', *Principes* 12, 39-103.

Braun, A. (1968), 'Cultivated Palms of Venezuela—Part II', *Principes* 12, 111-36.

Braun, A. (1984), 'More Venezuelan Palms', *Principes* 28(2), 73-84.

Broschat, T.K. (1984), 'Nutrient Deficiency Symptoms in Five Species of Palms Grown as Foliage Plants', *Principes* 28, 6-14.

Brown, F.B.H.. (1931), 'Flora of Southeastern Polynesia—1. Monocotyledons', *Bernice P. Bishop Museum Bulletin* 84, 117-21.

Buckley, R. & H. Harries (1984), 'Self-Sown Wild-Type Coconuts from Australia', *Biotropica* 16(2), 148-51.

Bullock, S.H. (1980), 'Dispersal of a Desert Palm by Opportunistic Frugivores', *Principes* 24, 29-32.

Burret, M. (1939), 'Palmae Gesammelt in Neu Guinea von L. Brass.', *Journal of the Arnold Arboretum* 20, 187-212.

Cárdenas, M. (1970), 'Palm Forests of the Bolivian High Andes', *Principes* 14, 50-4.

Corner, E.J.H. (1966), *The Natural History of Palms*, Weidenfield & Nicholson, London.

Dahlgren, B.E. & S.F. Glassman (1961), 'A Revision of the Genus *Copernicia* 1. South American Species', *Gentes Herbarium* 9(1), 3-39.

Dahlgren, B.E. & S.F. Glassman (1963), 'Revision of the Genus *Copernicia* 2. West Indian Species', *Gentes Herbarium* 9(2), 42-232.

Davis, T.A. (1971), 'Branching in *Chrysalidocarpus lutescens*', *Phytomorphology* 20(3), 199-209.

Davis, T.A. & T. Kuswara (1987), 'Observations on *Pigafetta filaris*', *Principes* 31, 127-37.

Doughty, S.C. , E.N. O'Rourke, E.P. Barrios & R.P. Mowers (1986), 'Germination Induction of Pygmy Date Palm Seed', *Principes* 30, 85-7.

Dowe, J.L. (1989), *Palms of the South-west Pacific*, Publication Fund: Palm & Cycad Societies of Australia.

Dransfield, J. (1972), 'The Genus *Johannesteijsmannia* H.E. Moore Jr.', *Gardens' Bulletin, Singapore* 26, 63-83.

Dransfield, J. (1974), 'Notes on the Palm Flora of Central Sumatra', *Reinwardtia* 8(4), 519-31.

Dransfield, J. (1974), 'Notes on *Caryota no* Becc. and other Malesian *Caryota* species', *Principes* 18, 87-93.

Dransfield, J. (1975), 'A Remarkable New *Nenga* From Sumatra', *Principes* 19, 27-35.

Dransfield, J. (1976), 'A Note on the Habitat of *Pigafetta filaris* in North Celebes', *Principes* 20, 48.

Dransfield (1977), 'A Dwarf *Livistona* (Palmae) from Borneo, *Kew Bulletin* 31(4), 759-62.

Dransfield, J. (1980), 'Systematic notes on *Pinanga* (Palmae) in Borneo', *Kew Bulletin* 34(4), 769-88.

Dransfield, J. (1980), 'Systematic notes on some Bornean Palmae', *Botanical Journal of the Linnean Society* 81, 39-42.

Dransfield, J. (1982), '*Clinostigma* in New Ireland', *Principes* 26, 73-6.

Dransfield, J. (1982), '*Pinanga cleistantha*, a new species with hidden flowers', *Principes* 26, 126-9.

Dransfield, J. (1984), 'The genus *Areca* (Palmae: Arecoideae) in Borneo', *Kew Bulletin* 39(1), 1-22.

Dransfield, J. & J.P. Mogea (1984), 'The Flowering behaviour of *Arenga* (Palmae: Caryotoideae)', *Botanical Journal of the Linnean Society* 88, 1-10.

Dransfield, J., Lee Shu-Kang & Wei Fa-Nan (1985), '*Guihaia*, a New Coryphoid Genus from China and Vietnam', *Principes* 29, 3-12.

Dransfield, J., A.K. Irvine & N.W. Uhl (1985), '*Oraniopsis appendiculata*, a Previously Misunderstood Queensland Palm', *Principes* 29, 56-63.

Dransfield, J. & N.W. Uhl (1986), 'An Outline of a Classification of Palms', *Principes* 30, 3-11.

Dransfield, J. & N.W. Uhl (1986), '*Ravenea* in the Comores', *Principes* 30, 156-60.

Endt, D. (1987), '*Rhopalostylis sapida* on Great Barrier Island, New Zealand', *Principes* 31, 165-8.

Esser, R.P. (1967), 'Nematode Pests and Associates of Five Species of Chamaedorea', *Principes* 11, 87-91.

Essig, F.B. (1977), 'A Preliminary Analysis of the Palm Flora of New Guinea and The Bismarck Archipelago', *Papua New Guinea Botany Bulletin* 9, 1-38.

Essig, F.B. (1978), 'A Revision of the Genus *Ptychosperma* Labill. (Arecaceae)', *Allertonia* 1(7), 415-78.

Essig, F.B. (1979), '*Ptychosperma hosinoi* growing in Hawaii', *Principes* 23, 174-5.

Essig, F.B. (1980) 'The genus *Orania* Zipp. (Arecaceae) in New Guinea', *Lyonia* 1(5), 211-33.

Essig, F.B. & B.E. Young (1980), 'Palm Collecting in Papua New Guinea. I. The Northeast', *Principes* 24, 14-28.

Essig, F.B. (1980), 'The genus *Orania* Zipp. (Arecaceae) in New Guinea', *Lyonia* 1(5), 211-31.

Essig, F.B. & B.E. Young (1981), 'Palm Collecting in Papua New Guinea II. The Sepik and the North Coast', *Principes* 25, 3-15.

Essig, F.B. & B.E. Young (1981), 'Palm Collecting in Papua New Guinea. III. Papua', *Principes* 25, 16-28.

Essig, F.B. (1982), 'A Synopsis of the Genus *Gulubia*', *Principes* 26, 159-73.

Essig, F.B. & B.E. Young (1985), 'A Reconsideration of *Gronophyllum* and *Nengella* (Arecoideae)', *Principes* 29 (3), 129-37.

Essig, F.B. (1987), 'A New Species of *Ptychosperma* (Palmae) from New Britain', *Principes* 31, 110-5.

Fernando, E.S. (1983), 'A Revision of the Genus *Nenga*', *Principes* 27, 55-70.

Fernando, E.S. (1983), 'The genus *Heterospathe* (Palmae: Arecoideae) in the Philippines', *Kew Bulletin* 45(2), 219-34.

Fisher, J.B. & J.H. Tsai (1979), 'Palm Tissue Culture Update', *Principes* 23, 178.

Fong, F.W. (1992), 'Perspectives for sustainable resource utilization and management of Nipa vegetation', *Economic Botany* 46(1), 45-54.

Fullington, J.G. (1987), '*Parajubaea*—an unsurpassed palm for cool, mild areas', *Principes* 31, 172-6.

Furtado, C.X. (1940), 'Palmae Malesicae—VIII, The Genus *Licuala* in the Malay Peninsula', *Gardens Bulletin, Singapore* 11, 31-73.

Furtado, C.X. (1949), 'Palmae Malesicae—X, The Malayan species of *Salacca*', *Gardens Bulletin, Singapore* 12, 378-403.

Furtado, C.X. (1970), 'Asian Species of *Hyphaene*', *Gardens Bulletin, Singapore* 25, 299-309.

Galeano-Garces, G. & R. Bernal-Gonzalez (1985), '*Aiphanes acaulis*, a New Species from Colombia', *Principes* 29, 20-22.

Galeano-Garces, G. (1986), 'Two New Species of Palmae From Colombia', *Brittonia* 38(1), 60-4.

Gentry, A.H. (1986), 'Notes on Peruvian Palms', *Annals of Missouri Botanic Garden* 73, 158-65.

Gibbons, M. (1993), 'Trekking on the *Trachycarpus* Trail', *Principes* 37, 19-25.

Gibbons, M. & T. Spanner (1993), 'In Search of *Trachycarpus nanus*', *Principes* 37, 64-72.

Gillett, G.W. (1971), '*Pelagodoxa* in the Marquesas Islands', *Principes* 15, 45-8.

Glassman, S.F. (1964), 'Two New Species of palms from Nicaragua', *Fieldiana: Botany* 31(2), 5-9.

Glassman, S.F. (1971), 'Rediscovery of *Syagrus werdermannii* Burret', *Fieldiana: Botany*, 34 (1), 1-10.

Glassman, S.F. (1971), 'Re-evaluation of *Syagrus loefgrenii* Glassman and *S. rachidii* Glassman', *Fieldiana: Botany* 34(2), 11-25.

Glassman, S.F. (1978), 'New Species of *Syagrus* from the State of Bahia (Brazil), with a Revisional Study of Closely Related Taxa', *Phytologia* 39(6), 401-23.

Glassman, S.F. (1979), 'Re-evaluation of the Genus *Butia* With a Description of a New Species', *Principes* 23(2) 65-79.

Glassman, S.F. (1987), 'Revisions of the Palm Genus *Syagrus* Mart. and Other Selected Genera in the *Cocos* Alliance', *Illinois Biological Monographs* 56, 1-231.

Gorman, M.L & S. Siwatibau (1975), 'The Status of *Neoveitchia storckii* (Wendl): A Species of Palm Tree Endemic to the Fijian Island of Viti Levu, *Biological Conservation* (8), 73-6.

Gruezo, W.Sm. & H.C. Harries (1984), 'Self-Sown Wild-Type Coconuts in the Philippines', *Biotropica* 16(2) 140-7.

Harries, H.C. (1978), 'The Evolution, Dissemination and Classification of *Cocos nucifera* L.', *The Botanical Review*, 44(3), 265-319.

Harries, H.C. (1981), 'Germination and Taxonomy of the Coconut', *Annals of Botany* 48, 873-83.

Harries, H.C. (1983), 'The Coconut Palm, The Robber Crab and Charles Darwin: April Fool or a Curious Case of Instinct?', *Principes* 27, 131-7.

Henderson, A. (1984), 'Observations on Pollination of *Cryosophila albida*', *Principes* 28, 120-6.

Henderson, A. (1984), 'The Native Palms of Puerto Rico', *Principes* 28, 168-172.

Henderson, A. (1985), 'Pollination of *Socratea exorrhiza* and *Iriartea ventricosa*', *Principes* 29, 64-71.

Hodel, D. (1980), 'Notes on *Pritchardia* in Hawaii', *Principes* 24, 65-81.

Hodel, D.R. (1985), 'A New *Pritchardia* from South Kona, Hawaii', *Principes* 29, 31-4.

Hodel, D.R. (1985), '*Gliocladium* and *Fusarium* Diseases of Palms', *Principes* 29, 85-8.

Hodel, D.R. (1991), 'The cultivated species of *Chamaedorea* with caespitose habit and pinnate leaves', *Principes* 35, 184-98.

Hodel, D.R. (1992), *Chamaedorea Palms, The Species and Their Cultivation*, Allen Press, Kansas.

Howard, F.W. (1980), 'Population Densities of *Myndus crudus* Van Duzee (Homoptera: Cixiidae) in Relation to Coconut Lethal Yellowing Distribution in Florida', *Principes* 24, 174-8.

Irvine, A. (1983), 'Wodyetia, A New Arecoid Genus from Australia', *Principes* 27, 158-67.

Johnson, D. (1971), 'Some Notes on Palms of the Genus *Copernicia*', *Principes* 15, 127-30.

Johnson, D. (1972), 'The Carnauba Wax palm (*Copernicia prunifera*) I. Botany', *Principes* 16, 16-19.

Johnson, D. (1987), 'Worldwide Endangerment of Useful Palms', *Principes* 31, 41.

Johnson, D. (1987), 'Conservation Status of Wild Palms in Latin America and the Caribbean', *Principes* 31, 96-7.

Jumelle, H. (1945), 'Flore De Madagascar—30 Famille Palmiers', *Flore de Madagascar* 1-186.

Kiew, R. (1976), 'The Genus *Iguanura* Bl. (Palmae)', *Gardens Bulletin, Singapore* 28, 191-226.

Kiew, R. (1978), 'New species and records of *Iguanura* (Palmae) from Sarawak and Thailand', *Kew Bulletin* 34(1), 143-5.

Kimnach, M. (1977), 'The Species of Trachycarpus', *Principes* 21, 155-60.

Kimnach, M. (1980), 'The "Fish-Tail" Palms', *Principes* 24, 125-7.

Koebernik, J. (1971), 'Germination of Palm Seed', *Principes* 15, 134-7.

Langlois, A.C. (1976), *Supplement to Palms of the World*, University Press, Florida.

MacKee, H.S., P. Morat & J.M. Veillon (1985), 'Palms in New Caledonia', *Principes* 29, 166-9.

Mahabale, T.S. (1976), 'The Origin of Coconut', *Palaeobotanist* 25, 238-48.

McCoy, R.E., M.E. Miller & D.S. Williams (1980), 'Lethal Yellowing in Texas *Phoenix* Palms', *Principes* 24, 179-180.

McCurrach, J.C. (1960), *Palms of the World*, Harper and Brothers, New York.

Merrill, E.D. (1922), *Enumeration of Philippine Plants—Palmae* 1, 142-72, Bureau of Science, Manila.

Mogea, J.P. (1980), 'The Flabellate-leaved Species of *Salacca* (Palmae)', *Reinwardtia* 9(4), 461-79.

Mogea, J.P. (1981), 'Notes on *Salacca wallichiana*', *Principes* 25, 120-3.

Mogea, J.P. (1982), '*Salacca zalacca*, The Correct Name for the Salak Palm', *Principes* 26, 70-2.

Mogea, J.P. (1986), 'A New Species in the Genus *Salacca*', *Principes* 30, 161-4.

Mohammed, S. (1981), 'Modernizing Date Production in Iraq', *Principes* 25, 73-7.

Moncur, M.W. & B.J. Watson (1987), 'Observations on the Floral Biology of the Monoecious Form of *Salacca zalacca*', *Principes* 31, 20-2.

Moore, Jr., H.E. & F.R. Fosberg (1956), 'Article 20, The Palms of Micronesia and the Bonin Islands', *Gentes Herbarium* 8(6), 423-78.

Moore, Jr., H.E. (1957), 'Article 21. *Veitchia*', *Gentes Herbarium* 8(7), 483-536.

Moore, Jr., H.E. (1957), 'Article 22. *Neoveitchia*', *Gentes Herbarium* 8(7), 537-40.

Moore, Jr., H.E. (1957), 'Article 23. *Reinhardtia*', *Gentes Herbarium* 8(7), 541-76.

Moore, Jr., H.E. (1965), '*Ptychococcus lepidotus*—A New Species from New Guinea', *Principes* 9, 11-12.

Moore, Jr., H.E. (1965), 'A New Species of *Arenga* from Borneo', *Principes* 9, 100-3.

Moore, Jr., H.E. (1966), '*Chamaedorea metallica*—A New Species from Cultivation', *Principes* 10, 44-50.

Moore, Jr., H.E. (1966), 'New Palms from the Pacific', *Principes* 10, 85-99.

Moore, Jr., H.E. (1967), 'The Genus *Gastrococos* (Palmae-Cocoideae)', *Principes* 11, 114-21.

Moore, Jr., H.E. (1969), 'Satakentia—A New Genus of Palmae-Arecoideae', *Principes* 13, 3-12.

Moore, Jr., H.E. (1970), 'The Genus *Rhopaloblaste* (Palmae)', *Principes* 14, 75-92.

Moore, Jr., H.E. (1971), 'The Genus *Synechanthus* (Palmae)', *Principes* 15, 10-18.

Moore, Jr., H.E. (1971), 'Notes on *Sabal* in Cultivation', *Principes* 15, 69-73.

Moore, Jr., H.E. (1971), 'Additions and Corrections to "An Annotated Checklist of Cultivated Palms"', *Principes* 15, 102-6.

Moore, Jr., H.E. (1973), 'The Major Groups of Palms and Their Distribution', *Gentes Herbarium* 11(2), 27-140.

Moore, Jr., H.E. (1978), 'The Genus *Hyophorbe* (Palmae)', *Gentes Herbarium* 11(4), 212-45.

Moore, Jr., H.E. (1978), 'New Genera and Species of Palmae from New Caledonia', *Gentes Herbarium* 11(4), 291-309.

Moore, Jr., H.E. (1979), 'Endangerment at the Specific and Generic Levels in Palms', *Principes* 23, 47-64.

Moore, Jr., H.E. (1980), 'New Genera and Species of Palmae from New Caledonia, II', *Gentes Herbarium* 12(1) 17-24.

Moore, Jr., H.E. & N.W. Uhl (1982), 'Major Trends of Evolution in Palms', *The Botanical Review* 48(1), 1-69.

Morgan, P. R.&J. Dransfield (1984), '*Livistona exigua*, A Rare Bornean Palm Refound', *Principes* 28, 3-5.

Morton, J.F. (1988), 'Notes on Distribution, Propagation and Products of Borassus Palms (Arecaceae)', *Economic Botany* 42(3) 420-41.

Nauman, C.E. & R.W. Sanders (1991), 'An Annotated Key to the Cultivated Species of *Coccothrinax*', *Principes* 35, 27-46.

Odetola, J.A. (1987), 'Studies on Seed Dormancy, Viability, and Germination in Ornamental Palms', *Principes* 31, 24-30.

Okita, Y. & J.L. Hollenberb (1981), *The Miniature Palms of Japan*, Weatherhill, Tokyo.

Otedoh, M.O. (1982), 'Revision of the Genus *Raphia* Beauv. (Palmae)', *Journal of Nigerian Institute for Oil Palm Research* 6(22), 145-89.

Parham, J.W. (1972), *Plants of the Fiji Islands*, Government Printer, Suva.

Pingitore, E.J. (1982), 'Rare Palms in Argentina', *Principes* 26, 9-18.

Quero, H.J. (1980), *Coccothrinax readii*, A New Species From the Peninsula of Yucatan, Mexico', *Principes* 24, 118-24.

Quero, H.J. (1981), '*Pseudophoenix sargentii* in the Yucatan Peninsula, Mexico', *Principes* 25, 63-72.

Quero, H.J. (1982), '*Opsiandra gomez-pompae*, A New Species from Oaxaca, Mexico', *Principes* 26, 144-9.

Quero, H.J. (1991), '*Sabal gretheriae*, a New Species of Palm from the Yucatan Peninsula, Mexico', *Principes* 35, 219-24.

Rauwerdink, J.B. (1986), 'An Essay on Metroxylon, the Sago Palm', *Principes* 30, 165-80.

Read, R.W. (1966), '*Coccothrinax inaguensis*—A New Species from the Bahamas', *Principes* 10, 29-35.

Read, R.W. (1966), '*Coccothrinax jamaicensis*—The Jamaican Silver Thatch', *Principes* 10, 133-41.

Read, R.W. (1968), 'A Study of *Pseudophoenix*', *Gentes Herbarium* 10(2), 169-213.

Read, R.W. (1969), '*Colpothrinax cookii*—A New Species from Central America', *Principes* 13, 13-22.

Read, R.W. (1969), 'Some Notes on *Pseudophoenix* and a Key to the Species', *Principes* 13, 77-79.

Read, R.W. (1979), 'Live Storage of Palm Pollen', *Principes* 23, 33-35.

Read, R.W., T.A. Zanoni & M. Mejia (1987), 'Reinhardtia paiewonskiana (Palmae), A New Species for the West Indies', Brittonia 39(1), 20-25.

Russell, T.A. (1965), 'The Raphia Palms of West Africa', Kew Bulletin 19(2), 173-95.

Shuey, A.G. & R.P. Wunderlin (1977), The Needle Palm: Rhapidophyllum hystrix', Principes 21, 47-59.

Smith, D. (1986), 'Brief Comments on Palms Suited to Indoor Culture', Principes 30, 32-34.

Stevenson, G.B. (1974), Palms of South Florida, published by the author.

Tomlinson, P.B. (1979), 'Systematics and Ecology of the Palmae', Annual Revue of Ecology & Systematics 10: 85-107.

Tomlinson, P.B. (1990), The Structural Biology of Palms, Clarendon Press, Oxford.

Uhl, N.W. & J. Dransfield (1987), Genera Palmarum, Allen Press, Kansas.

Whitmore, T.C. (1970), 'Taxonomic Notes on Some Malayan Palms', Principes 14, 123-27.

Whitmore, T.C. (1973), Palms of Malaya, Oxford University Press, Kuala Lumpur.

Young, B.E. (1985), 'A New Species of Gronophyllum (Palmae) from Papua New Guinea', Principes 29, 138-41.

Yuri S.J.A. (1987), 'Propagation of Chilean Wine Palm (Jubaea chilensis) by means of In Vitro Embryo Culture', Principes 31, 183-86.

Zona, S. (1985), 'A New Species of Sabal (Palmae) from Florida', Brittonia 37(4), 366-68.

Zona, S. (1990), 'A Monograph of Sabal (Arecacaea: Coryphoideae)', Aliso 12(4), 583-666.

Zona, S. (1991), 'Notes don Roystonea in Cuba', Principes 35, 225-3.

INDEX

Frequently used Synonyms